A New Concept for Tuning Design Weights in Survey Sampling

A New Concept for Tuning Design Weights in Survey Sampling

Jackknifing in Theory and Practice

by

Sarjinder Singh
Texas A&M University-Kingsville, TX, USA

Stephen A. Sedory
Texas A&M University-Kingsville, TX, USA

Maria del Mar Rueda
University of Granada, Spain

Antonio Arcos
University of Granada, Spain

Raghunath Arnab
University of Botswana and University of KwaZulu-Natal, S. Africa

AMSTERDAM · BOSTON · HEIDELBERG · LONDON
NEW YORK · OXFORD · PARIS · SAN DIEGO
SAN FRANCISCO · SINGAPORE · SYDNEY · TOKYO

Academic Press is an imprint of Elsevier

Academic Press is an imprint of Elsevier
125 London Wall, London, EC2Y 5AS, UK
525 B Street, Suite 1800, San Diego, CA 92101-4495, USA
225 Wyman Street, Waltham, MA 02451, USA
The Boulevard, Langford Lane, Kidlington, Oxford OX5 1GB, UK

Copyright © 2016 Elsevier Ltd. All rights reserved.

No part of this publication may be reproduced or transmitted in any form or by any means, electronic or mechanical, including photocopying, recording, or any information storage and retrieval system, without permission in writing from the publisher. Details on how to seek permission, further information about the Publisher's permissions policies and our arrangements with organizations such as the Copyright Clearance Center and the Copyright Licensing Agency, can be found at our website: www.elsevier.com/permissions.

This book and the individual contributions contained in it are protected under copyright by the Publisher (other than as may be noted herein).

Notices
Knowledge and best practice in this field are constantly changing. As new research and experience broaden our understanding, changes in research methods, professional practices, or medical treatment may become necessary.

Practitioners and researchers must always rely on their own experience and knowledge in evaluating and using any information, methods, compounds, or experiments described herein. In using such information or methods they should be mindful of their own safety and the safety of others, including parties for whom they have a professional responsibility.

To the fullest extent of the law, neither the Publisher nor the authors, contributors, or editors, assume any liability for any injury and/or damage to persons or property as a matter of products liability, negligence or otherwise, or from any use or operation of any methods, products, instructions, or ideas contained in the material herein.

ISBN: 978-0-08-100594-1

British Library Cataloguing in Publication Data
A catalogue record for this book is available from the British Library

Library of Congress Cataloging-in-Publication Data
A catalog record for this book is available from the Library of Congress

For information on all Academic Press publications
visit our website at http://store.elsevier.com/

Contents

Preface		**xi**
1	**Problem of estimation**	**1**
	1.1 Introduction	1
	1.2 Estimation problem and notation	1
	1.3 Modeling of jumbo pumpkins	5
	1.4 The concept of jackknifing	8
	1.5 Jackknifing the sample mean	11
	1.6 Doubly jackknifed sample mean	14
	1.7 Jackknifing a sample proportion	16
	1.8 Jackknifing of a double suffix variable sum	17
	1.9 Frequently asked questions	17
	1.10 Exercises	18
2	**Tuning of jackknife estimator**	**27**
	2.1 Introduction	27
	2.2 Notation	27
	2.3 Tuning with a chi-square type distance function	28
	2.4 Tuning with dell function	41
	2.5 An important remark	48
	2.6 Exercises	48
3	**Model assisted tuning of estimators**	**63**
	3.1 Introduction	63
	3.2 Model assisted tuning with a chi-square distance function	63
	3.3 Model assisted tuning with a dual-to-empirical log-likelihood (dell) function	69
	3.4 Exercises	74
4	**Tuned estimators of finite population variance**	**85**
	4.1 Introduction	85
	4.2 Tuned estimator of finite population variance	85
	4.3 Tuning with a chi-square distance	88
	4.4 Tuning of estimator of finite population variance with a dual-to-empirical log-likelihood (dell) function	98

	4.5	Alternative tuning with a chi-square distance	107
	4.6	Alternative tuning with a dell function	113
	4.7	Exercises	118
5	**Tuned estimators of correlation coefficient**		**139**
	5.1	Introduction	139
	5.2	Correlation coefficient	139
	5.3	Tuned estimator of correlation coefficient	140
	5.4	Exercises	151
6	**Tuning of multicharacter survey estimators**		**165**
	6.1	Introduction	165
	6.2	Transformation on selection probabilities	165
	6.3	Tuning with a chi-square distance function	166
	6.4	Tuning of the multicharacter estimator of population total with dual-to-empirical log-likelihood function	179
	6.5	Exercises	191
7	**Tuning of the Horvitz–Thompson estimator**		**199**
	7.1	Introduction	199
	7.2	Jackknifed weights in the Horvitz–Thompson estimator	199
	7.3	Tuning with a chi-square distance function while using jackknifed sample means	200
	7.4	Tuning of the Horvitz–Thompson estimator with a displacement function	208
	7.5	Exercises	214
8	**Tuning in stratified sampling**		**219**
	8.1	Introduction	219
	8.2	Stratification	219
	8.3	Tuning with a chi-square distance function using stratum-level known population means of an auxiliary variable	221
	8.4	Tuning with dual-to-empirical log-likelihood function using stratum-level known population means of an auxiliary variable	232
	8.5	Exercises	242
9	**Tuning using multiauxiliary information**		**257**
	9.1	Introduction	257
	9.2	Notation	257
	9.3	Tuning with a chi-square distance function	258
	9.4	Tuning with empirical log-likelihood function	267
	9.5	Exercises	275

10	**A brief review of related work**		**283**
	10.1	Introduction	**283**
	10.2	Calibration	**283**
	10.3	Jackknifing	**289**

Bibliography   **291**
Author Index   **297**
Subject Index   **299**

Preface

This monograph, "A New Concept for Tuning Design Weights in Survey Sampling: Jackknifing in Theory and Practice," introduces a fresh survey methodology that can be used to estimate population parameters and that, with the help of a computer, yet requiring minimal effort, leads to the construction of confidence intervals. The most important motivation for using the newly tuned estimation survey methodology is that it simplifies the estimation of the variance of those estimators used for estimating population parameters such as the mean, the total, the finite population variance, and the correlation coefficient, under various sampling schemes. The main ideas derive from the standard survey sampling methods of jackknifing, calibration, and imputation, but lead to an important modification of these existing methods. This new methodology should prove helpful to all government organizations, private organizations, and academic institutions that engage in survey sampling. The proposed new concept of tuning design weights leads to a computer friendly estimation survey methodology. Thus, it encourages development of a new statistical analysis software package in a language such as R, SAS, FORTRAN, or C++. Recall that the sampling design and estimation process has two main aspects: a selection process that includes the rules and operations by which some members of the population are included in the sample, and an estimation process for computing the sample statistics, which are sample estimates of population values. A good sample design and estimation methodology requires the balance of several important criteria: goal orientation, measurability, practicality, economy, time, simplicity, and reliability.

One need not take our word concerning the virtues of the new concept of tuning design weights which leads to the newly tuned estimation survey methodology developed here. One can easily test/apply these methods by executing the R codes that are provided. The authors are confident that one will find the performance of the developed theory amazing. In the 1940s (Cochran, 1940, ratio estimator), survey methodologists began looking to the future for a reliable, trustworthy, and easily applicable estimation methodology. It appears that for students currently doing a Ph.D. in survey sampling, the future is here.

In Chapter 1, we discuss the problem of estimation in survey sampling. The *Statistical Jumbo Pumpkin Model* (SJPM) is developed, which can produce very light to very heavy pumpkins, and is naturally correlated with their known circumferences. R code, used for generating the SJPM, is provided. A population of pumpkins is generated using the SJPM and is displayed. A sample of pumpkins is taken from the generated population and is also displayed. The idea of jackknifing a sample statistic is discussed. Examples of jackknifing a sample mean and a sample proportion are provided. The effect of double jackknifing on a sample mean is discussed. The concept of

Professor W.G. Cochran (1909–1980)
http://isi.cbs.nl/Nlet/images/N87_CochranWG.jpg

jackknifing a sum of doubly subscripted variables is also introduced. At the end of the chapter, a few unsolved exercises, which include jackknifing of sample geometric mean, sample harmonic mean, and sample median, are provided for practice and further investigation.

In Chapter 2, we introduce the new concept of tuning design weights, which, in turn, leads to a computer friendly tuning methodology. The proposed methodology results in an estimator that is equivalent to the linear regression estimator of the population mean or total. The newly tuned estimation methodology is able to efficiently and effectively estimate the variance of the estimator of the population mean. The estimation of variance of the estimator is a key feature of the proposed newly tuned estimation methodology. In Chapter 2, we restrict ourselves to the use of a simple random sampling (SRS) scheme. We show that under an SRS scheme, the linear regression estimator is a result of tuning the jackknife sample mean estimator with the chi-squared type distance function. Tuning of the sample jackknife estimator under a dual-to-empirical log-likelihood (dell) function leads to a newly tuned dell-estimator of the population mean. The coverage by the newly tuned estimators, under the chi-square distance and the dell function, is investigated for the SJPM. The R codes for studying the reliability of the tuned estimators under the chi-squared distance function and the dell function are also provided. Numerical illustrations for both methods are also presented. At the end of the chapter, a few unsolved exercises are given for practice and further investigation. The tuning of a nonresponse in sampling theory is addressed in one of the unsolved exercises. The tuning of a sensitive variable, in the context of estimating the population mean of a sensitive variable, is also introduced through an unsolved exercise. In addition, tuned estimators of geometric mean and harmonic mean are also suggested for further investigation.

In Chapter 3, the problem of estimating a population mean with the help of the newly tuned model assisted estimation methodology is developed by assuming that the auxiliary information is available at the unit level in the population. The newly tuned model assisted estimator does an excellent job for small samples in the case of both the chi-squared type distance function, and the dual-to-empirical log-likelihood (dell) function, insofar as the weight of pumpkins is concerned. The R codes for studying the reliability of the tuned model assisted estimators for the case of the chi-squared and the dell functions are also provided. At the end of the chapter, a few unsolved model assisted exercises are given for practice and further investigation. A new model assisted tuning of a nonresponse along with the concept of imputation is addressed in one of the exercises.

In Chapter 4, we present the problem of estimating the finite population variance with the help of the newly tuned estimation methodology. The newly tuned model assisted estimator of finite population variance is studied under both the chi-squared and the dual-to-empirical log-likelihood (dell) functions. The R code for studying the reliability of estimating the finite population variance of the weight of pumpkins, using circumference as an auxiliary variable, is provided. Here, it is suggested that only a slice of a pumpkin be removed from the sample instead of a complete pumpkin. Such a proposed method of jackknifing has been called partial jackknifing. An alternative tuned estimator of the finite population variance and the finite population variance estimator are also introduced and studied. At the end of the chapter, a few interesting, unsolved exercises for estimating the finite population variance are given for practice and further investigation. The exercises include new model assisted tuned estimators of finite population variance, as well as a few new tuning constraints that make use of jackknifed sample geometric mean and jackknifed sample harmonic mean.

In Chapter 5, the problem of tuning the estimator of the correlation coefficient using empirical log-likelihood style techniques is considered. In practice, it is difficult to estimate the variance of the estimator of the correlation coefficient. However, the proposed newly tuned method leads to a very successful estimator of variance of the estimator of the correlation coefficient when auxiliary information is available. A numerical illustration is also provided to estimate the correlation between the weight and circumference of pumpkins. The R code, used in the simulation study of reliability of the estimators, is also provided. At the end of the chapter, several exercises involving estimation of the correlation coefficient are provided for practice and further investigation. In addition, exercises leading toward tuned estimators of the ratio of two population means, two population variances, and the population regression coefficient are included for further investigation.

Until this point, we have considered only the simple random sampling (SRS) scheme at the selection stage of the sample. In Chapter 6, the problem of estimating a population total using the concept of a multi-character survey is addressed. The newly tuned multi-character survey estimators that we introduce are studied for the probability proportional to size and with replacement (PPSWR) sampling scheme. Simulation results with the relevant R code are reported. The code is also utilized to compute numerical illustrations. At the end of the chapter, a few unsolved exercises for further investigation are provided.

In Chapter 7, we consider the tuning of the estimator of population total using probability proportional to size and without replacement (PPSWOR) sampling scheme. A new linear regression type estimator under PPSWOR sampling scheme is proposed. The proposed tuned estimator is free of second order inclusion probabilities and provides a simple and safe solution to the important problem of estimating variance. The tuned regression type estimator and an estimator obtained by optimizing a displacement function are tested through simulation studies. The R code used in the simulation study of the reliability of the estimators is also provided. The tuned estimators are also supported by two numerical illustrations that consider the problem of estimating the weights of pumpkins using their known average circumference as a tuning variable and their known top size as an auxiliary variable at the selection stage. At the end of the chapter, a few exercises concerning the estimation of population parameters under the PPSWOR sampling scheme are provided for further investigation.

In Chapter 8, we consider the problem of tuning the estimators of population mean in stratified random sampling design, using both the chi-squared type, and the dual-to-empirical log-likelihood (dell) type distance functions, when the population mean of the auxiliary variable at the stratum level is available or known. The tuned estimators are supported with a simulation that considers the problem of estimating the average weights of pumpkins. Here, the pumpkins are divided into three different strata, consisting of *Sumbo*, *Mumbo*, and *Jumbo* pumpkins, based on their known circumferences. The adjusted tuned estimator of variance of the proposed estimator is found to be very effective in the case of small samples. The R code used in the simulation study for the case of chi-squared distance and dual to log-likelihood functions is also provided. Numerical illustrations for determining the average weight of the *Sumbo*, *Mumbo*, and *Jumbo* pumpkins using proportional allocation and newly tuned stratified sampling estimation methodology, along with R code, are presented. In the exercises, we consider the situation in which the pooled mean of the auxiliary variable across all strata is known, but the means at stratum level are unknown. An idea for extending the work to multistage stratified random sampling is introduced in an exercise at the end of the chapter.

In Chapter 9, we introduce the idea of using multiauxiliary information for semituning the estimators of population mean and finite population variance under simple random sampling. The use of multiauxiliary information for fully tuning estimators is addressed in the exercises. A new set of three tuning constraints is also introduced. These constraints make use of jackknifed sample arithmetic mean, jackknifed sample geometric mean, and jackknifed sample harmonic mean of three different auxiliary variables.

In Chapter 10, a very short review of the relevant literature is given. The concept of calibration and jackknifing is discussed in layman language.

Author and selected subject indexes have been also included at the end.

Further studies

Tuning of estimators of any parameter, such as median, mode, distribution function, for any sampling design, such as small area estimation, post-stratification, adaptive clustering, panel-surveys, systematic sampling, dual or multiple frame surveys, and longitudinal surveys etc., can also be investigated, if needed.

Acknowledgments

The authors would like to thank Mr. Steven Mathews, Academic Press, for his guidance and help during the preparation of final version of the monograph. The authors would like to thank Mr. Paul A. Rios, Graphic Designer, Marketing and Communications, Texas A&M University-Kingsville for his kind help in designing the cover page and creating the pumpkin in Figure 1.1. The authors are also thankful to a student at TAMUK, Lane Christiansen, for his help in going through the entire manuscript. We duly acknowledge the use of R-software to the, "R Development Core Team (2009). R: A language and environment for statistical computing. R Foundation for Statistical Computing, Vienna, Austria. ISBN 3-900051-07-0, http://www.R-project.org." Permission from Daniel Breze (Director) and Liliana Pinkasovych (Editorial Officer) International Statistical Institute, The Netherlands, to print the picture of Professor W.G. Cochran is duly acknowledged. Support for this research for two of the authors, Maria del Mar Rueda and Antonio Arcos, was provided in part by grant MTM2009-10055 from MEC, and is also acknowledged. The authors are also thankful to Dr. Glyn Jones and the reviewers for their recommendations to the publisher. Special thanks are due to Mr. Poulouse Joseph, Project Manager, Elsevier for his great patience during the editing process.

Finally, the authors would like to thank their department colleagues, their family members, and their other friends who always inspired them to complete this monograph.

Sarjinder Singh
Stephen A. Sedory
Maria del Mar Rueda
Antonio Arcos
Raghunath Arnab

Remark: The first author would like to mention that the first draft of this monograph was written while he was seeking employment and staying at the home of one of his friends, Mr. Sher Singh Sidhu, Melbourne, Australia, for a period of about 5–6 months during 2007. Thanks are also due to Mr. Sher Singh Sidhu, his wife Mrs Jaine Sidhu, and their two children Amy Sidhu and Shane Sidhu. It would not have been possible to write such a book if the first author had no support from this great family that is from his village in India.

Problem of estimation

1.1 Introduction

In this chapter, we discuss the problem of estimation, the benefits of sampling, differences between study and auxiliary variables, and basic scientific statistical terms and notation. We discuss the creation of a statistical jumbo pumpkin model (SJPM) and jackknifing of sample mean and proportion. We also introduce the ideas of double jackknifing and jackknifing of a double suffix variable. Frequently asked question are answered. At the end, we suggest jackknifed estimators of several parameters, such as skewness, kurtosis, geometric mean, harmonic mean, and median; these are provided in the form of unsolved exercises.

1.2 Estimation problem and notation

In this section, we discuss the real-life problem of estimation and the need for survey sampling. A farmer may be interested in estimating the total yield of a crop, say pumpkins from his field. A social scientist may be interested in estimating the total number of drug users in a country. A mahout may be interested in knowing the average diet of an elephant. A student may be interested in estimating a school's average GPA per semester. An employer may be interested in estimating his employees' total income. A minister may be interested in knowing the total unemployment rate in a country. There is no end to estimation problems in everyday life.

In statistical language the term *population* is applied to any finite or infinite collection of individuals or units. It has displaced the older term *universe* and is essentially synonymous with *aggregate*. The study population, or target population, refers to the collection of units about which one wants to make some inference or extract some information. The sample population is the population of units from which samples are drawn. Ideally, the two should coincide, but this is not always possible or convenient. One may be interested in the average weight of a certain variety of farm-grown pumpkin; the collection of all such pumpkins is the target population. However, samples can only be taken from those farms to which an organization or individual conducting a survey potentially has access; the collection of these pumpkins constitutes the sample population. Although both of these populations are finite, they are very large and, for purposes of computation and simplicity of mathematical expressions, are typically treated as infinite. We will denote the population under discussion by Ω, and its size by N.

A sample is a selected subset of the population that will be used for making inference. The organization or listing of the units used to actually draw the sample is referred to as the frame. One would like this subset to be representative of the target population, and methods for increasing the likelihood of this happening are an important part of the theory of survey sampling. We will denote a sample by s and its size by n. For example, a collection of a few pumpkins that are ready to eat may form a random and representative sample from the target population of interest. A census is a special case sample. If the entire population is taken as a sample, then the sampling survey is called a *census survey*. Table 1.1 shows some of the major differences between a *sample* survey and a *census*.

The variable of interest, or the variable about which we want to draw some inference, is called a study variable. The value of the ith unit is generally denoted by y_i. For example, the weight of a ripe pumpkin may be a study variable. A variable that has a direct or indirect relationship to the study variable is called an auxiliary variable. The value of an auxiliary variable for the ith unit is generally denoted by x_i or z_i, etc. The weight of a pumpkin may depend on its circumference, its top size, the fertilizer used, and irrigation, and so on. These could be regarded as auxiliary variables. The main differences between a study variable and an auxiliary variable are listed in Table 1.2.

Any numerical quantity or value obtained from all units in a population is called a parameter. A parameter is an unknown and a fixed quantity. It is generally denoted by θ. Any function of population values is also considered to be a parameter. Mathematically, suppose that a population Ω consists of N units and that the value of the ith unit is y_i. Then any function of all y_i values is a parameter, that is,

$$\text{Parameter} = \theta = f(y_1, y_2, \ldots, y_N) \tag{1.1}$$

Table 1.1 Difference between a sample and a census

Factor	Sample	Census
Cost	Less	More
Efforts	Less	More
Time required	Less	More
Accuracy of measurements	More	Less

Table 1.2 Difference between a study and auxiliary variable

Factor	Study variable	Auxiliary variable
Cost	More	Less
Effort	More	Less
Availability	Current surveys or experiments	Current or past survey, books, or journals, etc.
Interest to an investigator	More	Less
Error in measurements	More	Less
Sources of error	More	Fewer
Notation	Y	X, Z, etc.

For example, if y_i denotes the weight of the ith pumpkin, then the average weight of all the pumpkins produced on a farm is a parameter, the population mean $(\overline{Y})$, and is given by

$$\text{Population mean} = \overline{Y} = \frac{1}{N}(y_1 + y_2 + \cdots + y_N) \tag{1.2}$$

In short, a parameter is a fixed and unknown value that applies to an entire population. We wish to estimate this parameter as precisely as possible by taking a sample from the population of interest.

Any numerical quantity or value that is obtained from a sample is called a *statistic*. A *statistic* is known from a given sample and varies from sample to sample. Suppose that a sample s consists of n units and that the value of the ith unit in the sample is denoted by y_i. Any function of sample values, y_i, may be considered as an estimator of a population parameter, that is,

$$\text{Estimator} = \hat{\theta} = f(y_1, y_2, \ldots, y_n) \tag{1.3}$$

In general, $\hat{\theta}$ in a formula form as a function of y_i values in a sample is considered as an estimator of a parameter θ. The numerical value of $\hat{\theta}$ obtained from that formula forming the estimator is called a *statistic* or an *estimate* of the parameter θ.

For example, if y_i denotes the measured weight of the ith pumpkin in a sample, then the natural estimator of the population mean, $\theta = \overline{Y}$, is the sample mean, $\hat{\theta} = \bar{y}_n$, given by

$$\text{Sample mean} = \bar{y}_n = \frac{1}{n}(y_1 + y_2 + \cdots + y_n) \tag{1.4}$$

A sample can be selected from a population in many ways. We will assume that the reader is familiar with various sample selection schemes, such as those found in Brewer and Hanif (1983), Cochran (1977), Thompson (1997), Lohr (2010), and Singh (2003).

Assume it is possible to define two statistics $\hat{\theta}_1$ and $\hat{\theta}_2$ (functions of sample values only), between which the parameter θ is expected to fall with a probability of $(1 - \alpha)$, that is,

$$P(\hat{\theta}_1 < \theta < \hat{\theta}_2) = (1 - \alpha) \tag{1.5}$$

Then, the interval $(\hat{\theta}_1, \hat{\theta}_2)$ is called a $(1-\alpha)100\%$ confidence interval estimate of θ. In other words, if we imagine such intervals being constructed from every possible sample, then the proportion of these intervals that actually contain θ is $(1-\alpha)$.

Let y_i, $i = 1, 2, \ldots, N$ denote the value of the ith unit in a population Ω. The population mean is defined as

$$\overline{Y} = \frac{1}{N}(y_1 + y_2 + \cdots + y_N) = \frac{1}{N}\sum_{i=1}^{N} y_i \qquad (1.6)$$

and the population total is given by

$$Y = (y_1 + y_2 + \cdots + y_N) = \sum_{i=1}^{N} y_i = N\overline{Y} \qquad (1.7)$$

The rth order central population moment is defined as

$$\mu_r = \frac{1}{2N(N-1)}\sum_{i \neq j=1}^{N}(y_i - y_j)^r = \frac{1}{N-1}\sum_{i=1}^{N}(y_i - \overline{Y})^r \qquad (1.8)$$

where $r = 2, 3, \ldots$ is an integer.

If $r = 2$, then μ_2 represents the second-order population moment,

$$\mu_2 = S_y^2 = \frac{1}{2N(N-1)}\sum_{i \neq j=1}^{N}(y_i - y_j)^2 = \frac{1}{N-1}\sum_{i=1}^{N}(y_i - \overline{Y})^2 \qquad (1.9)$$

which is referred to as the population mean squared error.

Note that the population variance is defined as

$$\sigma_y^2 = \frac{1}{N}\sum_{i=1}^{N}(y_i - \overline{Y})^2 = \frac{(N-1)}{N}S_y^2 \approx S_y^2 \qquad (1.10)$$

for large values of N.

Let y_i, $i = 1, 2, \ldots, n$ denote the value of the ith unit selected in the sample s. The sample mean is given by

$$\overline{y}_n = \frac{1}{n}\sum_{i=1}^{n} y_i = \frac{1}{n}\sum_{i \in s} y_i \qquad (1.11)$$

The sample variance s_y^2 is defined as

$$s_y^2 = \frac{1}{n-1}\sum_{i=1}^{n}(y_i - \overline{y})^2 = \frac{1}{2n(n-1)}\sum_{i \neq j=1}^{n}(y_i - y_j)^2 \qquad (1.12)$$

or equivalently,

$$s_y^2 = \frac{1}{n-1}\sum_{i \in s}(y_i - \overline{y}_n)^2 = \frac{1}{2n(n-1)}\sum_{i \neq j \in s}(y_i - y_j)^2 \qquad (1.13)$$

where $i \neq j \in s$ means that both the ith and jth units are included in the sample s. Note that both $(y_1 - y_2)^2$ and $(y_2 - y_1)^2$ appear in the summation.

1.3 Modeling of jumbo pumpkins

On a pumpkin farm, the size of pumpkins may vary from very small to very large when considering their weights and circumferences. Similar to pumpkins on a farm, we propose here a statistical model that is able to produce very small to very large values of an output variable, say Y, for a given value of an input variable, say X. We named such a model the SJPM. We take the output value from such a model as weight (lbs) of a pumpkin and input value as circumference (inches) of a pumpkin.

To develop the SJPM, we begin with an assumed deterministic relationship between the weight (lbs), say $M(X)$, and circumference (inches), say X, of a pumpkin, modeled as

$$M(X) = 5.5 e^{0.047X - 0.0001X^2} \tag{1.14}$$

where $M(X)$ is the weight of a pumpkin in the range of 20.59–1124.11 lbs, and X is the circumference of a pumpkin in the range of 30–190 in. (Figure 1.1).

A graphical representation of the Deterministic Jumbo Pumpkin Model (DJPM) is given in Figure 1.2.

The proposed SJPM is an extension of the DJPM as

$$y_i = M(x_i) \exp(e_i) \tag{1.15}$$

Figure 1.1 Jackknifing a Jumbo Pumpkin.

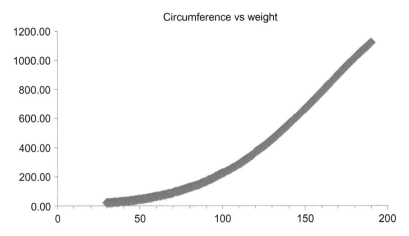

Figure 1.2 Deterministic Jumbo Pumpkin Model.

where $e_i \sim N(0, 2)$, that is, e_i is normally distributed with a mean of 0 and a standard deviation of 2.

A simulated population of $N = 10,000$ pumpkins from the SJPM is shown in Figure 1.3.

The population mean weight of the population generated earlier is $\overline{Y} = 3038.765$ lbs with a standard deviation of $\sigma_y = 20740.06$ lbs, and the population

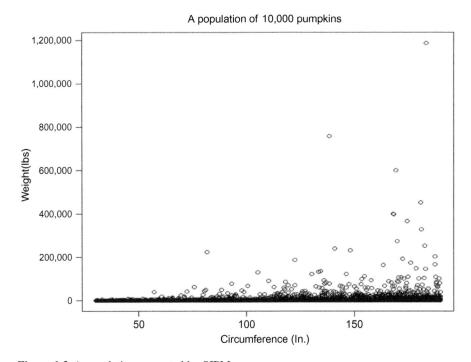

Figure 1.3 A population generated by SJPM.

Problem of estimation

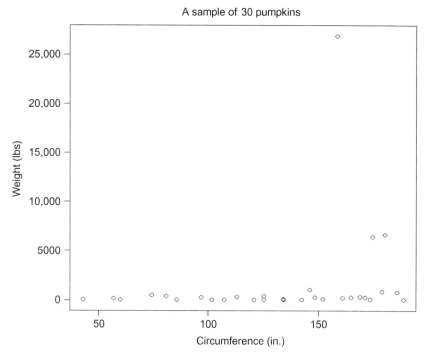

Figure 1.4 Sample representation from SJPM.

mean circumference is $\bar{X} = 109.9673$ in. with a standard deviation of $\sigma_x = 46.17808$ in.

A graphical representation of a sample s of $n = 30$ pumpkins selected with a simple random and without replacement sampling (SRSWOR) scheme from the preceding population is shown in Figure 1.4.

As computed from this particular sample s of $n = 30$ pumpkins, we have sample mean weight $\bar{y}_n = 1559.569$ lbs with a sample standard deviation of $s_y = 5056.591$ lbs, and a sample mean circumference $\bar{x}_n = 131.8182$ in. with a sample standard deviation of $s_x = 41.9988$ in.

1.3.1 R code

The following R code, **PUMPKIN1.R**, implements the proposed SJPM.

```
# R CODE USED FOR SJPM MODEL
# PROGRAM PUMPKIN1.R
set.seed(123456)
N<-10000
x<-runif(N, min=30, max=190)
m<-5.5*(exp(0.047*x - 0.0001*x*x))
z<-rnorm(N, 0, 2)
```

```
y<-m*exp(z)
mean(x)->XB; mean(y)->YB
sqrt(var(x))->Sx; sqrt(var(y))->Sy
cbind(XB,YB, Sx, Sy)
plot(x,y,main="A population of 10000 pumpkins",
xlab="Circumference (Inches)",ylab="Weight (lbs)")
#------------------------------------
n<-30
us<-sample(N,n)
xs<-x[us]; ys<-y[us]
mean(xs)->xm; mean(ys)->ym
sqrt(var(xs))->sx; sqrt(var(ys))->sy
cbind(xm,ym, sx, sy)
plot(xs,ys,main="A sample of 30 pumpkins",
xlab="Circumference (Inches)",ylab="Weight (lbs)")
```

1.4 The concept of jackknifing

The Quenouille (1956) method of bias reduction, popularly known as the jackknife procedure, has been successfully applied for estimating the variance of estimators. Tukey (1958) was the first to use the jackknife technique to estimate the variance of an estimator. Among others, Arnab and Singh (2006) have used it to estimate the variance of the ratio estimator in the presence of nonresponse. Farrell and Singh (2010) discussed the importance of jackknifing in survey sampling. Here we review the concept of jackknifing in a way that should be understandable to the interested layperson.

Suppose $\hat{\theta}_1, \hat{\theta}_2, \ldots, \hat{\theta}_n$ are independently distributed random variables with

$$E(\hat{\theta}_i) = \theta \text{ for all } i \in s \tag{1.16}$$

Suppose $\hat{\theta}$ is an estimator of a parameter θ, which is based on a sample of size n defined as

$$\hat{\theta} = \frac{1}{n}\sum_{i=1}^{n} \hat{\theta}_i \tag{1.17}$$

Then the estimator of the variance of the estimator $\hat{\theta}$ takes the form

$$\hat{v}(\hat{\theta}) = \frac{1}{n(n-1)}\sum_{i=1}^{n}(\hat{\theta}_i - \hat{\theta})^2 \tag{1.18}$$

For a given sample s of size n, let $\hat{\theta}_{(j)}$ denote the value of the estimator based on the sample of size $(n-1)$ that excludes the jth unit, that is,

Problem of estimation

$$\hat{\theta}_{(j)} = \frac{1}{n-1} \sum_{i(\neq j)=1}^{n} \hat{\theta}_i \qquad (1.19)$$

Finally, let the jackknifed estimator of θ be given by

$$\hat{\theta}_{\text{Jack}} = \frac{1}{n} \sum_{j=1}^{n} \hat{\theta}_{(j)} \qquad (1.20)$$

Then the standard jackknife estimator of variance of the estimator $\hat{\theta}$ is given by

$$\hat{v}_{\text{Jack}}(\hat{\theta}) = \frac{n-1}{n} \sum_{j=1}^{n} \left(\hat{\theta}_{(j)} - \hat{\theta}_{\text{Jack}} \right)^2 \qquad (1.21)$$

Sometimes it is more convenient to use another form of the jackknife estimator of the variance of $\hat{\theta}$, given by

$$\hat{v}_{\text{Jack}}(\hat{\theta}) = \frac{n-1}{n} \sum_{j=1}^{n} \left(\hat{\theta}_{(j)} - \hat{\theta} \right)^2 \qquad (1.22)$$

For example, assume

$$\hat{\theta} = \bar{y}_n = \frac{1}{n} \sum_{i=1}^{n} y_i \qquad (1.23)$$

is an estimator of the population mean $\bar{Y}$ under simple random and with replacement sampling (SRSWR) scheme.
Then, we have

$$\hat{\theta}(j) = \bar{y}_n(j) = \frac{1}{n-1} \sum_{i(\neq j)=1}^{n} y_i \qquad (1.24)$$

denote the estimator of the population mean $\bar{Y}$ obtained by dropping the jth unit from the sample.
Clearly, we can write

$$\begin{aligned}
\hat{\theta}(j) = \bar{y}_n(j) &= \frac{1}{n-1} \sum_{i(\neq j)=1}^{n} y_i = \frac{1}{n-1} \left[\sum_{i=1}^{n} y_i - y_j \right] = \frac{1}{n-1} [n\bar{y}_n - y_j] \\
&= \frac{1}{n-1} [n\bar{y}_n - (y_j - \bar{y}_n) - \bar{y}_n] = \frac{1}{n-1} [(n-1)\bar{y}_n - (y_j - \bar{y}_n)] \\
&= \bar{y}_n - \frac{1}{n-1} (y_j - \bar{y}_n)
\end{aligned} \qquad (1.25)$$

Also from Equation (1.20) we have

$$\hat{\theta}_{\text{Jack}} = \frac{1}{n}\sum_{j=1}^{n}\bar{y}_n(j) = \frac{1}{n}\sum_{j=1}^{n}\left[\frac{1}{n-1}\sum_{i(\neq j)=1}^{n}y_i\right]$$

$$= \frac{1}{n(n-1)}\sum_{j=1}^{n}\sum_{i(\neq j)=1}^{n}y_i = \frac{n(n-1)}{n(n-1)}\bar{y}_n = \bar{y}_n = \hat{\theta}$$
(1.26)

Therefore the jackknife estimator of variance of $\hat{\theta} = \bar{y}_n$ is given by

$$\hat{v}_{\text{Jack}}(\bar{y}_n)_{\text{srswr}} = \frac{(n-1)}{n}\sum_{j=1}^{n}(\bar{y}_n(j) - \bar{y}_n)^2$$

$$= \frac{(n-1)}{n}\sum_{j=1}^{n}\left[\left\{\bar{y}_n - \frac{1}{(n-1)}(y_j - \bar{y}_n)\right\} - \bar{y}_n\right]^2$$
(1.27)

$$= \frac{s_y^2}{n}$$

where

$$s_y^2 = \frac{1}{(n-1)}\sum_{i=1}^{n}(y_i - \bar{y}_n)^2$$
(1.28)

It is easy to verify that the expected value of $\hat{v}_{\text{Jack}}(\bar{y}_n)_{\text{srswr}}$ is

$$E\left[\hat{v}_{\text{Jack}}(\bar{y}_n)_{\text{srswr}}\right] = E\left[\frac{s_y^2}{n}\right] = \frac{\sigma_y^2}{n}$$
(1.29)

So under the SRSWR scheme, $\hat{v}_{\text{Jack}}(\bar{y}_n)_{\text{srswr}}$ is an unbiased estimator of variance of the sample mean given by

$$V(\bar{y}_n)_{\text{srswr}} = \frac{\sigma_y^2}{n}$$
(1.30)

where

$$\sigma_y^2 = \frac{1}{N}\sum_{i=1}^{N}(y_i - \bar{Y})^2$$
(1.31)

is called the finite population variance.

Problem of estimation

The jackknife technique provides a good estimate of variance under SRSWR scheme, but for other sampling schemes we need to adjust it to obtain a truly unbiased estimator of variance. For example, under the SRSWOR scheme, the variance of the sample mean $\bar{y}_n$ is given by

$$V(\bar{y}_n)_{\text{srswor}} = \frac{(1-f)}{n} S_y^2 \qquad (1.32)$$

where $f = n/N$ is called the finite population correction factor and

$$S_y^2 = \frac{1}{N-1} \sum_{i=1}^{N} (y_i - \bar{Y})^2 \qquad (1.33)$$

is called the population mean squared error.

Thus, for the SRSWOR scheme, an unbiased jackknife estimator of the variance of the sample mean, $V(\bar{y}_n)_{\text{srswor}}$, is given by

$$\hat{v}_{\text{Jack}}(\bar{y}_n)_{\text{srswor}} = \frac{(1-f)(n-1)}{n} \sum_{j=1}^{n} (\bar{y}_n(j) - \bar{y}_n)^2 \qquad (1.34)$$

Observe the difference between jackknifing a sample mean for the SRSWR scheme versus the SRSWOR scheme. The estimator of variance needs to be adjusted by a factor of $(1-f)$ for the SRSWOR scheme to obtain an unbiased estimator of variance. Note that it is not always possible to adjust the jackknife estimator of variance to make it unbiased for other sampling schemes available in the literature. A simple example is the case of a two-phase sampling scheme described by Sitter (1997). In the next section, we will investigate jackknifing the sample mean in detail.

1.5 Jackknifing the sample mean

Consider a sample s of n units is taken with the SRSWR scheme from a population. The values of the study variable y_i, $i = 1, 2, \ldots, n$, are noted, such that $E(y_i) = \bar{Y}$, $V(y_i) = \sigma_y^2$ and $\text{Cov}(y_i, y_j) = 0$ for $i \neq j$.

Let

$$\bar{y}_n = \frac{1}{n} \sum_{i=1}^{n} y_i \qquad (1.35)$$

denote the sample mean.

Obviously, the jackknife estimator of the population mean is given by

$$\bar{y}_{\text{Jack}} = \frac{1}{n} \sum_{j=1}^{n} \bar{y}_n(j) \qquad (1.36)$$

where

$$\bar{y}_n(j) = \frac{n\bar{y}_n - y_j}{n-1} \tag{1.37}$$

Taking the expected value of $\bar{y}_n(j)$ in Equation (1.37) we find

$$E\{\bar{y}_n(j)\} = E\left[\frac{n\bar{y}_n - y_j}{n-1}\right] = \left[\frac{n\bar{Y} - \bar{Y}}{n-1}\right] = \bar{Y} \tag{1.38}$$

The variance of the jth jackknife estimator, $\bar{y}_n(j)$, is given by

$$\begin{aligned}
V\{\bar{y}_n(j)\} &= V\left[\frac{n\bar{y}_n - y_j}{n-1}\right] \\
&= \frac{1}{(n-1)^2}\left[n^2 V(\bar{y}_n) + V(y_j) - 2n\mathrm{Cov}(\bar{y}_n, y_j)\right] \\
&= \frac{1}{(n-1)^2}\left[n^2 \frac{\sigma_y^2}{n} + \sigma_y^2 - 2n\frac{\sigma_y^2}{n}\right] \\
&= \frac{\sigma_y^2}{(n-1)}
\end{aligned} \tag{1.39}$$

The covariance between the two jackknife estimators $\bar{y}_n(j)$ and $\bar{y}_n(k)$ is given by

$$\begin{aligned}
\mathrm{Cov}\{\bar{y}_n(j), \bar{y}_n(k)\} &= \mathrm{Cov}\left[\frac{n\bar{y}_n - y_j}{(n-1)}, \frac{n\bar{y}_n - y_k}{(n-1)}\right] \\
&= \frac{n^2}{(n-1)^2}\mathrm{Cov}(\bar{y}_n, \bar{y}_n) - \frac{n}{(n-1)^2}\mathrm{Cov}(\bar{y}_n, y_k) \\
&\quad - \frac{n}{(n-1)^2}\mathrm{Cov}(\bar{y}_n, y_j) + \frac{1}{(n-1)^2}\mathrm{Cov}(y_j, y_k) \\
&= \frac{n^2}{(n-1)^2}\left(\frac{\sigma_y^2}{n}\right) - \frac{n}{(n-1)^2}\left(\frac{\sigma_y^2}{n}\right) \\
&\quad - \frac{n}{(n-1)^2}\left(\frac{\sigma_y^2}{n}\right) + \frac{1}{(n-1)^2}(0) \\
&= \left[\frac{n}{(n-1)^2} - \frac{1}{(n-1)^2} - \frac{1}{(n-1)^2}\right]\sigma_y^2 \\
&= \left[\frac{n-2}{(n-1)^2}\right]\sigma_y^2
\end{aligned} \tag{1.40}$$

Problem of estimation

The variance of the jackknife estimator $\bar{y}_{\text{Jack}}$ is

$$V(\bar{y}_{\text{Jack}}) = V\left[\frac{1}{n}\sum_{j=1}^{n}\bar{y}_n(j)\right]$$

$$= \frac{1}{n^2}\left[\sum_{j=1}^{n}V\{\bar{y}_n(j)\} + \sum_{j(\neq k)=1}^{n}\text{Cov}\{\bar{y}_n(j), \bar{y}_n(k)\}\right]$$

$$= \frac{1}{n^2}\left[\sum_{j=1}^{n}\frac{\sigma_y^2}{(n-1)} + \sum_{j\neq k=1}^{n}\frac{(n-2)}{(n-1)^2}\sigma_y^2\right] \quad (1.41)$$

$$= \frac{1}{n^2}\left[\sum_{j=1}^{n}\frac{1}{(n-1)} + \sum_{j\neq k=1}^{n}\frac{(n-2)}{(n-1)^2}\right]\sigma_y^2$$

$$= \frac{1}{n^2}\left[\frac{n}{(n-1)} + \frac{n(n-1)(n-2)}{(n-1)^2}\right]\sigma_y^2 = \frac{\sigma_y^2}{n}$$

Thus, the jackknife estimator $\bar{y}_{\text{Jack}}$ remains unbiased with variance

$$V(\bar{y}_{\text{Jack}}) = \frac{\sigma_y^2}{n} \quad (1.42)$$

Alternatively, the average jackknife estimator is given by

$$\bar{y}_{\text{Jack}} = \frac{1}{n}\sum_{j=1}^{n}\bar{y}_n(j) = \frac{1}{n}\sum_{j=1}^{n}\left[\frac{n\bar{y}_n - y_j}{n-1}\right] = \bar{y}_n \quad (1.43)$$

It is clear that

$$V(\bar{y}_{\text{Jack}}) = V(\bar{y}_n) = \frac{\sigma_y^2}{n} \quad (1.44)$$

For the SRSWR scheme, the jackknife estimator of the variance of the sample mean $\bar{y}_n$ is given by

$$\hat{v}(\bar{y}_{\text{Jack}}) = \frac{(n-1)}{n}\sum_{j=1}^{n}\{\bar{y}_n(j) - \bar{y}_{\text{Jack}}\}^2 \quad (1.45)$$

1.6 Doubly jackknifed sample mean

Note that

$$\bar{y}_{\text{Jack}} = \frac{1}{n}\sum_{j=1}^{n}\bar{y}_n(j)$$

$$= \frac{1}{n}\sum_{j=1}^{n}\left\{\frac{n\bar{y}_n - y_j}{n-1}\right\} \quad (1.46)$$

$$= \frac{1}{n}\left\{\frac{n^2\bar{y}_n - n\bar{y}_n}{n-1}\right\}$$

$$= \bar{y}_n$$

Also note that

$$\hat{v}(\bar{y}_{\text{Jack}}) = \frac{(n-1)}{n}\sum_{j=1}^{n}\{\bar{y}_n(j) - \bar{y}_{\text{Jack}}\}^2$$

$$= \frac{(n-1)}{n}\sum_{j=1}^{n}\left\{\frac{n\bar{y}_n - y_j}{n-1} - \bar{y}_n\right\}^2$$

$$= \frac{(n-1)}{n}\sum_{j=1}^{n}\left\{\frac{(n\bar{y}_n - y_j) - (n-1)\bar{y}_n}{n-1}\right\}^2 \quad (1.47)$$

$$= \frac{(n-1)}{n}\sum_{j=1}^{n}\left\{\frac{y_j - \bar{y}_n}{n-1}\right\}^2$$

$$= \frac{1}{n(n-1)}\sum_{j=1}^{n}(y_j - \bar{y}_n)^2$$

$$= \frac{s_y^2}{n}$$

Let

$$\bar{y}_{\text{Jack}}^{(j)} = \frac{n\bar{y}_{\text{Jack}} - \bar{y}_n(j)}{n-1} = \frac{\sum_{j \in s}\bar{y}_n(j) - \bar{y}_n(j)}{n-1} \quad (1.48)$$

Then we define the doubly jackknifed estimator of the population mean as

$$\bar{\bar{y}}_{\text{Jack}} = \frac{1}{n}\sum_{j \in s}\bar{y}_{\text{Jack}}^{(j)} \quad (1.49)$$

Problem of estimation

Now $\bar{\bar{y}}_{\text{Jack}}$ in Equation (1.49) can be written as

$$
\begin{aligned}
\bar{\bar{y}}_{\text{Jack}} &= \frac{1}{n}\sum_{j\in s}\bar{y}_{\text{Jack}}^{(j)} \\
&= \frac{1}{n}\sum_{j\in s}\left[\frac{\sum_{j\in s}\bar{y}_n(j) - \bar{y}_n(j)}{n-1}\right] \\
&= \frac{1}{n}\left[\frac{n\sum_{j\in s}\bar{y}_n(j) - \sum_{j\in s}\bar{y}_n(j)}{n-1}\right] \\
&= \frac{1}{n}\sum_{j\in s}\bar{y}_n(j) \\
&= \bar{y}_n
\end{aligned}
\qquad(1.50)
$$

We also define an estimator of the variance of the doubly jackknifed estimator as

$$
\hat{v}(\bar{\bar{y}}_{\text{Jack}}) = \frac{(n-1)^3}{n}\sum_{j\in s}\left\{\bar{y}_{\text{Jack}}^{(j)} - \bar{\bar{y}}_{\text{Jack}}\right\}^2 \qquad(1.51)
$$

Now Equation (1.51) can be written as

$$
\begin{aligned}
\hat{v}(\bar{\bar{y}}_{\text{Jack}}) &= \frac{(n-1)^3}{n}\sum_{j\in s}\left\{\bar{y}_{\text{Jack}}^{(j)} - \bar{\bar{y}}_{\text{Jack}}\right\}^2 \\
&= \frac{(n-1)^3}{n}\sum_{j\in s}\left\{\frac{\sum_{j\in s}\bar{y}_n(j) - \bar{y}_n(j)}{n-1} - \bar{y}_n\right\}^2 \\
&= \frac{(n-1)^3}{n}\sum_{j\in s}\left\{\frac{\sum_{j\in s}\left(\frac{n\bar{y}_n - y_j}{n-1}\right) - \left(\frac{n\bar{y}_n - y_j}{n-1}\right)}{n-1} - \bar{y}_n\right\}^2 \\
&= \frac{(n-1)^3}{n}\sum_{j\in s}\left\{\frac{n^2\bar{y}_n - 2n\bar{y}_n + y_j}{(n-1)^2} - \bar{y}_n\right\}^2 \\
&= \frac{s_y^2}{n}
\end{aligned}
\qquad(1.52)
$$

It shows if we rejackknife (or doubly jackknife) the sample mean, then there is no gain or loss so long as the estimation of variance is concerned.

1.7 Jackknifing a sample proportion

Consider a population Ω consisting of N units. Out of them N_A are the total number of units belonging to a group, say A, of the population. Obviously $P = N_A/N$ will be the proportion of units belonging to the group A in the entire population. Assume we selected a sample s of n units by the SRSWR scheme from the population Ω. Let y_i be an indicator variable in the sample such that $y_i = 1$ if the ith unit in the sample belongs to group A, and $y_i = 0$ otherwise. Then an unbiased estimator of the population proportion, P, is defined as

$$\hat{p} = \frac{1}{n}\sum_{i=1}^{n} y_i \tag{1.53}$$

The jth jackknifed sample proportion is given by

$$\hat{p}(j) = \frac{n\hat{p} - y_j}{n-1}, \quad j = 1, 2, \ldots, n \tag{1.54}$$

Obviously, an unbiased and jackknifed estimator of the population proportion, P, is given by

$$\hat{p}_{\text{Jack}} = \frac{1}{n}\sum_{j=1}^{n} \hat{p}(j) \tag{1.55}$$

The jackknife estimator of the variance of the estimator of sample proportion is given by

$$\hat{v}_{\text{Jack}}(\hat{p}) = \frac{(n-1)}{n}\sum_{j=1}^{n} \{\hat{p}(j) - \hat{p}\}^2 \tag{1.56}$$

Note that in this case $y_i^2 = 1$ if ith unit belongs to the group A, and $y_i^2 = 0$ otherwise. The estimator (1.56) can be written as

$$\begin{aligned}
\hat{v}_{\text{Jack}}(\hat{p}) &= \frac{(n-1)}{n}\sum_{j=1}^{n} \{\hat{p}(j) - \hat{p}\}^2 \\
&= \frac{(n-1)}{n}\sum_{j=1}^{n}\left[\frac{n\hat{p} - y_j}{n-1} - \hat{p}\right]^2 = \frac{(n-1)}{n}\sum_{j=1}^{n}\left[\frac{n\hat{p} - y_j - (n-1)\hat{p}}{n-1}\right]^2 \\
&= \frac{(n-1)}{n}\sum_{j=1}^{n}\left[\frac{n\hat{p} - y_j - n\hat{p} + \hat{p}}{n-1}\right]^2 = \frac{(n-1)}{n}\sum_{j=1}^{n}\left[\frac{-y_j + \hat{p}}{n-1}\right]^2 \\
&= \frac{1}{n(n-1)}\sum_{j=1}^{n}[-y_j + \hat{p}]^2 = \frac{1}{n(n-1)}\sum_{j=1}^{n}\left[y_j^2 + \hat{p}^2 - 2\hat{p}y_j\right] \\
&= \frac{1}{n(n-1)}\left[\sum_{j=1}^{n} y_j^2 + n\hat{p}^2 - 2\hat{p}\sum_{i=1}^{n} y_i\right] \\
&= \frac{1}{n(n-1)}\left[n\hat{p} + n\hat{p}^2 - 2n\hat{p}^2\right] = \frac{1}{(n-1)}\left[\hat{p} - \hat{p}^2\right] \\
&= \frac{\hat{p}(1-\hat{p})}{n-1}
\end{aligned} \tag{1.57}$$

Obviously,

$$E(\hat{v}_{\text{Jack}}(\hat{p})) = \frac{P(1-P)}{n} \quad (1.58)$$

Thus, the jackknife estimator of variance, $\hat{v}_{\text{Jack}}(\hat{p})$, is an unbiased estimator of the variance of the sample proportion for the SRSWR scheme. A recent contribution in estimating population proportions by a method of calibration can be had from Martinez, Arcos, Martinez, and Singh (2015).

1.8 Jackknifing of a double suffix variable sum

Let V_{ij}, $i \neq j = 1, 2, \ldots, n$, be a random variable, and let

$$V = \sum_{i \neq j} \sum_{\in s} V_{ij} \quad (1.59)$$

Note that

$$\sum_{i \neq j} \sum_{\in s} \left[\frac{V - V_{ij}}{n(n-1) - 1} \right] = \frac{1}{n(n-1) - 1} [n(n-1)V - V] = V \quad (1.60)$$

These results will remain useful in this monograph.

1.9 Frequently asked questions

Q1: Why did the authors choose this particular artificial Statistical Jumbo Pumpkin Model (SJPM)?
A1: It is the authors' choice!

Q2: Why did the authors not use real data?
A2: Real data has a lot of issues!

Q3: Is this SJPM unique?
A3: No, it was chosen by the authors. One can change the parameters, such as the population size and standard deviation, to any other suitable values.

Q4: Why R?
A4: R is free; anybody can use it.

Q5: Can people ask the authors questions?
A5: Sure!

1.10 Exercises

Exercise 1.1 Discuss the concept of jackknifing a sample. How can it be used to estimate the variance of the sample mean?

Exercise 1.2 Run the program PUMPKIN1.R to generate a sample of 50 pumpkins from the artificial SJPM. Plot weight versus circumference of the pumpkins, and comment on the shape of the graph.

Exercise 1.3 Imagine a situation where you can apply an SRSWOR scheme to estimate the population mean of a study variable. Create such a population using a linear or nonlinear model with (or without) the use of an auxiliary variable. From the population you generated, select a sample of 30 units using an SRSWOR scheme, and apply the concept of jackknifing to estimate the variance of the sample mean of the study variable. Construct the 95% confidence interval estimate of the population mean, and interpret your findings.

Exercise 1.4 (a) Let

$$\hat{\mu}_3 = \frac{1}{(n-1)} \sum_{i=1}^{n} (y_i - \bar{y}_n)^3 \qquad (1.61)$$

be an estimator of the third moment about the population mean $\bar{Y}$ defined as

$$\mu_3 = \frac{1}{N} \sum_{i=1}^{N} (y_i - \bar{Y})^3 \qquad (1.62)$$

Consider

$$\hat{\mu}_{3(\text{Jack})} = c \sum_{j=1}^{n} (\bar{y}_n(j) - \bar{y}_n)^3 \qquad (1.63)$$

where $\bar{y}_n(j) = \dfrac{n\bar{y}_n - y_j}{n-1}$ has its usual meaning. Determine c such that $\hat{\mu}_{3(\text{Jack})}$ can be used as an estimator of the third moment μ_3.

(b) Generate a sample of 50 pumpkins using the program PUMPKIN1.R from the artificial SJPM. Calculate the value of

$$\hat{\mu}_3 = \frac{1}{(n-1)} \sum_{i=1}^{n} (y_i - \bar{y}_n)^3 \qquad (1.64)$$

and

$$\hat{\mu}_{3(\text{Jack})} = c \sum_{j=1}^{n} (\bar{y}_n(j) - \bar{y}_n)^3 \qquad (1.65)$$

for the value of c you determined in (a). Comment on your findings.

Exercise 1.5 (a) Let

$$\hat{\mu}_r = \frac{1}{(n-1)} \sum_{i=1}^{n} (y_i - \bar{y}_n)^r \tag{1.66}$$

where $r \geq 2$, be an estimator of the rth central moment about the population mean $\bar{Y}$ defined as

$$\mu_r = \frac{1}{N} \sum_{i=1}^{N} (y_i - \bar{Y})^r \tag{1.67}$$

Let

$$\hat{\beta}_1 = \frac{\hat{\mu}_3}{\hat{\mu}_2^{3/2}} \tag{1.68}$$

be an estimator of the population coefficient of skewness defined as

$$\beta_1 = \frac{\mu_3}{\mu_2^{3/2}} \tag{1.69}$$

Let

$$\hat{\beta}_{1(\text{Jack})} = c \sum_{j=1}^{n} (\bar{y}_n(j) - \bar{y}_n)^3 \left[\sum_{j=1}^{n} (\bar{y}_n(j) - \bar{y}_n)^2 \right]^{-(3/2)} \tag{1.70}$$

Determine the constant c such that $\hat{\beta}_{1(\text{Jack})}$ can be considered as a jackknife estimator of β_1.

(b) Generate a sample of 50 pumpkins using the program PUMPKIN1.R from the artificial SJPM. Calculate the value of $\hat{\beta}_1$ and $\hat{\beta}_{1(\text{Jack})}$ for the value of c you determined in (a). Determine a numerical value of c such that $\hat{\beta}_{1(\text{Jack})} = \hat{\beta}_1$, and comment on your findings.

Exercise 1.6 (a) Let

$$\hat{\mu}_r = \frac{1}{(n-1)} \sum_{i=1}^{n} (y_i - \bar{y}_n)^r \tag{1.71}$$

where $r \geq 2$, be an estimator of the rth central moment about the population mean $\bar{Y}$ defined as

$$\mu_r = \frac{1}{N} \sum_{i=1}^{N} (y_i - \bar{Y})^r \tag{1.72}$$

Let

$$\hat{c}_y = \sqrt{\hat{\mu}_2}/\bar{y}_n \qquad (1.73)$$

be an estimator of the population coefficient of variation defined as

$$c_y = \sqrt{\mu_2}/\bar{Y} \qquad (1.74)$$

Let

$$\hat{c}_{y(\text{Jack})} = k\sqrt{\sum_{j=1}^{n}(\bar{y}_n(j)-\bar{y}_n)^2 \Big/ \sum_{j=1}^{n}\bar{y}_n(j)} \qquad (1.75)$$

Determine the constant k such that $\hat{c}_{y(\text{Jack})}$ can be considered as a jackknife estimator of c_y.

(b) Take a sample of 40 pumpkins using the program PUMPKIN1.R from the artificial SJPM. Calculate the value of $\hat{c}_y$ and $\hat{c}_{y(\text{Jack})}$ for the value of k you determined in (a). Compute a value of k such that $\hat{c}_y = \hat{c}_{y(\text{Jack})}$.

Exercise 1.7 (a) Let

$$\hat{\mu}_r = \frac{1}{(n-1)}\sum_{i=1}^{n}(y_i - \bar{y}_n)^r \qquad (1.76)$$

where $r \geq 2$, be an estimator of the rth central moment about the population mean $\bar{Y}$ defined as

$$\mu_r = \frac{1}{N}\sum_{i=1}^{N}(y_i - \bar{Y})^r \qquad (1.77)$$

Let

$$\hat{\beta}_2 = \hat{\mu}_4/\hat{\mu}_2^2 \qquad (1.78)$$

be an estimator of the population coefficient of kurtosis defined as

$$\beta_2 = \mu_4/\mu_2^2 \qquad (1.79)$$

Let

$$\hat{\beta}_{2(\text{Jack})} = c\sum_{j=1}^{n}(\bar{y}_n(j)-\bar{y}_n)^4 \left[\sum_{j=1}^{n}(\bar{y}_n(j)-\bar{y}_n)^2\right]^{-2} \qquad (1.80)$$

Determine the constant c such that $\hat{\beta}_{2(\text{Jack})}$ can be considered as a jackknife estimator of β_2.

(b) Take a sample of 20 pumpkins using the program PUMPKIN1.R from the SJPM. Calculate $\hat{\beta}_2$ and $\hat{\beta}_{2(\text{Jack})}$ for the value of c found in (a). Determine c such that $\hat{\beta}_{2(\text{Jack})} = \hat{\beta}_2$.

(c) Extend the preceding suggested method of jackknifing to estimate the value of kurtosis to the concept of two-phase sampling. Compare your method with the estimator of kurtosis suggested by Gamrot (2012).

Exercise 1.8 (a) For a sample s of n observations taken by the SRSWR scheme, consider the average uncorrected total sum of squares (TSS) given by

$$\text{TSS} = \frac{1}{n}\sum_{i=1}^{n} y_i^2 \tag{1.81}$$

After dropping the jth unit from the sample, the jackknifed average of the uncorrected TSS is defined as

$$\text{TSS}(j) = \frac{n(\text{TSS}) - y_j^2}{n-1}, \quad \text{for } j = 1, 2, \ldots, n \tag{1.82}$$

Expand the following squared term in terms of the central moments about the sample mean:

$$(n-1)\sum_{j=1}^{n}\left[\text{TSS}(j) - \frac{1}{n}\sum_{j=1}^{n}\text{TSS}(j)\right]^2 \tag{1.83}$$

Compare your findings with

$$\hat{\mu}_4 + \frac{(n-1)}{n}\hat{\mu}_2^2 + 4\bar{y}_n^2\hat{\mu}_2 + 4\bar{y}_n\hat{\mu}_3 - \frac{2(n-1)}{n}\hat{\mu}_2^3 \tag{1.84}$$

where

$$\hat{\mu}_r = \frac{1}{(n-1)}\sum_{i=1}^{n}(y_i - \bar{y}_n)^r \tag{1.85}$$

with r being any nonnegative integer, and the sample mean being $\bar{y}_n = n^{-1}\sum_{i=1}^{n} y_i$.

(b) Generate a sample of 30 pumpkins, using the program PUMPKIN1.R from the artificial SJPM. Calculate the numerical values of

$$(n-1)\sum_{j=1}^{n}\left[\text{TSS}(j) - \frac{1}{n}\sum_{j=1}^{n}\text{TSS}(j)\right]^2 \tag{1.86}$$

and

$$\hat{\mu}_4 + \frac{(n-1)}{n}\hat{\mu}_2^4 + 4\bar{y}_n^2\hat{\mu}_2 + 4\bar{y}_n\hat{\mu}_3 - \frac{2(n-1)}{n}\hat{\mu}_2^3 \qquad (1.87)$$

Are they equal? Justify your answer.

Exercise 1.9 Let

$$\hat{\mu}_r = \frac{1}{(n-1)}\sum_{i=1}^{n}(y_i - \bar{y}_n)^r \qquad (1.88)$$

where $r \geq 2$ is an integer, be an estimator of the rth central moment about the population mean $\bar{Y}$, given by

$$\mu_r = \frac{1}{N}\sum_{i=1}^{N}(y_i - \bar{Y})^r \qquad (1.89)$$

Consider

$$\hat{\mu}_{r(\text{Jack})} = c\sum_{j=1}^{n}(\bar{y}_n(j) - \bar{y}_n)^r \qquad (1.90)$$

where $\bar{y}_n(j) = \frac{n\bar{y}_n - y_j}{n-1}$ has its usual meaning. Determine c such that $\hat{\mu}_{r(\text{Jack})}$ can be used as a jackknife estimator of the rth central moment μ_r.

Hint: Finucan, Galbraith, and Stone (1974).

Exercise 1.10 Consider a farmer growing organic pumpkins and chemically treated pumpkins. A buyer took a random sample of n organic pumpkin and another random sample of m treated pumpkins, both using SRSWR schemes. Let $\bar{y}_n$ and $\bar{y}_m$ be the sample mean weights of the first and second samples, respectively.

Let

$$\bar{y}_{\text{pooled}} = \frac{n\bar{y}_n + m\bar{y}_m}{n+m} \qquad (1.91)$$

be the pooled estimator of the pooled population mean weight, $\bar{Y}$, of both types of pumpkins on the farm.

Let

$$\bar{y}_{\text{pooled}}(j,k) = \frac{(n-1)\bar{y}_n(j) + (m-1)\bar{y}_m(k)}{n+m-2} \qquad (1.92)$$

be a pooled jackknifed estimator of the population mean after dropping the jth organic pumpkin and the kth treated pumpkin from the pooled sample of $(m+n)$ pumpkins, for $j = 1, 2, \ldots, n$ and $k = 1, 2, \ldots, m$. If possible, develop a jackknife estimator of variance of the pooled estimator of the pooled population mean. If not, discuss the limitations of jackknifing in Equation (1.92). Support your findings by simulation.

Exercise 1.11 Geometric mean

Consider a pumpkin farmer who is facing a huge infestation of insects on his farm. The growth of the number of insects seems to be following an exponential distribution. Let $y_1, y_2, \ldots, y_n$ be the number of insects observed by the farmer on his n random visits to his farm. An estimator of the geometric mean of the number of insects on the farm is given by

$$\hat{G} = \left(\prod_{i=1}^{n} y_i \right)^{1/n} \tag{1.93}$$

Let

$$\hat{G}(j) = \left(\prod_{i(\neq j)=1}^{n} y_i \right)^{1/(n-1)}, \quad j = 1, 2, \ldots, n \tag{1.94}$$

denote the jackknife estimator of the population geometric mean G after the jth unit is dropped from the sample. Then the average jackknife geometric mean estimator is given by

$$\hat{G}_{\text{Jack}} = \frac{1}{n} \sum_{j=1}^{n} \hat{G}(j) \tag{1.95}$$

Assume an estimator of the variance of the jackknifed sample geometric mean is given by

$$\hat{v}(\hat{G}_{\text{Jack}}) = \sum_{j=1}^{n} c_j \left(\hat{G}(j) - \hat{G}_{\text{Jack}} \right)^2 \tag{1.96}$$

Determine, if possible, the values of weights c_j such that $\hat{v}(\hat{G}_{\text{Jack}})$ can be considered as an estimator of the variance of the geometric mean estimator. Support your views with a simulation study.

Exercise 1.12 Harmonic mean

A pumpkin farmer has a pumpkin pie factory and several vehicles that deliver his product to several destinations within the United States. Let $y_1, y_2, \ldots, y_n$ be the delivery times (in minutes) required by a random sample of n workers using those vehicles. An estimator of the harmonic mean delivery time is given by

$$\hat{H} = n \left(\sum_{i=1}^{n} y_i^{-1} \right)^{-1} \tag{1.97}$$

Let

$$\hat{H}(j) = (n-1)\left(\sum_{i(\neq j)\in s}^{n} y_i^{-1}\right)^{-1}, \quad j=1,2,\ldots,n \qquad (1.98)$$

denote the jackknife estimator of the population harmonic mean H where the jth unit is dropped from the sample. Then the average jackknife estimator of the harmonic mean is given by

$$\hat{H}_{\text{Jack}} = \frac{1}{n}\sum_{j=1}^{n}\hat{H}(j) \qquad (1.99)$$

Assume an estimator of the variance of the averaged jackknifed sample harmonic mean is given by

$$\hat{v}(\hat{H}_{\text{Jack}}) = \sum_{j=1}^{n} c_j \left(\hat{H}(j) - \hat{H}_{\text{Jack}}\right)^2 \qquad (1.100)$$

Determine, if possible, the values of weights c_j such that $\hat{v}(\hat{H}_{\text{Jack}})$ can be considered as an estimator of the variance of the harmonic mean estimator. Support your findings by a simulation study.

Exercise 1.13 (a) Suppose that a pumpkin farmer is interested in selling his pumpkins by weight. Let $y_1, y_2, \ldots, y_n$ be the prices in dollars for the n pumpkins, each pumpkin randomly selected by a buyer. Let $y^{(1)} \leq y^{(2)} \leq \cdots \leq y^{(n)}$ denote the n prices in ascending order. An estimator of the median price of pumpkins is given by

$$\hat{M}_y = \begin{cases} \text{Value of } y \text{ at the } \frac{n+1}{2}\text{th position if } n \text{ is odd} \\ \text{Average of } y \text{ values at } \frac{n}{2}\text{th and } \left(\frac{n}{2}+1\right)\text{th positions if } n \text{ is even} \end{cases} \qquad (1.101)$$

Let

$$\hat{M}_y(j) = \begin{cases} \text{Value of } y \text{ at the } \frac{n}{2}\text{th position if } (n-1) \text{ is odd} \\ \text{Average of } y \text{ values at } \frac{n-1}{2}\text{th} \\ \text{and } \left(\frac{n+1}{2}\right)\text{th positions if } (n-1) \text{ is even} \end{cases}, \quad j=1,2,\ldots,n$$

$$(1.102)$$

denote the jackknife estimator of the population median M_y after dropping the jth unit from the sample. Then, the average jackknife estimator of the median is given by

$$\hat{M}_{y(\text{Jack})} = \frac{1}{n} \sum_{j=1}^{n} \hat{M}_y(j) \tag{1.103}$$

Assume an estimator of the variance of the jackknifed sample median is given by

$$\hat{v}(\hat{M}_{y(\text{Jack})}) = \sum_{j=1}^{n} c_j (\hat{M}_y(j) - \hat{M}_{y(\text{Jack})})^2 \tag{1.104}$$

Determine, if possible, the values of weights c_j such that $\hat{v}(\hat{M}_{y(\text{Jack})})$ can be considered as an estimator of the variance of the sample median. Support your findings with a simulation study.

(b) Extend the results to estimate the first and third quartiles. Suggest a jackknifed estimator of Bowley's coefficient of skewness. Compare your estimator with the estimator by Singh, Solanki, and Singh (2015) through a simulation study.

Exercise 1.14 Consider a farmer who is growing organic pumpkins and chemically treated pumpkins. Let $\bar{Y}_1$ be the population mean weight (lbs) of all N_1 organic pumpkins and $\bar{Y}_2$ be the population mean weight (lbs) of all N_2 treated pumpkins. Then the pooled population mean weight of both types of pumpkins is given by

$$\bar{Y} = \frac{N_1 \bar{Y}_1 + N_2 \bar{Y}_2}{N_1 + N_2} \tag{1.105}$$

Let $\sigma_{y_1}^2 = N_1^{-1} \sum_{i=1}^{N_1} (y_{1i} - \bar{Y}_1)^2$ and $\sigma_{y_2}^2 = N_2^{-1} \sum_{i=1}^{N_2} (y_{2i} - \bar{Y}_2)^2$ be the population variances of the organic and treated pumpkins, respectively. The pooled population variance of the weight of both types of pumpkins is given by

$$\sigma_y^2 = \frac{1}{N_1 + N_2} \left[N_1 \{ \sigma_{y_1}^2 + (\bar{Y}_1 - \bar{Y})^2 \} + N_2 \{ \sigma_{y_2}^2 + (\bar{Y}_2 - \bar{Y})^2 \} \right] \tag{1.106}$$

The farmer selected two independent samples of sizes n_1 and n_2 of the organic and treated pumpkins with the SRSWR scheme. Let $\bar{y}_1$ be the sample mean weight (lbs) of the n_1 organic pumpkins and $\bar{y}_2$ be the sample mean weight (lbs) of the n_2 treated pumpkins. Then the pooled sample mean weight of both types of pumpkins is given by

$$\bar{y} = \frac{n_1 \bar{y}_1 + n_2 \bar{y}_2}{n_1 + n_2} \tag{1.107}$$

Let $s_{y_1}^2 = (n_1 - 1)^{-1} \sum_{i=1}^{n_1} (y_{1i} - \bar{y}_1)^2$ and $s_{y_2}^2 = (n_2 - 1)^{-1} \sum_{i=1}^{n_2} (y_{2i} - \bar{y}_2)^2$ be the sample variances of weight of the organic and treated pumpkins, respectively. Using this sample information, suggest a jackknifed estimator of the pooled population variance. Support your estimator through a simulation study or otherwise.

Exercise 1.15 Expand the following formula:

$$\frac{(n-1)}{2n} \sum_{i \neq j} \sum_{\in s} [\bar{y}_n(i) - \bar{y}_n(j)]^2 \tag{1.108}$$

where $\bar{y}_n(i)$ and $\bar{y}_n(j)$ are, respectively, the jackknifed sample means after removing the ith unit and jth unit.

Tuning of jackknife estimator 2

2.1 Introduction

In this chapter, we introduce a new methodology for tuning the jackknife technique in survey sampling that helps to estimate the population mean/total and that also estimates the variance of the resultant estimator. This new methodology is supported with a model used to estimate the average weight of pumpkins with the help of known circumference as an auxiliary variable. The possibility of tuning in cases of nonresponse, sensitive variable, geometric mean, and harmonic mean are also discussed at the end of the chapter, in the form of exercises.

2.2 Notation

Let y_i and x_i, $i = 1, 2, ..., N$ be the values of the study variable and auxiliary variable, respectively, of the ith unit in the population Ω. Here we consider the problem of estimating the population mean

$$\bar{Y} = N^{-1} \sum_{i=1}^{N} y_i \tag{2.1}$$

by assuming that the population mean

$$\bar{X} = N^{-1} \sum_{i=1}^{N} x_i \tag{2.2}$$

of the auxiliary variable is known.

Let (y_i, x_i), $i = 1, 2, ..., n$ be the values of the study variable and auxiliary variable of the ith unit in the sample s drawn using a simple random sampling (SRS) scheme.

Let

$$\bar{y}_n = n^{-1} \sum_{i \in s} y_i \tag{2.3}$$

and

$$\bar{x}_n = n^{-1} \sum_{i \in s} x_i \tag{2.4}$$

be the sample means for the study variable and the auxiliary variable, respectively.

2.3 Tuning with a chi-square type distance function

The newly tuned jackknife estimator of the population mean $\bar{Y}$ is defined as

$$\bar{y}_{\text{Tuned(cs)}} = \sum_{j \in s} \left[\left\{ (n-1)^2 \bar{w}_n(j) - (n-2) \right\} \bar{y}_n(j) \right] \qquad (2.5)$$

where

$$\bar{y}_n(j) = \frac{n\bar{y}_n - y_j}{n-1} \qquad (2.6)$$

is the jackknifed sample mean of the study variable and is obtained by removing the jth unit from the sample s, and

$$\bar{w}_n(j) = \frac{1 - w_j}{n - 1} \qquad (2.7)$$

is the tuned jackknifed weight of the calibrated weights w_j such that

$$\sum_{j \in s} w_j = 1 \qquad (2.8)$$

and

$$\sum_{j \in s} w_j x_j = \bar{X} \qquad (2.9)$$

Note that the calibration constraint (2.9) is due to Deville and Särndal (1992), and the constraint (2.8) is due to Owen (2001). Now, a set of newly tuned jackknife weights $\bar{w}_n(j)$ should satisfy the following two tuning constraints:

$$\sum_{j \in s} \bar{w}_n(j) = 1 \qquad (2.10)$$

and

$$\sum_{j \in s} \bar{w}_n(j) \bar{x}_n(j) = \frac{\bar{X} - n(2-n)\bar{x}_n}{(n-1)^2} \qquad (2.11)$$

where

$$\bar{x}_n(j) = \frac{n\bar{x}_n - x_j}{n-1} \qquad (2.12)$$

is the sample mean of the auxiliary variable obtained by removing the jth unit from the sample s.

Note the following results:

$$w_j = n\bar{w}_n - (n-1)\bar{w}_n(j) \tag{2.13}$$

$$x_j = n\bar{x}_n - (n-1)\bar{x}_n(j) \tag{2.14}$$

and

$$\bar{w}_n = \frac{1}{n}\sum_{j\in s} w_j = \frac{1}{n} \tag{2.15}$$

We consider tuning the calibrated weights $\bar{w}_n(j)$ by minimizing the following modified chi-square type distance function defined as

$$(2^{-1}n)\sum_{j\in s} q_j^{-1}\left(1 - (n-1)\bar{w}_n(j) - n^{-1}\right)^2 \tag{2.16}$$

where q_j are arbitrarily chosen weights, subject to the tuning constraints (2.10) and (2.11). Note that this is similar to the chi-square type distance function due to Deville and Särndal (1992), which, in particular, takes the form

$$(2^{-1}n)\sum_{j\in s} q_j^{-1}\left(w_j - n^{-1}\right)^2 \tag{2.17}$$

Obviously the Lagrange function is given by

$$L_1 = (2^{-1}n)\sum_{j\in s} q_j^{-1}\left(1 - (n-1)\bar{w}_n(j) - n^{-1}\right)^2$$
$$-\lambda_0\left\{\sum_{j\in s}\bar{w}_n(j) - 1\right\} - \lambda_1\left\{\sum_{j\in s} w_n(j)x_n(j) - (n-1)^{-2}(\bar{X} - n(2-n)\bar{x}_n)\right\} \tag{2.18}$$

where λ_0 and λ_1 are the Lagrange multiplier constants.
On setting

$$\frac{\partial L_1}{\partial \bar{w}_n(j)} = 0 \tag{2.19}$$

we have

$$\bar{w}_n(j) = \frac{1}{n}\left\{1 + \frac{1}{(n-1)^2}(q_j\lambda_0 + \lambda_1 q_j \bar{x}_n(j))\right\} \tag{2.20}$$

Using Equation (2.20) in Equations (2.10) and (2.11), a set of normal equations to find the optimum values of λ_0 and λ_1 is given by

$$\begin{bmatrix} \sum_{j \in s} q_j, & \sum_{j \in s} q_j \bar{x}_n(j) \\ \sum_{j \in s} q_j \bar{x}_n(j), & \sum_{j \in s} q_j \{\bar{x}_n(j)\}^2 \end{bmatrix} \begin{bmatrix} \lambda_0 \\ \lambda_1 \end{bmatrix} = \begin{bmatrix} 0 \\ (n-1)^2 \left\{ \dfrac{n(\bar{X} - n(2-n)\bar{x}_n)}{(n-1)^2} - \sum_{j \in s} \bar{x}_n(j) \right\} \end{bmatrix} \quad (2.21)$$

From this we determine that the newly tuned jackknife weights $\bar{w}_n(j)$ are given by

$$\bar{w}_n(j) = \frac{1}{n}\left[1 + \Delta_j \left\{ \frac{n(\bar{X} - n(2-n)\bar{x}_n)}{(n-1)^2} - \sum_{j \in s} \bar{x}_n(j) \right\} \right] \quad (2.22)$$

where

$$\Delta_j = q_j \left\{ \frac{\bar{x}_n(j)\left(\sum_{j \in s} q_j\right) - \sum_{j \in s} q_j \bar{x}_n(j)}{\left(\sum_{j \in s} q_j\right)\left\{\sum_{j \in s} q_j (\bar{x}_n(j))^2\right\} - \left\{\sum_{j \in s} q_j \bar{x}_n(j)\right\}^2} \right\} \quad (2.23)$$

Thus utilizing the chi-square (cs) type distance function (2.16), the newly tuned jackknife estimator (2.5) of the population mean becomes

$$\bar{y}_{\text{Tuned(cs)}} = \frac{(n-1)^2}{n}\left[\sum_{j \in s} \bar{y}_n(j) + \hat{\beta}_{\text{Tuned(cs)}} \left\{ \frac{n(\bar{X} - n(2-n)\bar{x}_n)}{(n-1)^2} - \sum_{j \in s} \bar{x}_n(j) \right\} \right] \\ - (n-2)\sum_{j \in s} \bar{y}_n(j) \quad (2.24)$$

or, equivalently

$$\bar{y}_{\text{Tuned(cs)}} = \bar{y}_n + \hat{\beta}_{\text{Tuned(cs)}}(\bar{X} - \bar{x}_n) \quad (2.25)$$

where

$$\hat{\beta}_{\text{Tuned(cs)}} = \frac{\left[\left(\sum_{j \in s} q_j\right)\left(\sum_{j \in s} q_j \bar{x}_n(j)\bar{y}_n(j)\right) - \left(\sum_{j \in s} q_j \bar{y}_n(j)\right)\left(\sum_{j \in s} q_j \bar{x}_n(j)\right)\right]}{\left[\left(\sum_{j \in s} q_j\right)\left(\sum_{j \in s} q_j (\bar{x}_n(j))^2\right) - \left(\sum_{j \in s} q_j \bar{x}_n(j)\right)^2\right]} \quad (2.26)$$

It can be easily seen that the estimator $\bar{y}_{\text{Tuned(cs)}}$ for the case $q_j = 1$ becomes

$$\bar{y}_{\text{Tuned(cs)}} = \bar{y}_n + \hat{\beta}_{\text{ols}}(\bar{X} - \bar{x}_n) \tag{2.27}$$

where

$$\hat{\beta}_{\text{ols}} = \frac{n\sum_{i=1}^{n} x_i y_i - \left(\sum_{i=1}^{n} x_i\right)\left(\sum_{i=1}^{n} y_i\right)}{n\sum_{i=1}^{n} x_i^2 - \left(\sum_{i=1}^{n} x_i\right)^2} \tag{2.28}$$

which is exactly the same linear regression estimator of population mean due to Hansen, Hurwitz, and Madow (1953), which was later rediscovered by Singh (2003). Therefore, it may be said that the proposed newly tuned estimation methodology is as efficient as the linear regression estimator for the choice of $q_j = 1$ and hence is always more efficient than the sample mean estimator. The major motivation and benefit of the proposed newly tuned estimation methodology is that it is computer friendly for estimating the variance of the resultant estimator through the doubly jackknifed method.

2.3.1 Problem of undercoverage

Let us recall that the main problem in survey sampling is estimation of the variance of an estimator of a population parameter. Assuming the reader's familiarity with standard survey sampling notation, let us now focus on the well-known linear regression estimator of the population mean $\bar{Y}$, defined as

$$\bar{y}_{\text{lr}} = \bar{y}_n + \hat{\beta}(\bar{X} - \bar{x}_n) \tag{2.29}$$

where $\hat{\beta} = \frac{s_{xy}}{s_x^2}$ is the estimator of the regression coefficient. Then for simple random and with replacement (SRSWR) sampling, the variance of the linear regression estimator may be approximated as

$$V(\bar{y}_{\text{lr}}) = \frac{1}{n}\sigma_y^2\left(1 - \rho_{xy}^2\right) \tag{2.30}$$

where $\sigma_y^2 = \frac{1}{N}\sum_{i=1}^{N}(y_i - \bar{Y})^2$ and $\rho_{xy} = \frac{\sigma_{xy}}{\sigma_x \sigma_y}$ is the population correlation coefficient, with $\sigma_{xy} = \frac{1}{N}\sum_{i=1}^{N}(y_i - \bar{Y})(x_i - \bar{X})$ being the population covariance between the two variables.

One way to estimate the variance in Equation (2.30) is to replace the parameters with their sample estimates to get an estimator of $V(\bar{y}_{lr})$ as

$$\hat{v}(\bar{y}_{lr}) = \frac{1}{n} s_y^2 \left(1 - r_{xy}^2\right) \tag{2.31}$$

where $r_{xy} = \dfrac{s_{xy}}{s_x s_y}$ is an estimator of the population correlation coefficient ρ_{xy}. One can use the estimator of variance in Equation (2.31) to construct a $(1 - \alpha)100\%$ confidence interval estimate as

$$\bar{y}_{lr} \pm \left[t_{\alpha/2}(\text{df} = n - 1)\right] \sqrt{\hat{v}(\bar{y}_{lr})} \tag{2.32}$$

We used degree of freedom (df) equal to $n - 1$, that is df $= n - 1$, to investigate the effect of the estimator of variance on the constructed confidence interval estimates by assuming that only one parameter, the population mean, is being estimated.

As is well known to the majority of survey statisticians, the interval estimate in Equation (2.32) provides very low coverage. We now provide an example of why the usual confidence interval based on the linear regression estimator gives very low coverage. Singh (2003) has also shown this in the following example. Assume a population consisting of five $(N = 5)$ units $A, B, C, D,$ and E, where two variables Y and X have been measured for each one of the units in the population.

Units	A	B	C	D	E
y_i	9	11	13	16	21
x_i	14	18	19	20	24

By selecting all possible samples of $n = 3$ units by using the simple random and without replacement (SRSWOR) scheme, Singh (2003) has shown that the ratio of approximate variance to the exact mean squared error of the linear regression estimator is given by

$$\text{Ratio} = \frac{V(\bar{y}_{lr})}{\text{Exact.MSE}(\bar{y}_{lr})} = \frac{0.230}{0.596} = 0.386 \tag{2.33}$$

In other words, the approximate variance $V(\bar{y}_{lr})$ of the linear regression estimator could be nearly as low as one-third of the exact mean squared error of the linear regression estimator.

We hope that the preceding discussion has made clear the seriousness of the problem of low coverage by a confidence interval estimator obtained using an estimator of the approximate variance of the linear regression estimator. The proposed tuned method of jackknifing is one of the ways to overcome this difficulty. In addition, for complex sampling designs, the computation and estimation of the variance of an estimator of a parameter becomes almost impossible whereas the jackknife technique sometimes could be helpful.

2.3.2 Estimation of variance and coverage

We investigate here an estimator of the variance of the estimator $\bar{y}_{\text{Tuned(cs)}}$, defined by

$$\hat{v}_{\text{Tuned(cs)}} = n(n-1)^3 \sum_{j \in s} (\bar{w}_n(j))^2 \left\{ \bar{y}_{(j)}^{\text{Tuned(cs)}} - \bar{y}_{\text{Tuned(cs)}} \right\}^2 \tag{2.34}$$

where each newly tuned doubly jackknifed estimator of the population mean is given by

$$\bar{y}_{(j)}^{\text{Tuned(cs)}} = \frac{n\bar{y}_{\text{Tuned(cs)}} - n\Big((n-1)^2 \bar{w}_n(j) - (n-2)\Big)\bar{y}_n(j)}{n-1} \tag{2.35}$$

for $j = 1, 2, \ldots, n$, with

$$\bar{y}_{\text{Tuned(cs)}} = \sum_{j \in s} \left[\Big((n-1)^2 \bar{w}_n(j) - (n-2)\Big) \bar{y}_n(j) \right] \tag{2.36}$$

It can be easily seen that the estimator of variance $\hat{v}_{\text{Tuned(cs)}}$ can be written as

$$\hat{v}_{\text{Tuned(cs)}} = \frac{(n-1)}{n} \left[\sum_{j=1}^{n} (\bar{y}_n(j) - \bar{y}_n)^2 \right.$$

$$+ 2(\bar{X} - \bar{x}_n) \left(\frac{n}{(n-1)^2} \sum_{j=1}^{n} \Delta_j (\bar{y}_n(j) - \bar{y}_n)^2 - \sum_{j=1}^{n} T_j (\bar{y}_n(j) - \bar{y}_n) \right)$$

$$+ (\bar{X} - \bar{x}_n)^2 \left(\sum_{j=1}^{n} T_j^2 - \frac{4n}{(n-1)^2} \sum_{j=1}^{n} \Delta_j T_j (\bar{y}_n(j) - \bar{y}_n) + \frac{n^2}{(n-1)^4} \sum_{j=1}^{n} \Delta_j^2 (\bar{y}_n(j) - \bar{y}_n)^2 \right)$$

$$+ \frac{2n(\bar{X} - \bar{x}_n)^3}{(n-1)^2} \left(\sum_{j=1}^{n} \Delta_j T_j^2 - \frac{n}{(n-1)^2} \sum_{j=1}^{n} \Delta_j^2 T_j (\bar{y}_n(j) - \bar{y}_n) \right) + \frac{n^2 (\bar{X} - \bar{x}_n)^4}{(n-1)^4} \sum_{j=1}^{n} \Delta_j^2 T_j^2 \right] \tag{2.37}$$

where

$$T_j = \sum_{j=1}^{n} \Delta_j \bar{y}_n(j) - n\Delta_j \bar{y}_n(j) \tag{2.38}$$

and Δ_j is given in Equation (2.23).

Clearly, as the sample size n increases, the sample mean $\bar{x}_n \to \bar{X}$. Therefore, the proposed jackknife estimator of variance, $\hat{v}_{\text{Tuned(cs)}}$, converges to the usual jackknife estimator of variance of sample mean $\bar{y}_n$ of the study variable given by

$$\hat{v}_{\text{jack}}(\bar{y}_n) = \frac{(n-1)}{n} \sum_{j=1}^{n} (\bar{y}_n(j) - \bar{y}_n)^2 \qquad (2.39)$$

Thus, the proposed jackknife estimator of variance $\hat{v}_{\text{Tuned(cs)}}$ seems reasonable so long as analytical thinking is concerned, although there is always a hope that someone can create a better estimator than it.

In the simulation study, we have taken $q_j = 1$. The coverage by the $(1-\alpha)100\%$ confidence interval estimates, obtained by this newly tuned jackknife estimator of population mean, is obtained by counting how many times the true population mean $\bar{Y}$ falls in the interval estimate given by

$$\bar{y}_{\text{Tuned(cs)}} \pm t_{\alpha/2}(\text{df} = n-1)\sqrt{\hat{v}_{\text{Tuned(cs)}}} \qquad (2.40)$$

We studied 90%, 95%, and 99% coverage by the newly tuned jackknife estimator of the population mean in Equation (2.40) and compared these to the usual estimator of variance of the linear regression estimator in Equation (2.32). These were computed by selecting 100,000 random samples from the Statistical Jumbo Pumpkin Model (SJPM). The results obtained for different sample sizes are as shown in Table 2.1.

Table 2.1 shows that for the population considered, when the sample size is small and the estimator $\hat{v}(\bar{y}_{\text{lr}})$ is used, the coverage by the usual linear regression estimator is less than expected. On the other hand, if the estimator $\hat{v}_{\text{Tuned(cs)}}$ is used, we note that coverage is much closer to the nominal coverage. In particular, for 9 pumpkins the 90% intervals cover the mean 89.44% of the time, for 13 pumpkins the 95% intervals have 95.79% actual coverage, and for 23 pumpkins the 99% intervals have 99.06% coverage. Thus, intervals desired from the newly tuned jackknife estimator of the population mean of the weight of the pumpkins shows quite good coverage if the sample size is small, which suggests good reliability of the newly tuned methodology in real practice.

2.3.3 R code

The following R code, **PUMPKIN21.R**, was used to study the coverage by the newly tuned jackknife estimator based on a chi-square type distance function.

```
#PROGRAM PUMPKIN21.R
set.seed(2013)
N<-10000
x<-runif(N, min=30, max=190)
m<-5.5*(exp(0.047*x-0.0001*x*x))
z<-rnorm(N,0,2)
y<-m*exp(z); mean(x)->XB; mean(y)->YB;
nreps<-100000
ESTP=rep(0,nreps)
EREG=rep(0,nreps)
```

Table 2.1 Performance of the newly tuned Jackknife estimators

Sample size (n)	CI using $\hat{v}_{\text{Tuned(cs)}}$ (proposed estimator)			CI using $\hat{v}(\bar{y}_{\text{lr}})$ (linear regression)		
	90% coverage	95% coverage	99% coverage	90% coverage	95% coverage	99% coverage
5	0.7408	0.7371	0.8646	0.3710	0.4195	0.5248
7	0.8346	0.8671	0.9160	0.4144	0.4586	0.5418
9	0.8944	0.9163	0.9467	0.4493	0.4904	0.5656
11	0.9266	0.9425	0.9626	0.4729	0.5149	0.5865
13	0.9466	0.9579	0.9726	0.4949	0.5361	0.6061
15	0.9594	0.9584	0.9797	0.5123	0.5505	0.6185
17	0.9669	0.9738	0.9830	0.5281	0.5660	0.6322
19	0.9740	0.9793	0.9861	0.5399	0.5774	0.6393
21	0.9791	0.9836	0.9893	0.5510	0.5913	0.6544
23	0.9820	0.9859	0.9906	0.5617	0.6024	0.6657
25	0.9849	0.9883	0.9923	0.5671	0.6096	0.6731
27	0.9870	0.9900	0.9932	0.5722	0.6162	0.6804
29	0.9878	0.9907	0.9938	0.5828	0.6259	0.6940
31	0.9896	0.9920	0.9947	0.5857	0.6312	0.7009
33	0.9912	0.9932	0.9956	0.5889	0.6337	0.7040
35	0.9923	0.9939	0.9961	0.5933	0.6399	0.7121
37	0.9925	0.9941	0.9962	0.5968	0.6432	0.7154
39	0.9936	0.9950	0.9968	0.6002	0.6470	0.7193
41	0.9943	0.9957	0.9972	0.6051	0.6516	0.7245

```
ci1.max=ci1.min=ci2.max=ci2.min=ci3.max=ci3.min=vESTP=ESTP
ci4.max=ci4.min=ci5.max=ci5.min=ci6.max=ci6.min=vREG=EREG
for ( n in seq(5,41,2))
  {
  for (r in 1: nreps)
  {
  us<-sample(N,n)
  xs<-x[us]; ys<-y[us]
  xmj<-(sum(xs)-xs)/(n-1); ymj<-(sum(ys)-ys)/(n-1)
delt<-n*sum(xmj^2)-(sum(xmj))^2
dif<-(n*XB-n*n*(2-n)*mean(xs))/((n-1)^2)-sum(xmj)
deltaj<-(xmj*n-sum(xmj))/delt
wbnj<-(1+deltaj*dif)/n
ESTI<-(((n-1)^2)*wbnj-(n-2))*ymj
ESTP[r]<-sum(ESTI)
EST_J<-(n*ESTP[r]-n*(((n-1)^2)*wbnj-(n-2))*ymj)/(n-1);
vESTP[r]<-n*((n-1)^3)*sum((wbnj^2)*((EST_J-ESTP[r])^2))
ci1.max[r]<-ESTP[r]+qt(0.95,n-1)*sqrt(vESTP[r])
ci1.min[r]<-ESTP[r]-qt(0.95,n-1)*sqrt(vESTP[r])
ci2.max[r]<-ESTP[r]+qt(0.975,n-1)*sqrt(vESTP[r])
ci2.min[r]<-ESTP[r]-qt(0.975,n-1)*sqrt(vESTP[r])
ci3.max[r]<-ESTP[r]+qt(0.995,n-1)*sqrt(vESTP[r])
ci3.min[r]<-ESTP[r]-qt(0.995,n-1)*sqrt(vESTP[r])
cov<-(sum(xs*ys)-sum(xs)*sum(ys)/n)/(n-1)
vx<-var(xs)
vy<-var(ys)
beta<-cov/vx
EREG[r]<-mean(ys) + beta*(XB-mean(xs))
corr<-cov/(vx*vy)^0.5
vREG[r]<-vy*(1-corr**2)/n
ci4.max[r]<-EREG[r]+qt(0.95,n-1)*sqrt(vREG[r])
ci4.min[r]<-EREG[r]-qt(0.95,n-1)*sqrt(vREG[r])
ci5.max[r]<-EREG[r]+qt(0.975,n-1)*sqrt(vREG[r])
ci5.min[r]<-EREG[r]-qt(0.975,n-1)*sqrt(vREG[r])
ci6.max[r]<-EREG[r]+qt(0.995,n-1)*sqrt(vREG[r])
ci6.min[r]<-EREG[r]-qt(0.995,n-1)*sqrt(vREG[r])
}
round(sum(ci1.min<YB & ci1.max>YB)/nreps, 4)->cov1
round(sum(ci2.min<YB & ci2.max>YB)/nreps, 4)->cov2
round(sum(ci3.min<YB & ci3.max>YB)/nreps, 4)->cov3
round(sum(ci4.min<YB & ci4.max>YB)/nreps, 4)->cov4
round(sum(ci5.min<YB & ci5.max>YB)/nreps, 4)->cov5
round(sum(ci6.min<YB & ci6.max>YB)/nreps, 4)->cov6
cat (n, cov1, cov2, cov3, cov4, cov5, cov6, '\n')
}
```

In the preceding R code, the variables cov4, cov5, and cov6 give the actual coverage of the nominal 90%, 95%, and 99% confidence interval when using the traditional linear regression estimator, assuming SRSWR sampling. By reexecuting the preceding R code, one may obtain results very similar to those given in Table 2.1.

2.3.4 Remark on tuning with a chi-square distance

We also consider tuning the jackknife weights $\bar{w}_n(j)$ so that the chi-square type distance function, defined as

$$(2^{-1}n)\sum_{j \in s} q_j^{-1}\left(1-(n-1)\bar{w}_n(j)-n^{-1}\right)^2 \tag{2.41}$$

is minimal subject only to the tuning constraint Equation (2.11), where q_j are some given weights.

Obviously, the Lagrange function can be taken as

$$L_0 = \sum_{j \in s} \frac{\left(1-(n-1)\bar{w}_n(j)-n^{-1}\right)^2}{(2q_j/n)} - \delta_0 \left\{ \sum_{j \in s} \bar{w}_n(j)\bar{x}_n(j) - \frac{(\bar{X}-n(2-n)\bar{x}_n)}{(n-1)^2} \right\} \tag{2.42}$$

where δ_0 is the Lagrange multiplier constant.

On setting:

$$\frac{\partial L_0}{\partial \bar{w}_n(j)} = 0 \tag{2.43}$$

we have

$$\bar{w}_n(j) = \frac{1}{n}\left\{1 + \frac{\delta_0}{(n-1)^2}q_j\bar{x}_n(j)\right\} \tag{2.44}$$

The newly tuned jackknife weights $\bar{w}_n(j)$ become

$$\bar{w}_n(j) = \frac{1}{n}\left[1 + \frac{q_j\bar{x}_n(j)}{\sum_{j \in s}q_j\{\bar{x}_n(j)\}^2}\left\{\frac{n(\bar{X}-n(2-n)\bar{x}_n)}{(n-1)^2} - \sum_{j \in s}\bar{x}_n(j)\right\}\right] \tag{2.45}$$

Under the chi-square (cs) type distance function and after eliminating the constraint (2.10), the newly tuned estimator (2.5) of the population mean becomes a modified generalized regression (greg) type estimator, namely,

$$\bar{y}_{\text{Tuned}(cs*)} = \frac{(n-1)^2}{n}\left[\sum_{j\in s}\bar{y}_n(j) + \hat{\beta}^*_{\text{Tuned}}\left\{\frac{n(\bar{X}-n(2-n)\bar{x}_n)}{(n-1)^2} - \sum_{j\in s}\bar{x}_n(j)\right\}\right]$$
$$-(n-2)\sum_{j\in s}\bar{y}_n(j)$$

(2.46)

or equivalently,

$$\bar{y}_{\text{Tuned}(cs*)} = \bar{y}_n + \hat{\beta}^*_{\text{Tuned}}(\bar{X}-\bar{x}_n) \tag{2.47}$$

where

$$\hat{\beta}^*_{\text{Tuned}} = \frac{\sum_{j\in s} q_j \bar{x}_n(j)\bar{y}_n(j)}{\sum_{j\in s} q_j(\bar{x}_n(j))^2} \tag{2.48}$$

If $q_j = (x_j - \bar{x}_n)/\bar{x}_n(j)$, then the estimator $\bar{y}_{\text{Tuned}(cs*)}$ in Equation (2.47) becomes exactly the linear regression estimator defined as

$$\bar{y}_{\text{Tuned}(cs*)} = \bar{y}_{\text{lr}} = \bar{y}_n + \left(\frac{s_{xy}}{s_x^2}\right)(\bar{x}-\bar{x}_n) \tag{2.49}$$

where $(n-1)s_{xy} = \sum_{j\in s}(x_j-\bar{x}_n)(y_j-\bar{y}_n)$. We remind the reader that the first bridge between the traditional linear regression and traditional greg estimators was built by Singh (2003).

If $q_j = 1/\bar{x}_n(j)$, then the estimator $\bar{y}_{\text{Tuned}(cs*)}$ in Equation (2.47) becomes exactly the ratio estimator, defined as

$$\bar{y}_{\text{Tuned}(cs*)} = \bar{y}_{\text{ratio}} = \bar{y}_n\left(\frac{\bar{X}}{\bar{x}_n}\right) \tag{2.50}$$

If $q_j = 1$, then the estimator (2.47) becomes the modified generalized regression estimator, defined as

$$\bar{y}_{\text{Tuned}(cs*)} = \bar{y}_{m(\text{greg})} = \bar{y}_n + \hat{\beta}_{m(\text{greg})}(\bar{X}-\bar{x}_n) \tag{2.51}$$

where

$$\hat{\beta}_{m(\text{greg})} = \frac{\bar{y}_n\left\{1+\dfrac{(n-1)s_{xy}}{n\{1+n(n-2)\}(\bar{x}_n\bar{y}_n)}\right\}}{\bar{x}_n\left\{1+\dfrac{(n-1)s_x^2}{n\{1+n(n-2)\}(\bar{x}_n^2)}\right\}} \tag{2.52}$$

The modified estimator in Equation (2.52) is a kind of Beale (1962) estimator of the regression coefficient. It seems that it is difficult to find a choice of q_j that reduces the estimator $\bar{y}_{\text{Tuned(cs*)}}$ in Equation (2.47) either to the exact product estimator or to the exact traditional generalized regression (greg) estimator due to Deville and Särndal (1992).

To examine the behavior of the modified greg estimator $(q_j = 1)$, we make the following change in the R code, **PUMPKIN21.R**.

deltaj<-xmj/sum(xmj ^2)

The three coverages, 90%, 95%, and 99%, were studied for the same sample sizes and the same number of iterations as in Section 2.3.2. The changes observed in the results are reported in Table 2.2.

The coverage by the newly tuned estimation methodology remains significantly lower than the nominal coverage when the tuned weights $\bar{w}_n(j)$ are computed with formula (2.45) instead of with Equation (2.22). Note that the modified greg estimator is far from the traditional greg estimator. Therefore the estimator $\bar{y}_{\text{Tuned(cs)}}$ is recommended so long as one is concerned about estimating the weight of a pumpkin using small samples. For large samples, the modified greg may perform just as well because better coverage is expected for the regression type estimator $\bar{y}_{\text{Tuned(cs)}}$. Here "better coverage" means coverage close to the nominal or anticipated coverage.

Table 2.2 Performance of the newly tuned jackknife estimator

Sample size (n)	90% coverage	95% coverage	99% coverage
5	0.4346	0.4834	0.5894
7	0.4580	0.4980	0.5839
9	0.4824	0.5202	0.5959
11	0.5030	0.5408	0.6120
13	0.5198	0.5580	0.6268
15	0.5342	0.5706	0.6392
17	0.5471	0.5836	0.6504
19	0.5572	0.5939	0.6580
21	0.5676	0.6063	0.6704
23	0.5790	0.6169	0.6812
25	0.5840	0.6230	0.6881
27	0.5892	0.6297	0.6947
29	0.5984	0.6397	0.7068
31	0.6016	0.6450	0.7139
33	0.6042	0.6479	0.7169
35	0.6087	0.6533	0.7250
37	0.6114	0.6564	0.7277
39	0.6147	0.6599	0.7321
41	0.6196	0.6648	0.7378

2.3.5 Numerical illustration

In the following example, we explain the computational steps involved in the construction of a confidence interval estimate with the tuned estimator.

Example 2.1 As an illustration of the previous estimator, we consider a particular sample of $n=7$ pumpkins drawn from the SJPM. The values of circumference (x) in inches and weight (y) in pounds are as follows:

x (in.)	122.0	67.0	106.5	98.0	115.2	132.0	101.1
y (lbs)	6400	800	3084	1042	4500	6700	2397

Use the regression type tuned estimator to construct a 95% confidence interval estimate of the average weight by assuming the population mean circumference, $\bar{X} = 105.40$ in., is known.

Solution. One can easily compute the following table:

$\bar{x}_n(j)$	$\bar{y}_n(j)$	Δ_j	$\bar{w}_n(j)$	$\bar{y}^{(j)}_{\text{Tuned(cs)}}$	$v(j)$
103.300	3082.16	−0.0366242	0.1434385	3491.560	0.7454596
112.466	4015.50	0.0890472	0.1414437	3649.638	462.5863682
105.883	3634.83	−0.0012077	0.1428763	3471.777	13.5906190
107.300	3975.16	0.01821420	0.1425680	3466.251	19.9488114
104.433	3398.83	−0.0210866	0.1431919	3466.257	20.1163339
101.633	3032.16	−0.059473	0.1438012	3454.925	37.6219900
106.783	3754.33	0.0111309	0.1426805	3482.646	4.5396352

where

$$v(j) = \{\bar{w}_n(j)\}^2 \left\{ y^{\text{Tuned(cs)}}_{(j)} - \bar{y}_{\text{Tuned(cs)}} \right\}^2$$

and

$$\hat{v}\left(\bar{y}_{\text{Tuned(cs)}}\right) = n(n-1)^3 \sum_{j \in s} v(j) = 848037.6$$

From the table values, we compute the tuned estimate of the average weight to be: $\bar{y}_{\text{Tuned(cs)}} = 3497.579$ and $SE\left(\bar{y}_{\text{Tuned(cs)}}\right) = 919.4746$. So, the 95% confidence interval estimate of the average pumpkin weight is 1247.706–5747.452 lbs. Here we used $t_{0.975}(6) = 2.447$. The confidence in the use of proposed estimator would increase with a large sample. Here $\varepsilon = |(\bar{x}_n/\bar{X}) - 1| = 0.0054215$ is close to zero, and we need it close to zero to get reliable results. This is a basic assumption made in survey sampling when applying ratio or regression type estimators.

2.3.6 R code used for illustration

We used the following R code, **PUMPKIN21EX.R**, to generate the preceding illustration.

#PROGRAM PUMPKIN21EX.R
```
n<-7
XB<-105.4
xs<-c(122,67,106.5,98,115.2,132,101.1)
ys<-c(6400,800,3084,1042,4500,6700,2367)
xmj<-(sum(xs) - xs)/(n-1)ymj<-(sum(ys) - ys)/(n-1)
delt<- n*sum(xmj^2) - (sum(xmj))^2
dif<- (n*XB-n*n*(2-n)*mean(xs))/((n-1)^2) - sum(xmj)
deltaj<-(xmj*n - sum(xmj) )/delt
wbnj<-(1+deltaj*dif )/n
ESTI<- (((n-1)^2)*wbnj - (n-2))*ymj
ESTP<- sum(ESTI)
EST_J<- (n*ESTP - n*(((n-1)^2)*wbnj-(n-2))*ymj)/(n-1)
nuj<-(wbnj^2)*((EST_J - ESTP)^2)
vESTP<-n*(n-1)^3*sum(nuj)
L<-ESTP-qt(0.975,n-1)*sqrt(vESTP)
U<-ESTP+qt(0.975,n-1)*sqrt(vESTP)
cbind(xmj,ymj,xmj^2,deltaj,wbnj,EST_J,nuj)
cat("Tuned estimate:", ESTP, "SE: ",vESTP^0.5 ,'\n')
cat("Confidence Interval:"," ", L,": ", U,'\n')
```

2.3.7 Problem of negative weights

It should be noted that with this method, individual weights may be negative, which may lead to a large number of rejections of samples or to a negative estimate of the average weight of pumpkins. To overcome this problem, we consider tuning the estimator with a dual-to-empirical log-likelihood (dell) distance function.

2.4 Tuning with dell function

Let w_j^* be positive calibrated weights constructed so that the following two constraints are satisfied:

$$\sum_{j \in s} w_j^* = 1 \qquad (2.53)$$

and

$$\sum_{j \in s} w_j^* \Phi_j = 0 \qquad (2.54)$$

where

$$\Phi_j = (x_j - \bar{X}) \qquad (2.55)$$

are called a *pivot*.

Note that constraints (2.53) and (2.54) are similar to those listed in Owen (2001). Let $\bar{w}_n^*(j)$ be the jackknife tuned weight such that

$$\bar{w}_n^*(j) = \frac{1 - w_j^*}{(n-1)} \qquad (2.56)$$

Note that

$$0 < \bar{w}_n^*(j) < \frac{1}{(n-1)} \qquad (2.57)$$

As before, now we define the newly tuned jackknife estimator of the population mean $\bar{Y}$ by

$$\bar{y}_{\text{Tuned(dell)}} = \sum_{j \in s} \left[(n-1)^2 \bar{w}_n^*(j) - (n-2) \right] \bar{y}_n(j) \qquad (2.58)$$

Here, for simplicity, we consider the optimization of a dell function defined by

$$\sum_{j \in s} \frac{\ln(1 - w_j^*)}{n} \qquad (2.59)$$

or equivalently, optimization of the log-likelihood function defined by

$$\sum_{j \in s} \frac{\ln(\bar{w}_n^*(j))}{n} \qquad (2.60)$$

such that the following two conditions are satisfied:

$$\sum_{j \in s} \bar{w}_n^*(j) = 1 \qquad (2.61)$$

and

$$\sum_{j \in s} \bar{w}_n^*(j) \Psi_j = 0 \qquad (2.62)$$

where

$$\Psi_j = \left\{ \bar{x}_n(j) - \frac{\bar{X} - n(2-n)\bar{x}_n}{(n-1)^2} \right\} \qquad (2.63)$$

The Lagrange function is given by

$$L_2 = \sum_{j \in s} \frac{\ln(\bar{w}_n^*(j))}{n} - \lambda_0^* \left\{ \sum_{j \in s} \bar{w}_n^*(j) - 1 \right\} - \lambda_1^* \left\{ \sum_{j \in s} \bar{w}_n^*(j)\Psi_j \right\} \qquad (2.64)$$

where λ_0^* and λ_1^* are Lagrange multiplier constants.

On setting:

$$\frac{\partial L_2}{\partial \bar{w}_n^*(j)} = 0 \qquad (2.65)$$

we have

$$\bar{w}_n^*(j) = \frac{1}{n(1 + \lambda_1^* \Psi_j)} \qquad (2.66)$$

Constraints (2.61) and (2.62) yield $\lambda_0^* = 1$, and λ_1^* is a solution to the nonlinear equation

$$\sum_{j \in s} \frac{\Psi_j}{1 + \lambda_1^* \Psi_j} = 0 \qquad (2.67)$$

Note that to have $\bar{w}_n^*(j) > 0$, we require $\lambda_1^* > 1/|\max(\Psi_j)|$ if $\max(\Psi_j) > 0$ and $\lambda_1^* < 1/|\min(\Psi_j)|$ if $\max(\Psi_j) < 0$.

Thus under the dell distance function, the newly tuned jackknife estimator (2.58) of the population mean becomes

$$\begin{aligned}\bar{y}_{\text{Tuned(dell)}} &= \sum_{j \in s} \left[(n-1)^2 \bar{w}_n^*(j) - (n-2) \right] \bar{y}_n(j) \\ &= \frac{1}{n} \sum_{j \in s} \left[\frac{(n-1)^2}{1 + \lambda_1^* \Psi_j} - n(n-2) \right] \bar{y}_n(j) \end{aligned} \qquad (2.68)$$

2.4.1 Estimation of variance and coverage

We suggest the following estimator of the variance of the preceding estimator $\bar{y}_{\text{Tuned(dell)}}$:

$$\hat{v}_{\text{Tuned(dell)}} = n(n-1)^3 \sum_{j \in s} \{\bar{w}_n^*(j)\}^2 \left\{ \bar{y}_{(j)}^{\text{Tuned(dell)}} - \bar{y}_{\text{Tuned(dell)}} \right\}^2 \qquad (2.69)$$

Note that for each sample the newly tuned dell doubly jackknifed estimator of the population mean is given by

$$\bar{y}_{(j)}^{\text{Tuned(dell)}} = \frac{n\bar{y}_{\text{Tuned(dell)}} - n\left((n-1)^2\bar{w}_n^*(j) - (n-2)\right)\bar{y}_n(j)}{n-1} \tag{2.70}$$

for $j = 1, 2, \ldots, n$, where

$$\bar{y}_{\text{Tuned(dell)}} = \sum_{j \in s}\left[\left((n-1)^2\bar{w}_n^*(j) - (n-2)\right)\bar{y}_n(j)\right] \tag{2.71}$$

The coverage by the $(1-\alpha)100\%$ confidence interval estimates, obtained by this newly tuned jackknife empirical log-likelihood estimator of population mean, is obtained by counting how many times the true population mean $\bar{Y}$ falls into the interval estimate given by

$$\bar{y}_{\text{Tuned(dell)}} \mp t_{\alpha/2}(\text{df} = n-1)\sqrt{\hat{v}_{\text{Tuned(dell)}}} \tag{2.72}$$

A very crude approximate value of λ_1^* is given by

$$\lambda_1^* \approx \frac{\sum_{j \in s}\Psi_j}{\sum_{j \in s}\Psi_j^2} = \frac{(\bar{x}_n - \bar{X})}{\frac{(n-1)}{n}s_x^2 + \frac{1}{(n-1)^2}(\bar{x}_n - \bar{X})^2} \tag{2.73}$$

The approximate value of λ_1^* in Equation (2.73) is obtained under the assumption that

$$-1 \leq \lambda_1^*\Psi_j \leq 1 \tag{2.74}$$

Solving the nonlinear equation (2.67) for λ_1^* by means of an iterative method may be computer time intensive, but use of the approximation for λ_1^* given in Equation (2.73) will speed things up considerably. However, because λ_1^* is a crude approximation, it may be preferable to find a fast subroutine for solving Equation (2.67).

We compared intervals with nominal 90%, 95%, and 99% coverage that were constructed using the newly tuned jackknife estimator of the population mean by selecting 100,000 random samples from the SJPM. The results obtained for different sample sizes are shown in Table 2.3. The results show that the coverage by the newly tuned jackknife empirical log-likelihood estimator of the population mean converges very quickly to the nominal coverage for moderate sample sizes.

The nominal 90%, 95%, and 99% coverages are estimated, shown as 89.39%, 95.79%, and 99.06%, respectively, for samples of sizes 9, 13, and 23 pumpkins. Thus, the newly tuned dell estimator of the population mean weight of the pumpkins gives intervals with coverage as good as that provided by intervals produced using the

Table 2.3 **Performance of the newly tuned dell**

Sample size (n)	90% coverage	95% coverage	99% coverage
5	0.7300	0.7787	0.8588
7	0.8329	0.8657	0.9151
9	0.8939	0.9159	0.9466
11	0.9264	0.9424	0.9625
13	0.9466	0.9579	0.9726
15	0.9594	0.9684	0.9796
17	0.9669	0.9738	0.9829
19	0.9740	0.9793	0.9861
21	0.9791	0.9836	0.9893
23	0.9820	0.9859	0.9906
25	0.9849	0.9883	0.9923
27	0.9870	0.9900	0.9932
29	0.9878	0.9907	0.9938
31	0.9896	0.9920	0.9947
33	0.9912	0.9932	0.9956
35	0.9923	0.9939	0.9961
37	0.9925	0.9941	0.9962
39	0.9936	0.9950	0.9968

proposed regression type estimator under a chi-square distance measure. For larger samples, the approximated crude value of λ_1^* converges to zero, indicating that the newly tuned dell estimator of the mean should be close to the sample mean estimator in the case of a large sample. The coverages based on the empirically tuned estimator are nearly the same as those based on the linear regression estimator $\bar{y}_{\text{Tuned(cs)}}$ given in Equation (2.25), but differ from those based on the generalized regression type (greg) estimator of the population mean given by $\bar{y}_{\text{Tuned(cs*)}}$ in Equation (2.47).

2.4.2 R code

The following R code, **PUMPKIN22.R**, was used to study the coverage based on the newly tuned dell function.

```
# PROGRAM: PUMPKIN22.R
set.seed(2013)
N<-10000
x<-runif(N, min=30, max=190)
m<-5.5*(exp(0.047*x - 0.0001*x*x))
z<-rnorm(N, 0, 2)
y<-m*exp(z)
mean(x)->XB; mean(y)->YB
nreps<-100000
```

```
ESTP=rep(0,nreps)
ci1.max=ci1.min=ci2.max=ci2.min=ci3.max=ci3.min=vESTP=ESTP
for (n in seq(5,41,2))
  {
  for (r in 1:nreps)
    {
    us<-sample(N,n)
    xs<-x[us]; ys<-y[us]
    xmj<-(sum(xs) - xs)/(n-1); ymj<-(sum(ys) - ys)/(n-1)
    shj<- xmj - (XB-n*(2-n)*mean(xs))/((n-1)^2)
    aphj<-sum(shj)/sum(shj^2)
    wbnj<-(1/n)*(1/(1+aphj*shj))
    ESTI<- n*(((n-1)^2)*wbnj - (n-2))*ymj
    ESTP[r]<- mean(ESTI)
    EST_J<-(n*ESTP[r] - ESTI)/(n-1)
    vj<-(wbnj^2)*((EST_J - ESTP[r])^2)
    vESTP[r]<-n*(n-1)^3*sum(vj)
    ci1.max[r]<- ESTP[r]+qt(0.95,n-1)*sqrt(vESTP[r])
    ci1.min[r]<- ESTP[r]-qt(0.95,n-1)*sqrt(vESTP[r])
    ci2.max[r]<- ESTP[r]+qt(0.975,n-1)*sqrt(vESTP[r])
    ci2.min[r]<- ESTP[r]-qt(0.975,n-1)*sqrt(vESTP[r])
    ci3.max[r]<- ESTP[r]+qt(0.995,n-1)*sqrt(vESTP[r])
    ci3.min[r]<- ESTP[r]-qt(0.995,n-1)*sqrt(vESTP[r])
    }
  round(sum(ci1.min<YB & ci1.max>YB)/nreps,4)->cov1
  round(sum(ci2.min<YB & ci2.max>YB)/nreps,4)->cov2
  round(sum(ci3.min<YB & ci3.max>YB)/nreps,4)->cov3
  cat(n, cov1,cov2,cov3,'\n')
  }
```

2.4.3 Numerical illustration

In the following example, we explain the computational steps involved in the construction of a confidence interval estimate with the dell estimator.

Example 2.2 Consider the following sample of $n=7$ pumpkins, where x and y represent the circumference (in.) and weight (lbs) of pumpkins:

x	122.0	67.0	106.5	98.0	115.2	132.0	101.1
y	6400	800	3084	1042	4500	6700	2397

Construct the 95% confidence interval estimate of the average weight by assuming the population mean circumference $\bar{X} = 105.40$ in. is known.

Solution. One can easily compute the following table:

$\bar{x}_n(j)$	$\bar{y}_n(j)$	Ψ_j	$\bar{w}_n^*(j)$	$\bar{y}_{(j)}^{\text{Tuned(dell)}}$	$v(j)$
103.3000	3082.167	−2.65555556	0.1434374	3491.785	0.7071573
112.4667	4015.500	6.51111111	0.1414542	3647.949	452.016458
105.8833	3634.833	−0.07222222	0.1428729	3472.384	13.0285075
107.3000	3975.167	1.34444444	0.1425652	3466.807	19.3324075
104.4333	3398.833	−1.52222222	0.1431892	3466.721	19.6103742
101.6333	3032.167	−4.32222222	0.1438039	3454.655	38.2248218
106.7833	3754.333	0.82777778	0.1426772	3483.235	4.2289600

where

$$v(j) = n(n-1)^3 \left(\bar{w}_n^*(j)\right)^2 \left\{\bar{y}_{(j)}^{\text{Tuned(dell)}} - \bar{y}_{\text{Tuned(dell)}}\right\}^2$$

The tuned estimate of the average weight is $\bar{y}_{\text{Tuned(dell)}} = 3497.648$ and $SE\left(\bar{y}_{\text{Tuned(dell)}}\right) = 909.5542$. Thus the 95% confidence interval estimate of the average weight of each pumpkin is 1272.049–5723.247 lbs.

2.4.4 R code used for illustration

We used the following R code, **PUMPKIN22EX.R**, to solve the preceding numerical illustration.

```
# ILLUSTRATION CODE
# PUMPKIN22EX.R
n<-7
XB<-105.4
xs<-c(122,67,106.5,98,115.2,132,101.1)
ys<-c(6400,800,3084,1042,4500,6700,2367)
xmj<-(sum(xs) - xs)/(n-1)
ymj<-(sum(ys) - ys)/(n-1)
shj<- xmj - (XB-n*(2-n)*mean(xs))/((n-1)^2)
wbnj<-(1/n)*(1/(1+(sum(shj)/sum(shj^2))*shj))
ESTI<- n*(((n-1)^2)*wbnj - (n-2))*ymj
ESTP<- mean(ESTI)
EST_J<-(n*ESTP - ESTI)/(n-1)
vj<-(wbnj^2)*((EST_J - ESTP)^2)
vESTP<-n*(n-1)^3*sum(vj)
L<-ESTP-qt(0.975,n-1)*sqrt(vESTP)
U<-ESTP+qt(0.975,n-1)*sqrt(vESTP)
cbind(xmj,ymj,shj,wbnj,EST_J,vj)
cat("Tuned estimate:", ESTP, "SE: ",vESTP^0.5,'\n')
cat("Confidence Interval:"," ", L,"; ", U,'\n')
```

2.5 An important remark

Note that several other versions of the doubly jackknifed estimator of variance could also be considered, namely,

$$\hat{v}_{\text{Tuned(cs)}}^{\text{no-weight}} = \frac{(n-1)^3}{n} \sum_{j \in s} \left(\bar{y}_{(j)}^{\text{Tuned(cs)}} - \bar{y}_{\text{Tuned(cs)}} \right)^2 \qquad (2.75)$$

or

$$\hat{v}_{\text{Tuned(cs)}}^{\text{no-weight}} = \frac{(n-1)^3}{n} \sum_{j \in s} \left(\bar{y}_{(j)}^{\text{Tuned(cs)}} - \frac{1}{n} \sum_{j=1}^{n} \bar{y}_{(j)}^{\text{Tuned(cs)}} \right)^2 \qquad (2.76)$$

Several such versions could be compared before application to real data, if necessary.
In the next section we provide a few possible extensions or further studies, although these are marked as exercises. In the same way, similar extensions may be developed and investigated in the future.

2.6 Exercises

Exercise 2.1 Consider a newly tuned jackknifed estimator of the population mean $\bar{Y}$ defined by

$$\bar{y}_{\text{Tuned(cs)}} = \sum_{j \in s} \left[\left\{ (n-1)^2 \bar{w}_n(j) - (n-2) \right\} \bar{y}_n(j) \right] \qquad (2.77)$$

where

$$\bar{y}_n(j) = \frac{n\bar{y}_n - y_j}{n-1} \qquad (2.78)$$

is the sample mean of the study variable obtained by removing the jth unit from the sample s, and $\bar{w}_n(j)$ is the jackknife tuned weight constructed such that the following three constraints are satisfied:

$$\sum_{j \in s} \bar{w}_n(j) = 1 \qquad (2.79)$$

$$\sum_{j \in s} \bar{w}_n(j) \bar{x}_n(j) = \frac{\bar{X} - n(2-n)\bar{x}_n}{(n-1)^2} \qquad (2.80)$$

and

$$\sum_{j \in s} \bar{w}_n(j) \hat{\sigma}_x^2(j) = \frac{\sigma_x^2 - n(2-n)\hat{\sigma}_x^2}{(n-1)^2} \tag{2.81}$$

where

$$\bar{w}_n(j) = \frac{1-w_j}{n-1} \tag{2.82}$$

for arbitrary weights w_j with $\sum_{j \in s} w_j = 1$,

$$\hat{\sigma}_x^2(j) = \frac{n\hat{\sigma}_x^2 - (x_j - \bar{x}_n)^2}{n-1} \tag{2.83}$$

and

$$\hat{\sigma}_x^2 = n^{-1} \sum_{i \in s} (x_i - \bar{x}_n)^2 \tag{2.84}$$

Note that $\hat{\sigma}_x^2$ is the maximum likelihood estimator of the known finite population variance $\sigma_x^2 = N^{-1} \sum_{i \in \Omega} (x_i - \bar{X})^2$ of the auxiliary variable, and $\hat{\sigma}_x^2(j)$ is a partial jth jackknifed estimator of the variance obtained by dropping the jth squared deviation about the sample mean from the total sum of squares from the sample s of the auxiliary variable divided by $(n-1)$. Subject to the preceding three tuning constraints in Equations (2.79), (2.80), and (2.81), optimize each of the following distance functions:

$$D_1 = (2^{-1}n) \sum_{j \in s} q_j^{-1} \left(1 - (n-1)\bar{w}_n(j) - n^{-1}\right)^2 \tag{2.85}$$

$$D_2 = \sum_{j \in s} [\bar{w}_n(j) \ln(\bar{w}_n(j))], \quad 0 < \bar{w}_n(j) < 1/(n-1) \tag{2.86}$$

$$D_3 = \frac{1}{2} \sum_{j \in s} q_j^{-1} \left(\sqrt{1 - (n-1)\bar{w}_n(j)} - \sqrt{n^{-1}}\right)^2, \quad 0 < \bar{w}_n(j) < 1/(n-1) \tag{2.87}$$

$$D_4 = \sum_{j \in s} [-n^{-1} \ln(\bar{w}_n(j))], \quad 0 < \bar{w}_n(j) < 1/(n-1) \tag{2.88}$$

$$D_5 = \sum_{j \in s} \frac{(1 - (n-1)\bar{w}_n(j) - n^{-1})^2}{2(1 - (n-1)\bar{w}_n(j))}, \quad 0 < \bar{w}_n(j) < 1/(n-1) \tag{2.89}$$

$$D_6 = \frac{1}{n}\sum_{j\in s} \tanh^{-1}\left(\frac{\{\bar{w}_n(j)\}^2 - 1}{\{\bar{w}_n(j)\}^2 + 1}\right) \qquad (2.90)$$

$$D_7 = \frac{1}{2}\sum_{j\in s} \frac{(1-(n-1)\bar{w}_n(j) - n^{-1})^2}{q_j n^{-1}} + \frac{1}{2}\sum_{j\in s} \frac{\varphi_j\{\bar{w}_n(j)\}^2}{q_j n^{-1}(n-1)^{-2}} \qquad (2.91)$$

and

$$D_8 = \sum_{j\in s} q_j^{-1}\left[\bar{w}_n(j)\ln(\bar{w}_n(j)) - \bar{w}_n(j) + n^{-1}\right] \qquad (2.92)$$

where q_j are the weights chosen to form different types of estimators, φ_j is a penalty as in Farrell and Singh (2002a), and $\tanh^{-1}()$ is the hyperbolic tangent function as in Singh (2012). Also optimize each of the preceding distance functions subject to: (a) only two tuning constraints, (2.79) and (2.80); and (b) only one tuning constraint (2.80). Write code in any scientific language, e.g. FORTRAN, C++, R, or SAS, to study these distance functions. Discuss the nature of tuned weights in each situation. Construct 90%, 95%, and 99% confidence interval estimates in each situation by estimating the variance using the method discussed in the chapter. Alternatively, construct your own confidence interval estimates that you can claim to be better based on some scientific criterion.

Exercise 2.2 Let the population variance $\sigma_x^2 = N^{-1}\sum_{i\in\Omega}(x_i - \bar{X})^2$ of the auxiliary variable be known. Replace the constraint

$$\sum_{j\in s}\bar{w}_n(j)\hat{\sigma}_x^2(j) = \frac{\sigma_x^2 - n(2-n)\hat{\sigma}_x^2}{(n-1)^2} \qquad (2.93)$$

in Exercise 2.1 with a newly tuned constraint:

$$\sum_{j\in s}\bar{w}_n(j)(\bar{x}_n(j) - \bar{x}_n)^2 = \frac{(n-1)s_x^2 - S_x^2}{(n-1)^3} \qquad (2.94)$$

where $s_x^2 = \frac{n}{n-1}\hat{\sigma}_x^2$ and $S_x^2 = \frac{N}{N-1}\sigma_x^2$. Report any changes observed in the resulting estimator. Now again consider its replacement with a newly tuned constraint:

$$\sum_{j\in s}\bar{w}_n(j)s_x^2(j) = \frac{nS_x^2 - (n-1)\{(n-1) - n(n-2)\}s_x^2}{(n-1)^2(n-2)} \qquad (2.95)$$

where

$$s_x^2(j) = \frac{(n-1)^2 s_x^2 - n(x_j - \bar{x}_n)^2}{(n-1)(n-2)}$$

or equivalently

$$s_x^2(j) = \frac{(n-1)s_x^2 - (x_j - \bar{x}_n)^2 - (n-1)(\bar{x}_n(j) - \bar{x}_n)^2}{(n-2)}$$

and again report any changes observed in the resulting estimator.

Exercise 2.3 Assume that the value of the finite population correlation coefficient ρ_{xy} between the study variable y and the auxiliary variable x is known. In Exercise 2.1, consider an additional newly tuned constraint given by

$$\sum_{j \in s} \bar{w}_n(j) \, r(j) = \frac{1}{(n-1)} \left[\sum_{j=1}^n r(j) - \rho_{xy} \right] \qquad (2.96)$$

Note that the constraint (2.96) has been designed by eliminating the additive effect of partial jackknifing on the estimator of the correlation coefficient with the tuned weights $\bar{w}_n(j)$ from a calibration constraint:

$$\sum_{j \in s} w_j r(j) = \rho_{xy} \qquad (2.97)$$

The value of the partial jackknifed correlation coefficient $r(j)$ is given as

$$r(j) = r \frac{\left[\frac{(n-1)}{(n-2)} - \frac{n(x_j - \bar{x}_n)(y_j - \bar{y}_n)}{(n-1)(n-2)s_{xy}} \right]}{\sqrt{\frac{(n-1)}{(n-2)} - \frac{n(x_j - \bar{x}_n)^2}{(n-1)(n-2)s_x^2}} \sqrt{\frac{(n-1)}{(n-2)} - \frac{n(y_j - \bar{y}_n)^2}{(n-1)(n-2)s_y^2}}} \qquad (2.98)$$

with

$$(n-1)s_{xy} = \sum_{i \in s}(x_i - \bar{x}_n)(y_i - \bar{y}_n), \quad (n-1)s_x^2 = \sum_{i \in s}(x_i - \bar{x}_n)^2,$$

$$(n-1)s_y^2 = \sum_{i \in s}(y_i - \bar{y}_n)^2, \quad n\bar{x}_n = \sum_{i \in s} x_i, \quad n\bar{y}_n = \sum_{i \in s} y_i \text{ and } r = s_{xy}/(s_x s_y)$$

Report any changes observed in the resultant estimators.

Exercise 2.4 In Exercise 2.1, find the distribution of the dual-to-log-likelihood function $-2\sum_{j\in s}\ln(1-w_j)$ or equivalently $-2\sum_{j\in s}\ln(\bar{w}_n(j))$ when optimizing the dual-to-log-likelihood distance function:

$$D = \sum_{j\in s}\left[n^{-1}\ln(\bar{w}_n(j))\right], \quad 0 < \bar{w}_n(j) < 1/(n-1) \tag{2.99}$$

Exercise 2.5 Tuning of a nonresponse Consider a sample s_n of n units selected using an SRS scheme where the value of the study variable y_i and an auxiliary variable x_i are measured $i = 1, 2, \ldots, n$. Suppose that the responses y_i, $i = 1, 2, \ldots, r$, on the study variable are available for a subset $s_r \subset s_n$ of the sample while the remaining $(n-r)$ responses on the study variable are missing completely at random. The problem is to estimate the population mean $\bar{Y}$ of the study variable. Let $\bar{y}_r = r^{-1}\sum_{i\in s_r} y_i$ be the sample mean of the study variable corresponding to the set of responding units, let $\bar{x}_r = r^{-1}\sum_{i\in s_r} x_i$ be the sample mean of the auxiliary variable corresponding to the set of responding units, and let $\bar{x}_n = n^{-1}\sum_{i\in s_n} x_i$ be the sample mean of the auxiliary variable for the entire sample selected in the survey.

(a) Consider the newly tuned jackknife estimator of the population mean $\bar{Y}$ in the occurrence of nonresponse defined by

$$\bar{y}_{\text{NRTuned(cs)}} = \sum_{j\in s_r}\left[\left\{(r-1)^2\bar{w}_r(j) - (r-2)\right\}\bar{y}_r(j)\right] \tag{2.100}$$

where

$$\bar{y}_r(j) = \frac{r\bar{y}_r - y_j}{r-1} \quad \text{and} \quad \bar{w}_r(j) = \frac{1-w_j}{r-1}$$

are the usual jackknife estimators of the population mean and weight obtained by removing the jth unit from the responding sample s_r for any set of weights w_j with unit total. Here we consider $\bar{w}_r(j)$ as the jth tuned weight constructed such that the following two constraints are satisfied:

$$\sum_{j\in s_r}\bar{w}_r(j) = 1 \tag{2.101}$$

and

$$\sum_{j\in s_r}\bar{w}_r(j)\bar{x}_r(j) = \frac{\bar{x}_n - r(2-r)\bar{x}_r}{(r-1)^2} \tag{2.102}$$

Tuning of jackknife estimator

Consider tuning the weights $\bar{w}_r(j)$ so that the chi-square type distance function, defined as

$$(2^{-1}r)\sum_{j\in s_r} q_j^{-1}\left(1-(r-1)\bar{w}_r(j)-r^{-1}\right)^2 \tag{2.103}$$

is minimum, subject to the tuning constraints (2.101) and (2.102), where q_j is some choice of weights. Show that the newly tuned estimator of the population mean in the presence of nonresponse becomes

$$\bar{y}_{\text{NRTuned(cs)}} = \bar{y}_r + \hat{\beta}_{\text{NRTuned}}(\bar{x}_n - \bar{x}_r) \tag{2.104}$$

where

$$\hat{\beta}_{\text{NRTuned}} = \frac{\left(\sum_{j\in s_r} q_j\right)\left(\sum_{j\in s_r} q_j\bar{x}_r(j)\bar{y}_r(j)\right) - \left(\sum_{j\in s_r} q_j\bar{y}_r(j)\right)\left(\sum_{j\in s_r} q_j\bar{x}_r(j)\right)}{\left(\sum_{j\in s_r} q_j\right)\left(\sum_{j\in s_r} q_j(\bar{x}_r(j))^2\right) - \left(\sum_{j\in s_r} q_j\bar{x}_r(j)\right)^2}$$

Show that for $q_j = 1$ the estimator $\bar{y}_{\text{NRTuned(cs)}}$ becomes

$$\bar{y}_{\text{NRTuned(cs)}} = \bar{y}_r + \hat{\beta}_{\text{ols}}(\bar{x}_r - \bar{x}_n) \tag{2.105}$$

where

$$\hat{\beta}_{\text{ols}} = \frac{r\sum_{i=1}^{r} x_i y_i - \left(\sum_{i=1}^{r} x_i\right)\left(\sum_{i=1}^{r} y_i\right)}{r\sum_{i=1}^{r} x_i^2 - \left(\sum_{i=1}^{r} x_i\right)^2}$$

Consider an estimator of the variance of the estimator $\bar{y}_{\text{NRTuned(cs)}}$ defined by

$$\hat{v}_{\text{NRTuned(cs)}} = r(r-1)^3 \sum_{j\in s_r} \{\bar{w}_r(j)\}^2 \left\{\bar{y}_{(j)}^{\text{NRTuned(cs)}} - \bar{y}_{\text{NRTuned(cs)}}\right\}^2 \tag{2.106}$$

where each newly tuned jackknifed estimator of the population mean is given by

$$\bar{y}_{(j)}^{\text{NRTuned(cs)}} = \frac{r\bar{y}_{\text{NRTuned(cs)}} - r\left((r-1)^2 \bar{w}_r(j) - (r-2)\right)\bar{y}_r(j)}{r-1} \tag{2.107}$$

for $j = 1, 2, \ldots, r$.

Generate a population of reasonable size, and create an environment through a simulations process where nonresponse could happen. For various values of response rate and sample size, study the coverage by the $(1-\alpha)100\%$ confidence interval estimates obtained from this newly tuned estimator of the population mean by counting how many times out of 10,000 attempts, the true population mean $\bar{Y}$ falls within the interval estimates given by

$$\bar{y}_{\text{NRTuned(cs)}} \pm t_{\alpha/2}(\text{df}=?)\sqrt{\hat{v}_{\text{NRTuned(cs)}}} \qquad (2.108)$$

Suggest and justify your choice of degree of freedom (df).

(b) Consider a newly tuned dell estimator of the population mean $\bar{Y}$ defined by

$$\bar{y}_{\text{NRTuned(dell)}} = \sum_{j \in s_r}\left[(r-1)^2 \bar{w}_r^*(j) - (r-2)\right] \bar{y}_r(j) \qquad (2.109)$$

where $0 < \bar{w}_r^*(j) < 1/(r-1)$ are positive tuned weights constructed such that the following two constraints are satisfied:

$$\sum_{j \in s_r} \bar{w}_r^*(j) = 1 \qquad (2.110)$$

and

$$\sum_{j \in s_r} \bar{w}_r^*(j) \psi_j = 0 \qquad (2.111)$$

where

$$\psi_j = \bar{x}_r(j) - \frac{(\bar{x}_n - r(2-r)\bar{x}_r)}{(r-1)^2} \qquad (2.112)$$

Here, for simplicity, consider the optimization of a dual-to-log-likelihood distance function given by

$$\sum_{j \in s_r} \frac{\ln(1-w_j^*)}{r} \qquad (2.113)$$

or equivalently, optimization of a new log-likelihood function

$$\sum_{j \in s_r} \frac{\ln(\bar{w}_r^*(j))}{r} \qquad (2.114)$$

subject to the two tuning constraints (2.110) and (2.111).

Show that under the dual-to-log-likelihood distance function, the newly tuned estimator (2.109) of the population mean becomes

$$\bar{y}_{\text{NRTuned(dell)}} = \frac{(r-1)^2}{r} \sum_{j \in s_r} \frac{\bar{y}_r(j)}{1+\lambda_1^* \psi_j} - (r-2) \sum_{j \in s_r} \bar{y}_r(j) \qquad (2.115)$$

where λ_1^* is a solution to the nonlinear equation

$$\sum_{j \in s_r} \frac{\psi_j}{1+\lambda_1^* \psi_j} = 0 \qquad (2.116)$$

Consider an estimator of the variance of the estimator $\bar{y}_{\text{NRTuned(dell)}}$ defined by

$$\hat{v}_{\text{NRTuned(dell)}} = r(r-1)^3 \sum_{j \in s_r} \{\bar{w}_r^*(j)\}^2 \{\bar{y}_{(j)}^{\text{NRTuned(dell)}} - \bar{y}_{\text{NRTuned(dell)}}\}^2 \qquad (2.117)$$

Assume each newly tuned dell estimator of the population mean is given by

$$\bar{y}_{(j)}^{\text{NRTuned(dell)}} = \frac{r\bar{y}_{\text{NRTuned(dell)}} - r\{(r-1)^2 \bar{w}_r^*(j) - (r-2)\}\bar{y}_r(j)}{r-1} \qquad (2.118)$$

for $j = 1, 2, \ldots, r$.

Generate such a population of reasonable size and create an environment through a simulations process where nonresponse could happen. Study the coverage of the $(1-\alpha)100\%$ confidence interval estimates given by this newly tuned dell estimator of the population mean by counting how many times out of 10,000 attempts, the true population mean $\bar{Y}$ falls within the interval estimates

$$\bar{y}_{\text{NRTuned(dell)}} \mp t_{\alpha/2}(\text{df} = ?) \sqrt{\hat{v}_{\text{NRTuned(dell)}}} \qquad (2.119)$$

Suggest and justify your choice of degree of freedom (df).

Exercise 2.6 Tuning of a sensitive variable Consider the problem of estimating the population mean of a sensitive variable and the problem of estimating the variance of this estimator. Suppose that we select a sample s of n respondents by the SRS scheme from a population consisting of N units. Let y_i be the true response, for example income, of the ith respondent in the sample. The ith respondent selected in the sample is requested to draw two numbers S_1 and S_2 from two independent randomization devises, say R_1 and R_2, respectively, and report the scrambled response Z_i computed by

$$Z_i = \frac{S_1 y_i + S_2 - B}{A}, \quad i = 1, 2, \ldots, r \qquad (2.120)$$

where $E_R(S_1) = A$ and $E_R(S_2) = B$, and A and B are known. Also let $V_R(S_1) = \sigma_A^2$ and $V_R(S_2) = \sigma_B^2$ be known. Let r be the number of respondents in s_r, the subsample who responded to the sensitive question using the preceding randomization device, and $(n-r)$ the number of units in $s_{(n-r)}$, the subsample who refused to respond, and let $s_n = s_r \cup s_{(n-r)}$. Let x_i be an auxiliary variable correlated with the study variable y_i, and assume the values of the auxiliary variable $x_i, i = 1, 2, \ldots, n$ for all the n units in s_n are available.

(a) Consider a newly tuned jackknife estimator of the population mean $\bar{Y}$ given by

$$\bar{y}_{\text{STuned(cs)}} = \sum_{j \in s_r} \left[\{(r-1)^2 \bar{w}_r(j) - (r-2)\} \bar{Z}_r(j) \right] \qquad (2.121)$$

where

$$\bar{Z}_r(j) = \frac{r\bar{Z}_r - Z_j}{r-1} \quad \text{and} \quad \bar{w}_r(j) = \frac{1-w_j}{r-1} \tag{2.122}$$

are, the usual, the jackknife estimator of the population mean and jackknife weight obtained by removing the jth unit from the responding sample s_r for any set of weights w_j with unit total. Let $\bar{w}_r(j)$ be the jth tuned weight, chosen such that the following two constraints are satisfied:

$$\sum_{j \in s_r} \bar{w}_r(j) = 1 \tag{2.123}$$

and

$$\sum_{j \in s_r} \bar{w}_r(j) \bar{x}_r(j) = \frac{\bar{x}_n - r(2-r)\bar{x}_r}{(r-1)^2} \tag{2.124}$$

Consider tuning the weights $\bar{w}_r(j)$ such that the chi-square type distance function, defined as

$$(2^{-1}r) \sum_{j \in s_r} q_j^{-1} \left(1 - (r-1)\bar{w}_r(j) - r^{-1}\right)^2 \tag{2.125}$$

is minimum, subject to the tuning constraints (2.123) and (2.124), and where q_j is some choice of weights. Show that under the chi-square (cs) type distance function, the newly tuned estimator of the population mean of the sensitive variable becomes

$$\bar{y}_{\text{STuned(cs)}} = \bar{Z}_r + \hat{\beta}_{\text{STuned}} (\bar{x}_n - \bar{x}_r) \tag{2.126}$$

where

$$\hat{\beta}_{\text{STuned}} = \frac{\left[\left(\sum_{j \in s_r} q_j\right) \left(\sum_{j \in s_r} q_j \bar{x}_r(j) \bar{Z}_r(j)\right) - \left(\sum_{j \in s_r} q_j \bar{Z}_r(j)\right) \left(\sum_{j \in s_r} q_j \bar{x}_r(j)\right)\right]}{\left(\sum_{j \in s_r} q_j\right) \left(\sum_{j \in s_r} q_j (\bar{x}_r(j))^2\right) - \left(\sum_{j \in s_r} q_j \bar{x}_r(j)\right)^2}$$

is an estimator of the regression coefficient.

Consider an estimator of the variance of the estimator $\bar{y}_{\text{STuned(cs)}}$ defined by

$$\hat{v}\left(\bar{y}_{\text{STuned(cs)}}\right) = r(r-1)^3 \sum_{j \in s_r} \{\bar{w}_r(j)\}^2 \left\{\bar{y}_{(j)}^{\text{STuned(cs)}} - \bar{y}_{\text{STuned(cs)}}\right\}^2 \tag{2.127}$$

Note that each newly tuned jackknifed estimator of the population mean $\bar{Y}$ is given by

$$\bar{y}_{(j)}^{\text{STuned(cs)}} = \frac{r\bar{y}_{\text{STumed(cs)}} - r\left\{(r-1)^2 \bar{w}_r(j) - (r-2)\right\} \bar{Z}_r(j)}{r-1} \tag{2.128}$$

for $j = 1, 2, \ldots, r$.

Following Singh, Joarder, and King (1996) generate a population where a study variable could be sensitive in nature and also generate scrambling variables following them. Then for different sample sizes, study the coverage of the $(1-\alpha)100\%$ confidence interval estimates given by this newly tuned estimator of the population mean by counting how many times, out of 10,000 attempts, the true population mean $\bar{Y}$ falls within the interval estimates given by:

$$\bar{y}_{\text{STuned(cs)}} \mp t_{\alpha/2}(\text{df} = ?)\sqrt{\hat{v}\left(\bar{y}_{\text{STuned(cs)}}\right)} \tag{2.129}$$

Suggest and justify your choice of degree of freedom (df).

(b) Consider a newly tuned estimator of the population mean $\bar{Y}$ given by

$$\bar{y}_{\text{STuned(dell)}} = \sum_{j \in s_r} \left[(r-1)^2 \bar{w}_r^*(j) - (r-2)\right] \bar{Z}_r(j) \tag{2.130}$$

where $0 < \bar{w}_r^*(j) < 1/(r-1)$ are the positive tuned weights constructed such that the following two constraints are satisfied:

$$\sum_{j \in s_r} \bar{w}_r^*(j) = 1 \tag{2.131}$$

and

$$\sum_{j \in s_r} \bar{w}_r^*(j)\psi_j = 0 \tag{2.132}$$

where

$$\psi_j = \bar{x}_r(j) - \frac{(\bar{x}_n - r(2-r)\bar{x}_r)}{(r-1)^2} \tag{2.133}$$

Here, for simplicity, consider the optimization of a dual-to-log-likelihood function given by

$$\sum_{j \in s_r} \frac{\ln\left(1 - \bar{w}_j^*\right)}{r} \tag{2.134}$$

or equivalently, consider the optimization of a new log-likelihood function

$$\sum_{j \in s_r} \frac{\ln\left(\bar{w}_r^*(j)\right)}{r} \tag{2.135}$$

subject to the preceding two tuning constraints.

Show that under the dual-to-log-likelihood function, the newly tuned dell estimator (2.130) of the population mean becomes

$$\bar{y}_{\text{STuned(dell)}} = \frac{(r-1)^2}{r} \sum_{j \in s_r} \frac{\bar{Z}_r(j)}{1 + \lambda_1^* \psi_j} - (r-2)\sum_{j \in s_r} \bar{Z}_r(j) \tag{2.136}$$

where λ_1^* is a solution to the nonlinear equation:

$$\sum_{j \in s_r} \frac{\psi_j}{1 + \lambda_1^* \psi_j} = 0 \qquad (2.137)$$

Consider an estimator of the variance of the estimator $\bar{y}_{\text{STuned(dell)}}$ given by

$$\hat{v}_{\text{STuned(dell)}} = r(r-1)^3 \sum_{j \in s_r} \{\bar{w}_r^*(j)\}^2 \left\{ \bar{y}_{(j)}^{\text{STuned(dell)}} - \bar{y}_{\text{STuned(dell)}} \right\}^2 \qquad (2.138)$$

Assume that each newly tuned jackknifed empirical log-likelihood estimator of the population mean is given by

$$\bar{y}_{(j)}^{\text{STuned(dell)}} = \frac{r \bar{y}_{\text{STuned(dell)}} - r\left\{(r-1)^2 \bar{w}_r^*(j) - (r-2)\right\} \bar{Z}_r(j)}{r-1} \qquad (2.139)$$

for $j = 1, 2, \ldots, r$. Study the coverage by the $(1-\alpha)100\%$ confidence interval estimates constructed using this newly tuned empirical log-likelihood estimator of the population mean, by counting how many times out of 10,000 attempts the true population mean $\bar{Y}$ falls within the interval estimates given by

$$\bar{y}_{\text{STuned(dell)}} \mp t_{\alpha/2}(\text{df} = ?) \sqrt{\hat{v}\left(\bar{y}_{\text{STuned(dell)}}\right)} \qquad (2.140)$$

Suggest and justify your choice of degree of freedom (df).
Hint: Tracy and Singh (1999).

Exercise 2.7 Estimating geometric mean Consider the problem of estimation of population geometric mean defined as

$$G_y = \left(\prod_{i=1}^N y_i\right)^{1/N} \qquad (2.141)$$

Consider a tuned estimator of the population geometric mean G_y given by

$$\hat{G}_{\text{Tuned(cs)}} = \sum_{j \in s} \left[\left\{(n-1)^2 \bar{w}_n(j) - (n-2)\right\} \hat{G}_y(j)\right] \qquad (2.142)$$

where

$$\hat{G}_y(j) = \left(\prod_{i \neq j=1}^n y_i\right)^{1/(n-1)}, \quad j = 1, 2, \ldots, n \qquad (2.143)$$

is the jth jackknifed estimator of the geometric mean of the study variable obtained by removing the jth unit from the usual estimator of the geometric mean given by

$$\hat{G}_y = \left(\prod_{i=1}^n y_i\right)^{1/n} \tag{2.144}$$

The tuning weights $\bar{w}_n(j)$ in the estimator $\hat{G}_{\text{Tuned(cs)}}$ are obtained by minimizing the tuned chi-square type distance function

$$D = \frac{n}{2}\sum_{j\in s} q_j^{-1}\left[1 - (n-1)\bar{w}_n(j) - n^{-1}\right]^2 \tag{2.145}$$

subject to the following two tuning constraints:

$$\sum_{j\in s} \bar{w}_n(j) = 1 \tag{2.146}$$

and

$$\sum_{j\in s} \bar{w}_n(j)\left[\hat{G}_x(j)\right]^{(1-n)} = \frac{1}{(n-1)}\left[\sum_{j\in s}\left(\hat{G}_x(j)\right)^{(1-n)} - \bar{X}\left(\hat{G}_x\right)^{-n}\right] \tag{2.147}$$

where $\bar{X} = N^{-1}\sum_{i=1}^N x_i$ denotes the known population arithmetic mean of the auxiliary variable, and

$$\hat{G}_x(j) = \left(\prod_{i\neq j=1}^n x_i\right)^{1/(n-1)}, \quad j=1,2,\ldots,n \tag{2.148}$$

is the jth jackknifed estimator of the geometric mean of the auxiliary variable obtained by dropping jth unit from the usual estimator of the geometric mean of the auxiliary variable given by

$$\hat{G}_x = \left(\prod_{i=1}^n x_i\right)^{1/n} \tag{2.149}$$

Suggest a doubly tuned jackknife estimator of variance of the tuned estimator of the population geometric mean, G_y. Generate a population of 10,000 pumpkins from the SJPM. Investigate the nominal 90%, 95%, and 99% coverages by simulating 5000 interval estimates for different sample sizes in the range of 10–100. Comment on your findings.

Exercise 2.8 Estimating harmonic mean Consider the problem of estimating population harmonic mean defined as

$$H_y = \frac{N}{\sum_{i=1}^{N} \frac{1}{y_i}} \tag{2.150}$$

Consider a tuned estimator of the population harmonic mean H_y given by

$$\hat{H}_{\text{Tuned(cs)}} = \sum_{j \in s} \left[\left\{ (n-1)^2 \bar{w}_n(j) - (n-2) \right\} \hat{H}_y(j) \right] \tag{2.151}$$

where

$$\hat{H}_y(j) = \frac{(n-1)}{\sum_{i \neq j=1}^{n} \frac{1}{y_i}}, \quad j = 1, 2, \ldots, n \tag{2.152}$$

is the jth jackknifed estimator of the harmonic mean of the study variable obtained by dropping the jth unit from the usual estimator of the harmonic mean given by

$$\hat{H}_y = \frac{n}{\sum_{i=1}^{n} \frac{1}{y_i}} \tag{2.153}$$

The tuning weights $\bar{w}_n(j)$ in the estimator $\hat{H}_{\text{Tuned(cs)}}$ are obtained by minimizing the tuned chi-square type distance function

$$D = \frac{n}{2} \sum_{j \in s} q_j^{-1} \left[1 - (n-1) \bar{w}_n(j) - n^{-1} \right]^2 \tag{2.154}$$

subject to the following two tuning constraints:

$$\sum_{j \in s} \bar{w}_n(j) = 1 \tag{2.155}$$

and

$$\sum_{j \in s} \frac{\bar{w}_n(j) \hat{H}_x(j)}{n\hat{H}_x(j) - (n-1)\hat{H}_x} = \frac{1}{(n-1)} \left[\sum_{j \in s} \frac{\hat{H}_x(j)}{n\hat{H}_x(j) - (n-1)\hat{H}_x} - \frac{\bar{X}}{\hat{H}_x} \right] \tag{2.156}$$

where $\bar{X} = N^{-1} \sum_{i=1}^{N} x_i$ denotes the known population arithmetic mean of the auxiliary variable, and

$$\hat{H}_x(j) = \frac{n-1}{\sum_{i \neq j=1}^{n} \frac{1}{x_i}}, \quad j = 1, 2, \ldots, n \tag{2.157}$$

is the jth jackknifed estimator of the harmonic mean of the auxiliary variable obtained by dropping the jth unit from the usual estimator of the harmonic mean of the auxiliary variable given by

$$\hat{H}_x = \frac{n}{\sum_{i=1}^{n} \frac{1}{x_i}} \tag{2.158}$$

Suggest a doubly tuned jackknife estimator of variance of the tuned estimator of the population harmonic mean, H_y. Generate a population of 10,000 pumpkins from the SJPM. Investigate the nominal 90%, 95%, and 99% coverages by simulating 5000 interval estimates for different sample sizes in the range of 10–100. Comment on your findings.

Exercise 2.9 Consider a tuned estimator of the population mean $\bar{Y}$ as

$$\bar{y}_{\text{Tuned(cs)}} = \sum_{j \in s} \left[\left\{ (n-1)^2 \bar{w}_n(j) - (n-2) \right\} \bar{y}_n(j) \right] \tag{2.159}$$

where

$$\bar{y}_n(j) = \frac{n\bar{y}_n - y_j}{n-1} \tag{2.160}$$

is the sample mean of the study variable obtained by removing the jth unit from the sample s, and $\bar{w}_n(j)$ is the jackknife tuned weight constructed such that the following two constraints are satisfied:

$$\sum_{j \in s} \bar{w}_n(j) = 1 \tag{2.161}$$

and

$$\sum_{j \in s} \bar{w}_n(j) \left[\hat{G}_x(j) \right]^{(1-n)} = \frac{1}{(n-1)} \left[\sum_{j \in s} \left(\hat{G}_x(j) \right)^{(1-n)} - \bar{X} \left(\hat{G}_x \right)^{-n} \right] \tag{2.162}$$

where $\bar{X} = N^{-1}\sum_{i=1}^{N} x_i$ denotes the known population arithmetic mean of the auxiliary variable, and

$$\hat{G}_x(j) = \left(\prod_{i \neq j=1}^{n} x_i\right)^{1/(n-1)}, \quad j=1,2,\ldots,n \tag{2.163}$$

is the jth jackknifed estimator of the geometric mean of the auxiliary variable obtained by dropping jth unit from the usual estimator of the geometric mean of the auxiliary variable given by

$$\hat{G}_x = \left(\prod_{i=1}^{n} x_i\right)^{1/n} \tag{2.164}$$

Subject to the preceding two tuned constraints in Equations (2.161) and (2.162), optimize the following distance function:

$$D_1 = (2^{-1}n)\sum_{j \in s} q_j^{-1}\left(1-(n-1)\bar{w}_n(j)-n^{-1}\right)^2 \tag{2.165}$$

Suggest a doubly tuned jackknife estimator of variance of the tuned estimator of the population mean, $\bar{Y}$. Generate a population of 10,000 pumpkins from the SJPM. Investigate the nominal 90%, 95%, and 99% coverages by simulating 5000 interval estimates for different sample sizes in the range of 10–100. Comment on your findings.

Exercise 2.10 From the SJPM code listed in **PUMPKIN1.R**, we generated a random sample of seven pumpkins and noted their weights (y) and circumferences (x) as follows:

Weight (lbs)	4430	4060	2000	4100	5080	3790	2108
Circumference (in.)	112	101	98	110	120	104	94

(a) Construct the 95% confidence interval estimate of the average weight of the pumpkins by using the tuned estimator with chi-square type distance function.
(b) Construct the 95% confidence interval estimate of the average weight of the pumpkins by using the dell estimator constructed with a dual-to-log-likelihood (dell) distance function (given: $\bar{X} = 105.4$ in.).

Model assisted tuning of estimators

3

3.1 Introduction

In this chapter, we discuss the tuning of a jackknife estimator of population mean and the estimation of its variance through a model assisted technique. Model assisted tuning of nonresponse has been discussed in one of the exercises at the end.

3.2 Model assisted tuning with a chi-square distance function

The newly tuned jackknife estimator of the population mean $\bar{Y}$ is defined as

$$\bar{y}_{\text{MATuned(cs)}} = \sum_{j \in s} \left[\left\{ (n-1)^2 \bar{w}_n(j) - (n-2) \right\} \bar{y}_n(j) \right] \quad (3.1)$$

where

$$\bar{y}_n(j) = \frac{n\bar{y}_n - y_j}{n-1} \quad (3.2)$$

is the sample mean of the study variable obtained by removing the jth unit from the sample s, and

$$\bar{w}_n(j) = \frac{1 - w_j}{n-1} \quad (3.3)$$

is the tuned jackknife weight of the calibrated weights w_j such that

$$\sum_{j \in s} w_j = 1 \quad (3.4)$$

$$\sum_{j \in s} w_j x_j = \bar{X} \quad (3.5)$$

and

$$\sum_{j \in s} w_j \hat{y}_j = \overline{\hat{Y}} \quad (3.6)$$

A New Concept for Tuning Design Weights in Survey Sampling. http://dx.doi.org/10.1016/B978-0-08-100594-1.00003-6
Copyright © 2016 Elsevier Ltd. All rights reserved.

where $\overline{\hat{Y}} = N^{-1} \sum_{i \in \Omega} \hat{y}_i$ for $\hat{y}_i = f(x_i, \hat{\beta})$ is the predicted value of the study variable based on any linear or nonlinear model.

Now a set of newly tuned jackknifed weights $\bar{w}_n(j)$ should satisfy the following three tuning constraints:

$$\sum_{j \in s} \bar{w}_n(j) = 1 \tag{3.7}$$

$$\sum_{j \in s} \bar{w}_n(j) \bar{x}_n(j) = \frac{\bar{X} - n(2-n)\bar{x}_n}{(n-1)^2} \tag{3.8}$$

and

$$\sum_{j \in s} \bar{w}_n(j) \overline{\hat{y}}_n(j) = \frac{\overline{\hat{Y}} - n(2-n)\overline{\hat{y}}_n}{(n-1)^2} \tag{3.9}$$

where

$$\bar{x}_n(j) = \frac{n\bar{x}_n - x_j}{n-1} \text{ and } \overline{\hat{y}}_n(j) = \frac{n\overline{\hat{y}}_n - \hat{y}_j}{n-1}$$

are the jth jackknifed sample means of the auxiliary variable and the predicted values obtained by removing the jth unit from the sample s, and where $\bar{x}_n = n^{-1} \sum_{i \in s} x_i$ and $\overline{\hat{y}}_n = n^{-1} \sum_{i \in s} \hat{y}_i$.

We consider the chi-square type distance function defined as

$$(2^{-1}n) \sum_{j \in s} q_j^{-1} (w_j - n^{-1})^2 = (2^{-1}n) \sum_{j \in s} q_j^{-1} (1 - (n-1)\bar{w}_n(j) - n^{-1})^2 \tag{3.10}$$

to be optimized subject to tuning constraints (3.7), (3.8), and (3.9), where q_j is an arbitrary choice of weights.

The Lagrange function is given by

$$L_1 = (2^{-1}n) \sum_{j \in s} q_j^{-1} (1 - (n-1)\bar{w}_n(j) - n^{-1})^2 - \lambda_0 \left\{ \sum_{j \in s} \bar{w}_n(j) - 1 \right\}$$

$$- \lambda_1 \left\{ \sum_{j \in s} \bar{w}_n(j) \bar{x}_n(j) - (n-1)^{-2} (\bar{X} - n(2-n)\bar{x}_n) \right\} \tag{3.11}$$

$$- \lambda_2 \left\{ \sum_{j \in s} \bar{w}_n(j) \overline{\hat{y}}_n(j) - (n-1)^{-2} (\overline{\hat{Y}} - n(2-n)\overline{\hat{y}}_n) \right\}$$

where λ_0, λ_1, and λ_2 are Lagrange multiplier constants.
On setting

$$\frac{\partial L_1}{\partial \bar{w}_n(j)} = 0 \qquad (3.12)$$

we have

$$\bar{w}_n(j) = \frac{1}{n}\left[1 + \frac{1}{(n-1)^2}\{q_j\lambda_0 + \lambda_1 q_j \bar{x}_n(j) + \lambda_2 q_j \bar{\hat{y}}_n(j)\}\right] \qquad (3.13)$$

Using Equation (3.13) in Equations (3.7), (3.8), and (3.9), a set of normal equations for the optimum values of λ_0, λ_1, and λ_2 is given by

$$\begin{bmatrix} A, & B, & C \\ B, & D, & E \\ C, & E, & F \end{bmatrix} \begin{bmatrix} \lambda_0 \\ \lambda_1 \\ \lambda_2 \end{bmatrix} = \begin{bmatrix} 0 \\ G \\ H \end{bmatrix} \qquad (3.14)$$

where

$$A = \sum_{j \in s} q_j, \quad B = \sum_{j \in s} q_j \bar{x}_n(j), \quad C = \sum_{j \in s} q_j \bar{\hat{y}}_n(j), \quad D = \sum_{j \in s} q_j \{\bar{x}_n(j)\}^2,$$

$$E = \sum_{j \in s} q_j \{\bar{x}_n(j)\bar{\hat{y}}_n(j)\}, \quad G = (n-1)^2 \left\{ \frac{n(\bar{X} - n(2-n)\bar{x}_n)}{(n-1)^2} - \sum_{j \in s} \bar{x}_n(j) \right\},$$

$$F = \sum_{j \in s} q_j \{\bar{\hat{y}}_n(j)\}^2, \text{ and } H = (n-1)^2 \left\{ \frac{n(\hat{Y} - n(2-n)\bar{\hat{y}}_n)}{(n-1)^2} - \sum_{j \in s} \bar{\hat{y}}_n(j) \right\}.$$

The newly tuned jackknifed weights $\bar{w}_n(j)$ are given by

$$\bar{w}_n(j) = \frac{1}{n}\left[1 + \frac{1}{(n-1)^2}\{\hat{K}_{1j}G + \hat{K}_{2j}H\}\right] \qquad (3.15)$$

with

$$\hat{K}_{1j} = \frac{q_j}{\Delta}\left[(CE - BF) + (AF - C^2)\bar{x}_n(j) + (BC - AE)\bar{\hat{y}}_n(j)\right]$$

and

$$\hat{K}_{2j} = \frac{q_j}{\Delta}\left[(BE - CD) + (BC - AE)\bar{x}_n(j) + (AD - B^2)\bar{\hat{y}}_n(j)\right]$$

where

$$\Delta = ADF - AE^2 - B^2F - C^2D + 2BCE$$

Thus under the chi-square (cs) type distance function, the newly tuned jackknife estimator (3.1) of the population mean becomes

$$\bar{y}_{\text{MATuned(cs)}} = \frac{(n-1)^2}{n}\left[\sum_{j\in s}\bar{y}_n(j) + \hat{\beta}_1\left\{\frac{n(\bar{X} - n(2-n)\bar{x}_n)}{(n-1)^2} - \sum_{j\in s}\bar{x}_n(j)\right\}\right.$$
$$\left. + \hat{\beta}_2\left\{\frac{n(\hat{\bar{Y}} - n(2-n)\bar{\hat{y}}_n)}{(n-1)^2} - \sum_{j\in s}\bar{\hat{y}}_n(j)\right\}\right] - (n-2)\sum_{j\in s}\bar{y}_n(j)$$

(3.16)

where

$$\hat{\beta}_1 = \frac{1}{\Delta}\left[(CE - BF)\sum_{j\in s}q_j\bar{y}_n(j) + (AF - C^2)\sum_{j\in s}q_j\bar{x}_n(j)\bar{y}_n(j)\right.$$
$$\left. + (BC - AE)\sum_{j\in s}q_j\bar{\hat{y}}_n(j)\bar{y}_n(j)\right]$$

(3.17)

and

$$\hat{\beta}_2 = \frac{1}{\Delta}\left[(BE - CD)\sum_{j\in s}q_j\bar{y}_n(j) + (BC - AE)\sum_{j\in s}q_j\bar{x}_n(j)\bar{y}_n(j)\right.$$
$$\left. + (AD - B^2)\sum_{j\in s}q_j\bar{\hat{y}}_n(j)\bar{y}_n(j)\right]$$

(3.18)

are the partial regression coefficients in the newly tuned linear regression type estimator $\bar{y}_{\text{MATuned(cs)}}$ of the population mean.

3.2.1 Estimation of variance and coverage

An estimator of variance of the estimator $\bar{y}_{\text{MATuned(cs)}}$ is

$$\hat{V}_{\text{MATuned(cs)}} = n(n-1)^3 \sum_{j\in s}(\bar{w}_n^2(j))\left\{\bar{y}_{(j)}^{\text{MATuned(cs)}} - \bar{y}_{\text{MATuned(cs)}}\right\}^2$$

(3.19)

where each newly tuned doubly jackknifed estimator of the population mean is given by

$$\bar{y}_{(j)}^{\text{MATuned(cs)}} = \frac{n\bar{y}_{\text{MATuned(cs)}} - n\left\{(n-1)^2\bar{w}_n(j) - (n-2)\right\}\bar{y}_n(j)}{n-1}$$

(3.20)

for $j = 1, 2, \ldots, n$.

The coverage by the $(1-\alpha)100\%$ confidence interval estimates obtained by this newly tuned jackknife estimator of population mean is obtained by counting how many times the true population mean $\bar{Y}$ falls within the interval estimates given by

$$\bar{y}_{\text{MATuned(cs)}} \pm t_{\alpha/2}(\text{df}=n-1)\sqrt{\hat{v}_{\text{MATuned(cs)}}} \qquad (3.21)$$

Again, we used degree of freedom df $= n-1$ for our convenience. We studied 90%, 95%, and 99% coverage by the newly tuned jackknife estimator of the population mean by selecting 10,000 random samples from a finite population of $N = 70,000$ pumpkins generated by the Statistical Jumbo Pumpkin Model (SJPM). The results obtained for different sample sizes are shown in Table 3.1.

Table 3.1 shows that the coverage by the model assisted newly tuned doubly jackknifed estimator of population mean performs well in the case of small sample sizes in the range of 5–9. The nominal 90% coverage is approximated as 89.53% for a sample of 7 pumpkins, the nominal 95% coverage is approximated as 95.23% for a sample of 9 pumpkins, and the nominal 99% coverage is approximated as 96.51% for a sample of 9 pumpkins. The last column in Table 3.1 indicates the proportion of negative estimates obtained during the entire simulation process for different sample sizes.

To predict the weights of the pumpkins, we fit the nonlinear model to each sample as

$$y_i = ae^{bx_i + cx_i^2}e^{\varepsilon_i} \qquad (3.22)$$

where $\varepsilon_i \sim N(0,1)$. Therefore, model assisted tuning of the estimators works well for small sample sizes, as far as estimation of the weights of pumpkins is concerned, assuming the circumferences of the pumpkins are known. The newly tuned jackknife estimator of the population mean weight estimates works well even though the relation between the weight of pumpkins and their circumferences is not linear.

Table 3.1 **Performance of the newly tuned jackknife estimator**

Sample size (n)	90% coverage	95% coverage	99% coverage	Negative estimates
5	0.8192	0.8447	0.8821	0.0720
6	0.8570	0.8779	0.9112	0.0537
7	0.8953	0.9113	0.9329	0.0438
8	0.9230	0.9362	0.9524	0.0332
9	0.9434	0.9523	0.9651	0.0255

3.2.2 R code

The following R code, **PUMPKIN31.R**, was used to study the coverage by the newly tuned jackknife estimator based on a chi-square type distance function.

PROGRAM PUMPKIN31.R
```
set.seed(2013)
N<-70000
x<-runif(N, min=30, max=190)
m<-5.5*(exp(0.047*x - 0.0001*x*x))
z<-rnorm(N, 0, 2)
y<-m*exp(z)
mean(x)->XB; mean(y)->YB; nreps<-10000
ESTP=rep(0,nreps)
ci1.max=ci1.min=ci2.max=ci2.min=ci3.max=ci3.min=vESTP=ESTP
for (n in 5:9)
 {
 for (r in 1:nreps)
  {
  us<-sample(N,n)
  xs<-x[us]; ys<-y[us]
  model<-lm(log(ys)~xs+I(xs^2))
  exp(model$fitted)->yss
  exp(predict(model,newdata=data.frame(xs=x)))->ysp
  yssj<-(n*mean(yss) - yss)/(n-1)
  xmj<-(sum(xs) - xs)/(n-1); ymj<-(sum(ys) - ys)/(n-1)
  a<-n; b<-sum(xmj); c<- sum(yssj); d<- sum(xmj^2)
  e<- sum(xmj*yssj); f<- sum(yssj^2)
  g<- (XB-n*(2-n)*mean(xs)) - mean(xmj)*((n-1)^2)
  h<- (mean(ysp)-n*(2-n)*mean(yss))-mean(yssj)*((n-1)^2)
  delta<-a*d*f - a*e*e - b*b*f - c*c*d + 2*b*c*e
  landa1<-((c*e - b*f )+(a*f-c*c)*xmj+(b*c-a*e)*yssj)/delta
  landa2<-((b*e - c*d)+(b*c-a*e)*xmj+(a*d-b*b)*yssj)/delta
  wbnj<-(1/n) + (landa1*g + landa2*h)/((n-1)^2)
  ESTI<- n*(((n-1)^2)*wbnj - (n-2))*ymj
  ESTP[r]<- (mean(ESTI))
  EST_J<-(n*ESTP[r] - ESTI)/(n-1)
  vESTP[r]<-n*(n-1)^3*sum((wbnj^2)*((EST_J- ESTP[r])^2))
  ci1.max[r]<- ESTP[r]+qt(0.95,n-1)*sqrt(vESTP[r])
  ci1.min[r]<- ESTP[r]-qt(0.95,n-1)*sqrt(vESTP[r])
  ci2.max[r]<- ESTP[r]+qt(0.975,n-1)*sqrt(vESTP[r])
  ci2.min[r]<- ESTP[r]-qt(0.975,n-1)*sqrt(vESTP[r])
  ci3.max[r]<- ESTP[r]+qt(0.995,n-1)*sqrt(vESTP[r])
  ci3.min[r]<- ESTP[r]-qt(0.995,n-1)*sqrt(vESTP[r])
  }
 sum(ESTP <0,na.rm=T) + sum(ESTP=="NaN")->out
```

```
for (r in 1:nreps)
  if (ESTP[r]!="NaN") { if (ESTP[r] < 0) {
    ci1.max[r]<-NaN;ci1.min[r]<-NaN
    ci2.max[r]<-NaN;ci2.min[r]<-NaN
    ci3.max[r]<-NaN;ci3.min[r]<-NaN
                   }}
  round(sum(ci1.min<YB & ci1.max>YB,na.rm=T)/nreps,4)->cov1
  round(sum(ci2.min<YB & ci2.max>YB,na.rm=T)/nreps,4)->cov2
  round(sum(ci3.min<YB & ci3.max>YB,na.rm=T)/nreps,4)->cov3
  cat(n, round(out/nreps,4), cov1,cov2,cov3,'\n')
  }
}
```

3.3 Model assisted tuning with a dual-to-empirical log-likelihood (dell) function

Let w_j^* be positive calibrated weights such that the following three constraints are satisfied:

$$\sum_{j \in s} w_j^* = 1 \tag{3.23}$$

$$\sum_{j \in s} w_j^* \Phi_{1j} = 0 \tag{3.24}$$

and

$$\sum_{j \in s} w_j^* \Phi_{2j} = 0 \tag{3.25}$$

where

$$\Phi_{1j} = \left(x_j - \bar{X}\right) \tag{3.26}$$

and

$$\Phi_{2j} = \left(\hat{y}_j - \overline{\hat{Y}}\right) \tag{3.27}$$

Let $\bar{w}_n^*(j)$ be the tuned jackknife weights such that:

$$\bar{w}_n^*(j) = \frac{1 - w_j^*}{(n-1)} \tag{3.28}$$

Note that

$$0 < \bar{w}_n^*(j) < \frac{1}{(n-1)} \qquad (3.29)$$

Now we consider a newly tuned jackknife estimator of the population mean $\bar{Y}$ defined as

$$\bar{y}_{\text{MATuned(dell)}} = \sum_{j \in s} \left[(n-1)^2 \bar{w}_n^*(j) - (n-2)\right] \bar{y}_n(j) \qquad (3.30)$$

Here, for simplicity, we consider the optimization of the following dell function:

$$\sum_{j \in s} \frac{\ln\left(1 - w_j^*\right)}{n} \qquad (3.31)$$

or equivalently, optimization of a new log-likelihood function

$$\sum_{j \in s} \frac{\ln\left(\bar{w}_n^*(j)\right)}{n} \qquad (3.32)$$

such that the following three conditions are satisfied:

$$\sum_{j \in s} \bar{w}_n^*(j) = 1 \qquad (3.33)$$

$$\sum_{j \in s} \bar{w}_n^*(j) \Psi_{1j} = 0 \qquad (3.34)$$

and

$$\sum_{j \in s} \bar{w}_n^*(j) \Psi_{2j} = 0 \qquad (3.35)$$

where

$$\Psi_{1j} = \bar{x}_n(j) - \frac{(\bar{X} - n(2-n)\bar{x}_n)}{(n-1)^2} \qquad (3.36)$$

and

$$\Psi_{2j} = \bar{\hat{y}}_n(j) - \frac{\left(\bar{\hat{Y}} - n(2-n)\bar{\hat{y}}_n\right)}{(n-1)^2} \qquad (3.37)$$

The Lagrange function is given by

$$L_2 = \sum_{j \in s} \frac{\ln(\bar{w}_n^*(j))}{n} - \lambda_0^* \left\{ \sum_{j \in s} \bar{w}_n^*(j) - 1 \right\} \tag{3.38}$$

$$- \lambda_1^* \left\{ \sum_{j \in s} \bar{w}_n^*(j) \Psi_{1j} \right\} - \lambda_2^* \left\{ \sum_{j \in s} \bar{w}_n^*(j) \Psi_{2j} \right\}$$

where λ_0^*, λ_1^* and λ_2^* are Lagrange multiplier constants.

On setting

$$\frac{\partial L_2}{\partial \bar{w}_n^*(j)} = 0 \tag{3.39}$$

we have

$$n\bar{w}_n^*(j) = \frac{1}{1 + \lambda_1^* \Psi_{1j} + \lambda_2^* \Psi_{2j}} \tag{3.40}$$

Constraints (3.33), (3.34) and (3.35) yield $\lambda_0^* = 1$, and λ_1^* and λ_2^* are solutions to the following two nonlinear equations:

$$\sum_{j \in s} \frac{\Psi_{1j}}{1 + \lambda_1^* \Psi_{1j} + \lambda_2^* \Psi_{2j}} = 0 \tag{3.41}$$

and

$$\sum_{j \in s} \frac{\Psi_{2j}}{1 + \lambda_1^* \Psi_{1j} + \lambda_2^* \Psi_{2j}} = 0 \tag{3.42}$$

Under the dell distance function, the newly tuned jackknife estimator (3.30) of the population mean becomes

$$\bar{y}_{\text{MATuned(dell)}} = \frac{(n-1)^2}{n} \sum_{j \in s} \frac{\bar{y}_n(j)}{1 + \sum_{k=1}^{2} \lambda_k^* \Psi_{kj}} - (n-2) \sum_{j \in s} \bar{y}_n(j) \tag{3.43}$$

3.3.1 Estimation of variance and coverage

An estimator of the variance of the estimator $\bar{y}_{\text{MATuned(dell)}}$ is

$$\hat{v}_{\text{MATuned(dell)}} = n(n-1)^3 \sum_{j \in s} \{\bar{w}_n^*(j)\}^2 \left\{ \bar{y}_{(j)}^{\text{MATuned(dell)}} - \bar{y}_{\text{MATuned(dell)}} \right\}^2 \tag{3.44}$$

Note that the newly tuned empirical log-likelihood doubly jackknifed estimator of population mean is given by

$$\bar{y}^{\text{MATuned(dell)}}_{(j)} = \frac{n\bar{y}_{\text{MATuned(dell)}} - n\left((n-1)^2 \bar{w}_n^*(j) - (n-2)\right)\bar{y}_n(j)}{n-1} \quad (3.45)$$

for $j = 1, 2, \ldots, n$.

The coverage by the $(1-\alpha)100\%$ confidence interval estimates, provided by this jackknifed dell estimator of population mean, is obtained by counting how many times the true population mean $\bar{Y}$ falls within the interval estimates given by

$$\bar{y}_{\text{MATuned(dell)}} \mp t_{\alpha/2}(\text{df} = n-1)\sqrt{\hat{v}_{\text{MATuned(dell)}}} \quad (3.46)$$

Again, note the use of degree of freedom $\text{df} = n-1$. We studied coverage of the nominal 90%, 95%, and 99% intervals using 10,000 random samples selected from the SJPM. The results obtained for different sample sizes are shown in Table 3.2.

In particular, we see from Table 3.2 that using the dell estimator to construct nominal 90%, 95%, and 99% confidence intervals we obtained, respectively, 89.19% coverage with a sample of 8 pumpkins, 94.39% coverage with a sample of 9, and 99.02% coverage with a sample of 12. Thus, the newly tuned jackknifed dell estimator of population mean seems to work well for small sample sizes.

We used the following approximate values of λ_1^* and λ_2^* in computing the weights:

$$\lambda_1^* = \frac{D^*C^* - B^*E^*}{A^*C^* - B^{*2}} \quad (3.47)$$

and

Table 3.2 **Performance of the newly tuned model assisted dell estimator**

Sample size (n)	90% coverage	95% coverage	99% coverage
5	0.6464	0.7182	0.8338
6	0.7605	0.8081	0.8929
7	0.8474	0.8808	0.9342
8	0.8919	0.9181	0.9570
9	0.9298	0.9439	0.9717
10	0.9475	0.9590	0.9770
11	0.9675	0.9744	0.9852
12	0.9762	0.9820	0.9902
13	0.9863	0.9897	0.9948
14	0.8877	0.9901	0.9955
15	0.9911	0.9934	0.9957

$$\lambda_2^* = \frac{A^*E^* - B^*D^*}{A^*C^* - B^{*2}} \tag{3.48}$$

where

$$A^* = \sum_{j \in s} \Psi_{1j}^2, \ C^* = \sum_{j \in s} \Psi_{2j}^2, \ B^* = \sum_{j \in s} \Psi_{1j}\Psi_{2j}, \ D^* = \sum_{j \in s} \Psi_{1j}, \ E^* = \sum_{j \in s} \Psi_{2j}, \text{ and no}$$

doubt a better approximation may be used.

3.3.2 R code

To study the coverage of intervals obtained using the newly tuned jackknife estimator based on the dell distance function, the following R code, **PUMPKIN32.R**, was used.

PROGRAM PUMPKIN32.R
```
set.seed(2013)
N<-70000
x<-runif(N, min=30, max=190)
m<-5.5*(exp(0.047*x - 0.0001*x*x))
z<-rnorm(N, 0, 2)
y<-m*exp(z)
mean(x)->XB; mean(y)->YB
nreps<-10000
ESTP=rep(0,nreps)
ci1.max=ci1.min=ci2.max=ci2.min=ci3.max=ci3.min= ESTP
vESTP=ESTP
for (n in 5:15)
 {
 for (r in 1:nreps)
  {
  us<-sample(N,n)
  xs<-x[us]; ys<-y[us]
  model<-lm(log(ys)~xs+I(xs^2))
  exp(model$fitted)->yss
  exp(predict(model,newdata=data.frame(xs=x)))->ysp
  yssj<-(n*mean(yss) - yss)/(n-1)
  xmj<-(sum(xs) - xs)/(n-1); ymj<-(sum(ys) - ys)/(n-1)
  shi1<-xmj-(XB-n*(2-n)*mean(xs))/((n-1)^2)
  shi2<-yssj-(mean(ysp)-n*(2-n)*mean(yss))/((n-1)^2)
  a<-sum(shi1^2); b<-sum(shi1*shi2); c<- sum(shi2^2)
  d<- sum(shi1); e<- sum(shi2)
  l1<-(d*c-b*e)/(a*c-b*b);l2<-(a*e-b*d)/(a*c-b*b)
  wbnj<-(1/n)*(1/(1+l1*shi1+l2*shi2))
  ESTI<- n*(((n-1)^2)*wbnj - (n-2))*ymj
```

```
ESTP[r]<- mean(ESTI)
EST_J<-(n*ESTP[r] - ESTI)/(n-1)
vESTP[r]<-n*(n-1)^3*sum((wbnj^2)*((EST_J -ESTP[r])^2))
ci1.max[r]<- ESTP[r]+qt(0.95,n-1)*sqrt(vESTP[r])
ci1.min[r]<- ESTP[r]-qt(0.95,n-1)*sqrt(vESTP[r])
ci2.max[r]<- ESTP[r]+qt(0.975,n-1)*sqrt(vESTP[r])
ci2.min[r]<- ESTP[r]-qt(0.975,n-1)*sqrt(vESTP[r])
ci3.max[r]<- ESTP[r]+qt(0.995,n-1)*sqrt(vESTP[r])
ci3.min[r]<- ESTP[r]-qt(0.995,n-1)*sqrt(vESTP[r])
}
round(sum(ci1.min<YB & ci1.max>YB,na.rm=T)/nreps,4)->cov1
round(sum(ci2.min<YB & ci2.max>YB,na.rm=T)/nreps,4)->cov2
round(sum(ci3.min<YB & ci3.max>YB,na.rm=T)/nreps,4)->cov3
cat(n, cov1,cov2,cov3,'\n')
}
```

A researcher could also refer to Wu and Sitter (2001), Farrell and Singh (2002b, 2005), and Montanari and Ranalli (2005) for model calibration in survey sampling, although their approach is different than the one discussed in this chapter.

3.4 Exercises

Exercise 3.1 Consider a newly tuned jackknife estimator of the population mean $\bar{Y}$:

$$\bar{y}_{\text{MATuned(cs)}} = \sum_{j \in s} \left[\left\{ (n-1)^2 \bar{w}_n(j) - (n-2) \right\} \bar{y}_n(j) \right] \tag{3.49}$$

where

$$\bar{y}_n(j) = \frac{n\bar{y}_n - y_j}{n-1} \tag{3.50}$$

is the sample mean of the study variable obtained by removing the jth unit from the sample s, and $\bar{w}_n(j)$ is the jackknife tuned weight such that the following three constraints are satisfied:

$$\sum_{j \in s} \bar{w}_n(j) = 1 \tag{3.51}$$

$$\sum_{j \in s} \bar{w}_n(j)\bar{x}_n(j) = \frac{\bar{X} - n(2-n)\bar{x}_n}{(n-1)^2} \tag{3.52}$$

and

$$\sum_{j\in s} \bar{w}_n(j)\bar{\hat{y}}_n(j) = \frac{\bar{Y} - n(2-n)\bar{\hat{y}}_n}{(n-1)^2} \qquad (3.53)$$

where

$$\bar{w}_n(j) = \frac{1-w_j}{n-1} \qquad (3.54)$$

for arbitrary weights w_j such that $\sum_{j\in s} w_j = 1$. Assume that the predicted values $\hat{y}_i = f(x_i, \hat{\beta})$ for $i \in \Omega$ are known, $\bar{Y} = \frac{1}{N}\sum_{i\in\Omega}\hat{y}_i$, $\bar{\hat{y}}_n = \frac{1}{n}\sum_{i\in s}\hat{y}_i$ and $\bar{\hat{y}}_n(j) = \frac{n\bar{\hat{y}}_n - \hat{y}_j}{n-1}$ have their usual meanings. Subject to these three constraints, optimize each of the following distance functions:

$$D_1 = (2^{-1}n)\sum_{j\in s} q_j^{-1}\left(1 - (n-1)\bar{w}_n(j) - n^{-1}\right)^2 \qquad (3.55)$$

$$D_2 = \sum_{j\in s}\left[\bar{w}_n(j)\ln(\bar{w}_n(j))\right], \quad 0 < \bar{w}_n(j) < 1/(n-1) \qquad (3.56)$$

$$D_3 = \sum_{j\in s} q_j^{-1}\left(\sqrt{1 - (n-1)\bar{w}_n(j)} - \sqrt{n^{-1}}\right)^2, \quad 0 < \bar{w}_n(j) < 1/(n-1) \qquad (3.57)$$

$$D_4 = \sum_{j\in s}\left[-n^{-1}\ln(\bar{w}_n(j))\right], \quad 0 < \bar{w}_n(j) < 1/(n-1) \qquad (3.58)$$

$$D_5 = \sum_{j\in s}\frac{(1 - (n-1)\bar{w}_n(j) - n^{-1})^2}{2q_j(1 - (n-1)\bar{w}_n(j))}, \quad 0 < \bar{w}_n(j) < 1/(n-1) \qquad (3.59)$$

$$D_6 = \frac{1}{n}\sum_{j\in s}\tanh^{-1}\left(\frac{\{\bar{w}_n(j)\}^2 - 1}{\{\bar{w}_n(j)\}^2 + 1}\right) \qquad (3.60)$$

$$D_7 = \frac{1}{2}\sum_{j\in s}\frac{(1 - (n-1)\bar{w}_n(j) - n^{-1})^2}{q_j n^{-1}} + \frac{1}{2}\sum_{j\in s}\frac{\varphi_j\{\bar{w}_n(j)\}^2}{q_j n^{-1}(n-1)^{-2}} \qquad (3.61)$$

and

$$D_8 = \sum_{j\in s} q_j^{-1}\left[\bar{w}_n(j)\ln(\bar{w}_n(j)) - \bar{w}_n(j) + n^{-1}\right] \qquad (3.62)$$

where q_j are suitably chosen weights that form different types of estimators, φ_j is a penalty, and $\tanh^{-1}()$ is the hyperbolic tangent function. Also, optimize each one of the preceding eight distance functions subject to only one of the tuning constraints

(3.52) and (3.53). Write code in any scientific language, like R, FORTRAN, or C, to study the resulting estimators of population mean. Discuss the nature of tuned weights in each situation. In each case, investigate the nominal 90%, 95%, and 99% confidence interval estimates by estimating the variance using the method discussed in the chapter.

Exercise 3.2 Consider a newly tuned additional constraint for Exercise 3.1:

$$\sum_{j \in s} \bar{w}_n(j) s_x^2(j) = \frac{nS_x^2 - (n-1)\{(n-1) - n(n-2)\}s_x^2}{(n-1)^2(n-2)} \quad (3.63)$$

where

$$s_x^2(j) = \frac{(n-1)s_x^2 - n(n-1)^{-1}(x_j - \bar{x}_n)^2}{n-2} \quad (3.64)$$

and $S_x^2 = (N-1)^{-1} \sum_{i \in \Omega} (x_i - \bar{X})^2$ is known. Report any changes observed in the resultant estimator.

Exercise 3.3 Assume the predicted values $\hat{y}_i = f(x_i, \hat{\beta})$ for $i \in \Omega$ are known. Consider a newly tuned additional constraint for Exercise 3.1

$$\sum_{j \in s} \bar{w}_n(j) s_{\hat{y}}^2(j) = \frac{nS_{\hat{y}}^2 - (n-1)\{(n-1) - n(n-2)\}s_{\hat{y}}^2}{(n-1)^2(n-2)} \quad (3.65)$$

where

$$s_{\hat{y}}^2(j) = \frac{(n-1)s_{\hat{y}}^2 - n(n-1)^{-1}(\hat{y}_j - \bar{\hat{y}}_n)^2}{n-2} \quad (3.66)$$

$S_{\hat{y}}^2 = N(N-1)^{-1} \sigma_{\hat{y}}^2$ and $\sigma_{\hat{y}}^2 = N^{-1} \sum_{i \in \Omega} (\hat{y}_i - \bar{Y})^2$. Report any changes observed in the resultant estimator.

Exercise 3.4 Assume that the value of the finite population correlation coefficient ρ_{xy} between the study variable y and the auxiliary variable x is known. Let us consider

$$(n-1)s_{xy} = \sum_{i \in s}(x_i - \bar{x}_n)(y_i - \bar{y}_n), \ (n-1)s_x^2 = \sum_{i \in s}(x_i - \bar{x}_n)^2,$$

$$(n-1)s_y^2 = \sum_{i \in s}(y_i - \bar{y}_n)^2, \ n\bar{x}_n = \sum_{i \in s} x_i, \ n\bar{y}_n = \sum_{i \in s} y_i, \text{ and } r = s_{xy}/(s_x s_y).$$

Consider a newly tuned additional constraint for Exercise 3.1:

$$\sum_{j \in s} \bar{w}_n(j) r(j) = \rho_{xy} \quad (3.67)$$

where the value of the partially jackknifed correlation coefficient $r(j)$ after dropping some influence of the jth pair of the study and auxiliary variable values is given by

$$r_{(j)} = \frac{r\left\{\frac{(n-1)}{(n-2)} - \frac{n(x_j - \bar{x}_n)(y_j - \bar{y}_n)}{(n-1)(n-2)s_{xy}}\right\}}{\sqrt{\frac{(n-1)}{(n-2)} - \frac{n(x_j - \bar{x}_n)^2}{(n-1)(n-2)s_x^2}}\sqrt{\frac{(n-1)}{(n-2)} - \frac{n(y_j - \bar{y}_n)^2}{(n-1)(n-2)s_y^2}}} \tag{3.68}$$

Report any changes observed in the resultant estimators.

Exercise 3.5 Assume that the heteroscedasticity of the function $v(x_i) > 0$ is known in the linear model:

$$y_i = \beta x_i + e_i \tag{3.69}$$

such that $E(e_i|x_i) = 0$, $E(e_i^2|x_i) = \sigma^2 v(x_i)$, and $E(e_i e_j|x_i x_j) = 0$.

Consider the following additional tuning constraints:

$$\sum_{j \in s} \bar{w}_n(j)\hat{\sigma}_x^2(j) = \frac{\sigma_x^2 - n(2-n)\hat{\sigma}_x^2}{(n-1)^2} \tag{3.70}$$

and

$$\sum_{j \in s} \bar{w}_n(j) v_{s(x)}^*(j) = \frac{v_{\Omega(x)}^* - n(2-n)v_{s(x)}^*}{(n-1)^2} \tag{3.71}$$

where

$$v_{\Omega(x)}^* = \frac{1}{N}\sum_{i \in \Omega} v(x_i), \quad v_{s(x)}^* = \frac{1}{n}\sum_{i \in s} v(x_i), \quad v_{s(x)}^*(j) = \frac{nv_{s(x)}^* - v(x_j)}{n-1},$$

$$\hat{\sigma}_x^2(j) = \frac{n\hat{\sigma}_x^2 - (x_j - \bar{x}_n)^2}{(n-1)}, \quad \hat{\sigma}_x^2 = n^{-1}\sum_{i \in s}(x_i - \bar{x}_n)^2, \text{ and } \sigma_x^2 = N^{-1}\sum_{i \in \Omega}(x_i - \bar{X})^2 \text{ have their}$$

usual meanings. Repeat Exercise 3.1, incorporating the preceding constraints, and discuss your findings.

Hint: Stearns and Singh (2008).

Exercise 3.6 In Exercise 3.1, find the distribution of $-2\sum_{j \in s} \ln(1 - w_j)$, or equivalently $-2\sum_{j \in s} \ln(\bar{w}_n(j))$, while optimizing the dell distance function:

$$D = \sum_{j \in s} \left[n^{-1} \ln(\bar{w}_n(j))\right], \quad 0 < \bar{w}_n(j) < 1/(n-1) \tag{3.72}$$

Exercise 3.7 Model assisted tuning of nonresponse Consider a sample s_n on n units, selected using simple random sampling scheme, where the value of the study variable y_i and an auxiliary variable x_i are measured $i = 1, 2, \ldots, n$. Assume that the responses $s_r \subset s_n$ on the study variable y_i, $i = 1, 2, \ldots, r$ for the r units are available, and the remaining $(n-r)$ responses on the study variable are missing at random. Estimate the population mean $\bar{Y}$ of the study variable. Let $\bar{y}_r = r^{-1} \sum_{i \in s_r} y_i$ be the sample mean of the study variable corresponding to the set of responding units, let $\bar{x}_r = r^{-1} \sum_{i \in s_r} x_i$ be the sample mean of the auxiliary variable corresponding to the set of responding units, and let $\bar{x}_n = n^{-1} \sum_{i \in s_n} x_i$ be the sample mean of the auxiliary variable for the entire sample selected in the survey.

(a) Consider a newly tuned jackknife estimator of the population mean $\bar{Y}$, defined as

$$\bar{y}_{\text{NRMATuned(cs)}} = \sum_{j \in s_r} \left[\left\{ (r-1)^2 \bar{w}_r(j) - (r-2) \right\} \bar{y}_r(j) \right] \qquad (3.73)$$

where

$$\bar{y}_r(j) = \frac{r\bar{y}_r - y_j}{r-1} \quad \text{and} \quad \bar{w}_r(j) = \frac{1 - w_j}{n - 1}$$

have their usual meanings. The newly tuned jackknife weights $\bar{w}_r(j)$ should satisfy the following three tuning constraints:

$$\sum_{j \in s_r} \bar{w}_r(j) = 1 \qquad (3.74)$$

$$\sum_{j \in s_r} \bar{w}_r(j) \bar{x}_r(j) = \frac{\bar{x}_n - r(2-r)\bar{x}_r}{(r-1)^2} \qquad (3.75)$$

and

$$\sum_{j \in s_r} \bar{w}_r(j) \overline{\hat{y}}_r(j) = \frac{\overline{\hat{y}}_n - r(2-r)\overline{\hat{y}}_r}{(r-1)^2} \qquad (3.76)$$

where

$$\bar{x}_r(j) = \frac{r\bar{x}_r - x_j}{r-1} \quad \text{and} \quad \overline{\hat{y}}_r(j) = \frac{r\overline{\hat{y}}_r - \hat{y}_j}{r-1}$$

are the jth jackknife sample mean of the auxiliary variable and mean of predicted values obtained by removing the jth unit from the sample s_r such that $\bar{x}_r = r^{-1} \sum_{i \in s_r} x_i$ and $\overline{\hat{y}}_r = r^{-1} \sum_{i \in s_r} \hat{y}_i$ for $\hat{y}_i = f(x_i, \hat{\beta})$ are the predicted values of the study variable based on any linear or nonlinear model. Consider the tuning of the weights $\bar{w}_r(j)$ such that the chi-square type distance function defined as

$$(2^{-1}r)\sum_{j\in s_r} q_j^{-1}\left(1-(r-1)\bar{w}_r(j)-r^{-1}\right)^2 \tag{3.77}$$

is optimum subject to tuning constraints (3.74), (3.75), and (3.76), and q_j is a choice of weights. Show that under the chi-square (cs) type distance function, the newly tuned jackknife estimator (3.72) of the population mean becomes

$$\bar{y}_{\text{NRMATuned}(cs)} = \frac{(r-1)^2}{r}\left[\sum_{j\in s_r}\bar{y}_r(j) + \hat{\beta}_1\left\{\frac{r(\bar{x}_n - r(2-r)\bar{x}_r)}{(r-1)^2} - \sum_{j\in s_r}\bar{x}_r(j)\right\}\right. \\
\left. + \hat{\beta}_2\left\{\frac{n(\bar{\hat{y}}_n - r(2-r)\bar{\hat{y}}_r)}{(r-1)^2} - \sum_{j\in s_r}\bar{\hat{y}}_r(j)\right\}\right] - (r-2)\sum_{j\in s_r}\bar{\hat{y}}_r(j) \tag{3.78}$$

where

$$\hat{\beta}_1 = \frac{1}{\Delta}\left[(CE-BF)\sum_{j\in s_r}q_j\bar{y}_r(j) + (AF-C^2)\sum_{j\in s_r}q_j\bar{x}_r(j)\bar{y}_r(j)\right. \\
\left. + (BC-AE)\sum_{j\in s_r}q_j\bar{\hat{y}}_r(j)\bar{y}_r(j)\right] \tag{3.79}$$

and

$$\hat{\beta}_2 = \frac{1}{\Delta}\left[(BE-CD)\sum_{j\in s_r}q_j\bar{y}_r(j) + (BC-AE)\sum_{j\in s_r}q_j\bar{x}_r(j)\bar{y}_r(j)\right. \\
\left. + (AD-B^2)\sum_{j\in s_r}q_j\bar{\hat{y}}_r(j)\bar{y}_r(j)\right] \tag{3.80}$$

where

$$A = \sum_{j\in s_r}q_j,\ B = \sum_{j\in s_r}q_j\bar{x}_r(j),\ C = \sum_{j\in s_r}q_j\bar{\hat{y}}_r(j),\ D = \sum_{j\in s_r}q_j\{\bar{x}_r(j)\}^2,\ E = \sum_{j\in s_r}q_j\{\bar{x}_r(j)\bar{\hat{y}}_r(j)\},$$

$$G = (r-1)^2\left\{\frac{r(\bar{x}_n - r(2-r)\bar{x}_r)}{(r-1)^2} - \sum_{j\in s_r}\bar{x}_r(j)\right\},\qquad F = \sum_{j\in s_r}q_j\{\bar{\hat{y}}_r(j)\}^2,\qquad \text{and}$$

$$H = (r-1)^2\left\{\frac{r(\bar{\hat{y}}_n - r(2-r)\bar{\hat{y}}_r)}{(r-1)^2} - \sum_{j\in s_r}\bar{\hat{y}}_r(j)\right\}.$$

Consider an estimator of variance of the estimator $\bar{y}_{\text{NRMATuned}(cs)}$ as

$$\hat{v}_{\text{NRMATuned}(cs)} = r(r-1)^3\sum_{j\in s_r}\{\bar{w}_r(j)\}^2\left\{\bar{y}^{\text{NRMATuned}(cs)}_{(j)} - \bar{y}_{\text{NRMATuned}(cs)}\right\}^2 \tag{3.81}$$

Note that each newly tuned doubly jackknifed estimator of the population mean is given by

$$\bar{y}_{(j)}^{\text{NRMATuned(cs)}} = \frac{r\bar{y}_{\text{NRMATuned(cs)}} - r\left\{(r-1)^2 \bar{w}_r(j) - (r-2)\right\}\bar{y}_r(j)}{r-1} \qquad (3.82)$$

for $j = 1, 2, \ldots, r$. Simulate a population of 10,000 units. For various sample sizes and response rates, examine the coverage by the $(1-\alpha)100\%$ confidence interval estimates provided by this newly tuned jackknife estimator of the population mean by counting how many times, say out of 100,000 attempts, the true population mean $\bar{Y}$ falls within the interval estimates given by

$$\bar{y}_{\text{NRMATuned(cs)}} \mp t_{\alpha/2}(\text{df} = ?)\sqrt{\hat{v}_{\text{NRMATuned(cs)}}} \qquad (3.83)$$

where df $=$? stands for the degree of freedom to be determined.

(b) Consider a newly tuned jackknife dell estimator of the population mean $\bar{Y}$, defined as

$$\bar{y}_{\text{NRMATuned(dell)}} = \sum_{j \in s_r} \left[(r-1)^2 \bar{w}_r^*(j) - (r-2)\right] \bar{y}_r(j) \qquad (3.84)$$

where $\bar{w}_r^*(j)$ are the tuned positive weights such that the following three constraints are satisfied:

$$\sum_{j \in s_r} \bar{w}_r^*(j) = 1 \qquad (3.85)$$

$$\sum_{j \in s_r} \bar{w}_r^*(j) \Psi_{1j} = 0 \qquad (3.86)$$

and

$$\sum_{j \in s_r} \bar{w}_r^*(j) \Psi_{2j} = 0 \qquad (3.87)$$

where

$$\Psi_{1j} = \bar{x}_r(j) - \frac{(\bar{x}_n - r(2-r)\bar{x}_r)}{(r-1)^2} \qquad (3.88)$$

and

$$\Psi_{2j} = \bar{y}_r(j) - \frac{(\bar{y}_n - r(2-r)\bar{y}_r)}{(r-1)^2} \qquad (3.89)$$

Now consider the optimization of a dell function:

$$n^{-1} \sum_{j \in s_r} \ln\left(1 - w_j^*\right) \qquad (3.90)$$

or equivalently, optimization of the log-likelihood function

$$n^{-1} \sum_{j \in s_r} \ln(\bar{w}_n^*(j)) \qquad (3.91)$$

such that conditions (3.85), (3.86), and (3.87) are satisfied. Show that under the dell distance function, the newly tuned jackknife estimator (3.84) of the population mean becomes

$$\bar{y}_{\text{NRMATuned(dell)}} = \frac{(r-1)^2}{r} \sum_{j \in s_r} \frac{\bar{y}_r(j)}{1 + \sum_{k=1}^{2} \lambda_k^* \Psi_{kj}} - (r-2) \sum_{j \in s_r} \bar{y}_r(j) \qquad (3.92)$$

where λ_1^* and λ_2^* is a solution to the single parametric equation

$$\sum_{j \in s_r} \frac{\Psi_{1j} + \Psi_{2j}}{1 + \lambda_1^* \Psi_{1j} + \lambda_2^* \Psi_{2j}} = 0 \qquad (3.93)$$

Consider an estimator of variance of the estimator $\bar{y}_{\text{NRMATuned(dell)}}$ as

$$\hat{v}_{\text{NRMATuned(dell)}} = r(r-1)^3 \sum_{j \in s_r} \{\bar{w}_r^*(j)\}^2 \left\{ \bar{y}_{(j)}^{\text{NRMATuned(dell)}} - \bar{y}_{\text{NRMATuned(dell)}} \right\}^2 \qquad (3.94)$$

where each newly tuned doubly jackknifed estimator of the population mean is given by

$$\bar{y}_{(j)}^{\text{NRMATuned(dell)}} = \frac{r \bar{y}_{\text{NRMATuned(dell)}} - r\{(r-1)^2 \bar{w}_r^*(j) - (r-2)\} \bar{y}_r(j)}{r-1} \qquad (3.95)$$

for $j = 1, 2, \ldots, r$. Simulate a population of 10,000 units where nonresponse is expected. For various sample sizes and response rates, study the $(1-\alpha)100\%$ confidence interval estimates given by this newly tuned empirical log-likelihood estimator of population mean by counting how many times out of 10,000 attempts, the true population mean $\bar{Y}$ falls within the interval estimates given by

$$\bar{y}_{\text{NRMATuned(dell)}} \mp t_{\alpha/2}(\text{df} = ?) \sqrt{\hat{v}_{\text{NRMATuned(dell)}}} \qquad (3.96)$$

where df = ? stands for the degree of freedom to be determined. Comment on your findings in each situation.

Exercise 3.8 Consider a newly tuned jackknife estimator of the population mean $\bar{Y}$:

$$\bar{y}_{\text{MATuned(cs)}} = \sum_{j \in s} \left[\{(n-1)^2 \bar{w}_n(j) - (n-2)\} \bar{y}_n(j) \right] \qquad (3.97)$$

where

$$\bar{y}_n(j) = \frac{n \bar{y}_n - y_j}{n-1} \qquad (3.98)$$

is the sample mean of the study variable obtained by removing the jth unit from the sample s, and $\bar{w}_n(j)$ is the jackknife tuned weight. Assume that the predicted values $\hat{y}_i = g(x_i, \hat{\beta})$ for $i \in \Omega$ are known from any linear or nonlinear model. Let $\overline{\hat{Y}} = \dfrac{1}{N} \sum_{i \in \Omega} \hat{y}_i$ be the known population mean of the predicted values. Let $\hat{G}_{\hat{y}} = \left(\prod_{i=1}^{n} \hat{y}_i \right)^{1/n}$ be the geometric mean of the predicted values in the sample s. Let

$$\hat{G}_{\hat{y}}(j) = \left(\prod_{i \neq j=1}^{n} \hat{y}_i \right)^{1/(n-1)}, j = 1, 2, \ldots, n,$$ be the jth jackknifed sample geometric mean of the predicted values. Also let $\hat{G}_x = \left(\prod_{i=1}^{n} x_i \right)^{1/n}$ be the sample geometric mean of the sampled auxiliary variable x values. Let $\hat{G}_x(j) = \left(\prod_{i \neq j=1}^{n} x_i \right)^{1/(n-1)}, j = 1, 2, \ldots, n,$ be the jth jackknifed sample geometric mean of the auxiliary variable. To determine the tuned weights $\bar{w}_n(j)$, optimize the following distance functions:

$$D = (2^{-1}n) \sum_{j \in s} q_j^{-1} \left(1 - (n-1)\bar{w}_n(j) - n^{-1} \right)^2 \tag{3.99}$$

such that the following three constraints are satisfied:

$$\sum_{j \in s} \bar{w}_n(j) = 1 \tag{3.100}$$

$$\sum_{j \in s} \bar{w}_n(j) [\hat{G}_x(j)]^{(1-n)} = \frac{1}{(n-1)} \left[\sum_{j \in s} [\hat{G}_x(j)]^{(1-n)} - \bar{X}(\hat{G}_x)^{-n} \right] \tag{3.101}$$

and

$$\sum_{j \in s} \bar{w}_n(j) [\hat{G}_{\hat{y}}(j)]^{(1-n)} = \frac{1}{(n-1)} \left[\sum_{j \in s} [\hat{G}_{\hat{y}}(j)]^{(1-n)} - \overline{\hat{Y}}(\hat{G}_{\hat{y}})^{-n} \right] \tag{3.102}$$

Construct an estimator of variance of the estimator of population mean $\bar{Y}$ using double jackknifing technique discussed in the text. Write R code to simulate a population of your choice of 5000 units. For various sample sizes, study the nominal 90%, 95%, and 99% coverage by the $(1-\alpha)100\%$ confidence interval estimates over 10,000 attempts. Comment on your findings.

Exercise 3.9 Consider a newly tuned jackknife estimator of the population mean $\bar{Y}$ as

$$\bar{y}_{\text{MATuned(cs)}} = \sum_{j \in s} \left[\left\{ (n-1)^2 \bar{w}_n(j) - (n-2) \right\} \bar{y}_n(j) \right] \tag{3.103}$$

where

$$\bar{y}_n(j) = \frac{n\bar{y}_n - y_j}{n-1} \tag{3.104}$$

is the sample mean of the study variable obtained by removing the jth unit from the sample s, and $\bar{w}_n(j)$ is the jackknife tuned weight. Assume that the predicted values $\hat{y}_i = h(x_i, \hat{\beta})$ for $i \in \Omega$ are known. Let $\overline{\hat{Y}} = \frac{1}{N} \sum_{i \in \Omega} \hat{y}_i$ be the known population mean of the predicted values. Let $\hat{H}_{\hat{y}} = n \left(\sum_{i=1}^{n} \hat{y}_i^{-1} \right)^{-1}$ be the harmonic mean of the predicted values in the sample s. Let $\hat{H}_{\hat{y}}(j) = (n-1) \left(\sum_{i \neq j=1}^{n} \hat{y}_i^{-1} \right)^{-1}$, $j = 1, 2, ..., n$, be the jth jackknifed sample harmonic mean of the predicted values. Also let

$$\hat{H}_x = n \left(\sum_{i=1}^{n} x_i^{-1} \right)^{-1}$$

be the sample harmonic mean of the auxiliary variable. Let

$$\hat{H}_x(j) = (n-1) \left(\sum_{i \neq j=1}^{n} x_i^{-1} \right)^{-1}, \ j = 1, 2, ..., n,$$ be the jth jackknifed sample harmonic mean of the auxiliary variable. Find the tuned weights $\bar{w}_n(j)$, by optimizing the following distance function:

$$D = (2^{-1}n) \sum_{j \in s} q_j^{-1} \left(1 - (n-1)\bar{w}_n(j) - n^{-1} \right)^2 \tag{3.105}$$

such that the following three constraints are satisfied:

$$\sum_{j \in s} \bar{w}_n(j) = 1 \tag{3.106}$$

$$\sum_{j \in s} \frac{\bar{w}_n(j) \hat{H}_x(j)}{n\hat{H}_x(j) - (n-1)\hat{H}_x} = \frac{1}{(n-1)} \left[\sum_{j \in s} \frac{\hat{H}_x(j)}{n\hat{H}_x(j) - (n-1)\hat{H}_x} - \bar{X}(\hat{H}_x)^{-1} \right] \tag{3.107}$$

and

$$\sum_{j \in s} \frac{\bar{w}_n(j) \hat{H}_{\hat{y}}(j)}{n\hat{H}_{\hat{y}}(j) - (n-1)\hat{H}_{\hat{y}}} = \frac{1}{(n-1)} \left[\sum_{j \in s} \frac{\hat{H}_{\hat{y}}(j)}{n\hat{H}_{\hat{y}}(j) - (n-1)\hat{H}_{\hat{y}}} - \overline{\hat{Y}}(\hat{H}_{\hat{y}})^{-1} \right] \tag{3.108}$$

Construct an estimator of variance of the estimator of population mean using double jackknifing technique. Write R code to simulate a population of your choice. For various sample sizes, study the nominal 90%, 95%, and 99% coverages by $(1-\alpha)100\%$ confidence interval estimates over 10,000 attempts. Comment on your findings.

Exercise 3.10 In Exercise 3.8, replace the constraint (3.102) with

$$\sum_{j \in s} \bar{w}_n(j)\bar{\hat{y}}_n(j) = (n-1)^{-2}\left[\bar{\hat{Y}} - n(2-n)\bar{\hat{y}}_n\right] \qquad (3.109)$$

and report any changes in the results.

Exercise 3.11 In Exercise 3.9, replace the constraint (3.107) with

$$\sum_{j \in s} \bar{w}_n(j)\bar{x}_n(j) = (n-1)^{-2}[\bar{X} - n(2-n)\bar{x}_n] \qquad (3.110)$$

and report any changes in the results.

Tuned estimators of finite population variance

4.1 Introduction

In this chapter, we discuss tuning a jackknife estimator of finite population variance and then estimate the variance of this estimator. We discuss tuning of the weights under a chi-square (cs) distance function and also when using the dual to the empirical log-likelihood estimation method. At the end of the chapter, a few more unsolved exercises are given.

4.2 Tuned estimator of finite population variance

The newly tuned jackknife estimator of the finite population variance

$$S_y^2 = \{2N(N-1)\}^{-1} \sum\sum_{i \neq j \in \Omega} (y_i - y_j)^2 \qquad (4.1)$$

is defined as

$$\hat{\sigma}_{T(cs)}^2 = \sum\sum_{i \neq j \in s} \left[(n(n-1)-1)^2 \bar{w}(i,j) - (n(n-1)-2) \right] s_y^2(y_{(i)}, y_{(j)}) \qquad (4.2)$$

where

$$s_y^2(y_{(i)}, y_{(j)}) = \frac{2n(n-1)s_y^2 - (y_i - y_j)^2}{2n(n-1) - 2} \qquad (4.3)$$

is the sample variance of the study variable obtained by removing the square of the difference between the ith and jth units from the sum.
Note that

$$\sum\sum_{i \neq j \in s} s_y^2(y_{(i)}, y_{(j)}) = n(n-1)s_y^2 \qquad (4.4)$$

and

$$s_y^2 = \{2n(n-1)\}^{-1} \sum\sum_{i \neq j \in s} (y_i - y_j)^2 \qquad (4.5)$$

Table 4.1 **Squared differences**

	$(y_i - y_j)^2$				
	y_1	y_2	y_3	y_4	y_5
y_1	–	4	9	100	1681
y_2	4	–	25	64	1521
y_3	9	25	–	169	1936
y_4	100	64	169	–	961
y_5	1681	1521	1936	961	–

We illustrate this with the following example. Consider a sample consisting of $n=5$ units with values $y_1 = 15$, $y_2 = 17$, $y_3 = 12$, $y_4 = 25$, and $y_5 = 56$ Now, consider the symmetric matrix given in Table 4.1.
Obviously,

$$s_y^2 = \frac{1}{2n(n-1)} \sum_{i \neq j=1}^{n} \sum^{n} (y_i - y_j)^2 = \frac{12,940}{2 \times 5 \times 4} = 323.50$$

and

$$s_y^2(2,4) = \frac{12,876}{2 \times 5 \times 4 - 2} = 338.84$$

Clearly $s_y^2(2,4)$ is not a sample variance of $y_1 = 15$, $y_3 = 12$, and $y_5 = 56$. Therefore $s_y^2(2,4)$ is a partial jackknifed estimator of variance. It is easy to verify that the equality (4.4) olds (Singh & Grewal, 2012). Partial jackknifing is kind of like taking one slice out of a pumpkin at a time, instead of removing the whole pumpkin from the sample.

Let w_{ij} be a set of calibrated weights constructed so that the chi-square distance function defined by

$$\frac{1}{2} \sum_{i \neq j \in s} \sum \frac{(w_{ij} - 1/(n(n-1)))^2}{q_{ij}/(n(n-1))} \qquad (4.6)$$

is optimum subject to the two constraints:

$$\sum_{i \neq j \in s} \sum w_{ij} = 1 \qquad (4.7)$$

and

$$\frac{1}{2} \sum_{i \neq j \in s} \sum w_{ij} (x_i - x_j)^2 = S_x^2 \qquad (4.8)$$

where

$$S_x^2 = \{2N(N-1)\}^{-1} \sum\sum_{i \neq j \in \Omega} (x_i - x_j)^2 \qquad (4.9)$$

is the known population variance of the auxiliary variable.

Now let $\bar{w}(i,j)$ be the set of two-dimensional jackknife tuned weight determined by

$$w_{ij} = n(n-1)\bar{w} - (n(n-1)-1)\bar{w}(i,j) \qquad (4.10)$$

where

$$\bar{w} = \frac{1}{n(n-1)} \sum\sum_{i \neq j \in s} w_{ij} = \frac{1}{n(n-1)} \qquad (4.11)$$

Thus, we have

$$\bar{w}(i,j) = \frac{1 - w_{ij}}{n(n-1) - 1} \qquad (4.12)$$

which satisfies the following two tuning constraints:

$$\sum\sum_{i \neq j \in s} \bar{w}(i,j) = 1 \qquad (4.13)$$

and

$$\sum\sum_{i \neq j \in s} \bar{w}(i,j) s_x^2 (x_{(i)}, x_{(j)}) = \frac{S_x^2 - n(n-1)(2 - n(n-1))s_x^2}{(n(n-1) - 1)^2} \qquad (4.14)$$

where

$$s_x^2(x_{(i)}, x_{(j)}) = \frac{n(n-1)s_x^2 - 0.5(x_i - x_j)^2}{n(n-1) - 1} \qquad (4.15)$$

and

$$s_x^2 = \{2n(n-1)\}^{-1} \sum\sum_{i \neq j \in s} (x_i - x_j)^2 \qquad (4.16)$$

4.3 Tuning with a chi-square distance

Note that the chi-square distance function defined as

$$\frac{1}{2}\sum_{i\neq j\in s}\sum \frac{(w_{ij}-1/(n(n-1)))^2}{q_{ij}/(n(n-1))} \quad (4.17)$$

is equivalent to the chi-square function of the tuned weights $\bar{w}(i,j)$ expressed as

$$\frac{1}{2}\sum_{i\neq j\in s}\sum \frac{\{1-(n(n-1)-1)\bar{w}(i,j)-1/(n(n-1))\}^2}{q_{ij}/(n(n-1))} \quad (4.18)$$

We choose the newly tuned weights $\bar{w}(i,j)$ such that the chi-square distance in Equation 4.18 is optimal subject to the tuning constraints (4.13) and (4.14), where the q_{ij} are weights used to form various estimators.

Obviously, the Lagrange function in terms of tuned weights is given by

$$L_1 = \frac{1}{2}\sum_{i\neq j\in s}\sum \frac{\{1-(n(n-1)-1)\bar{w}(i,j)-1/(n(n-1))\}^2}{q_{ij}/(n(n-1))}$$

$$-\lambda_0\left\{\sum_{i\neq j\in s}\sum \bar{w}(i,j)-1\right\} - \lambda_1\left\{\sum_{i\neq j\in s}\sum \bar{w}(i,j)s_x^2(x_{(i)},x_{(j)})\right.$$

$$\left.-\frac{S_x^2-n(n-1)(2-n(n-1))s_x^2}{(n(n-1)-1)^2}\right\} \quad (4.19)$$

where λ_0 and λ_1 are the Lagrange multiplier constants.

On setting

$$\frac{\partial L_1}{\partial \bar{w}(i,j)} = 0 \quad (4.20)$$

we have

$$\bar{w}(i,j) = \frac{1}{n(n-1)}\left[1 + \frac{1}{(n(n-1)-1)^2}\{q_{ij}\lambda_0 + \lambda_1 q_{ij}s_x^2(x_{(i)},x_{(j)})\}\right] \quad (4.21)$$

On using Equation (4.21) in Equations (4.13) and (4.14), a set of normal equations used to find the optimum values of λ_0 and λ_1 is given by

$$\left[\begin{array}{cc} \sum\sum_{i\neq j\in s} q_{ij}, & \sum\sum_{i\neq j\in s} q_{ij}s_x^2(x_{(i)},x_{(j)}) \\ \sum\sum_{i\neq j\in s} q_{ij}s_x^2(x_{(i)},x_{(j)}), & \sum\sum_{i\neq j\in s} q_{ij}\{s_x^2(x_{(i)},x_{(j)})\}^2 \end{array}\right]\left[\begin{array}{c} \lambda_0 \\ \lambda_1 \end{array}\right]$$

$$=\left[\begin{array}{c} 0 \\ (n(n-1)-1)^2\left\{\dfrac{n(n-1)\{S_x^2 - n(n-1)(2-n(n-1))s_x^2\}}{(n(n-1)-1)^2}\right. \\ \left. -\sum\sum_{i\neq j\in s} s_x^2(x_{(i)},x_{(j)})\right\} \end{array}\right]$$

Note that

$$(n(n-1)-1)^2\left\{\dfrac{n(n-1)\{S_x^2 - n(n-1)(2-n(n-1))s_x^2\}}{(n(n-1)-1)^2} - \sum\sum_{i\neq j\in s} s_x^2(x_{(i)},x_{(j)})\right\}$$

$$=n(n-1)\left(S_x^2 - s_x^2\right)$$

Thus, the set of jackknife tuned weights $\bar{w}(i,j)$ is given by

$$\bar{w}(i,j) = \dfrac{1}{n(n-1)} + \dfrac{\Delta_{ij}}{(n(n-1)-1)^2}\left(S_x^2 - s_x^2\right) \qquad (4.22)$$

where

$$\Delta_{ij} = \dfrac{\left(\sum\sum_{i\neq j\in s} q_{ij}\right)q_{ij}s_x^2(x_{(i)},x_{(j)}) - q_{ij}\sum\sum_{i\neq j\in s} q_{ij}s_x^2(x_{(i)},x_{(j)})}{\left(\sum\sum_{i\neq j\in s} q_{ij}\right)\left(\sum\sum_{i\neq j\in s} q_{ij}\{s_x^4(x_{(i)},x_{(j)})\}\right) - \left\{\sum\sum_{i\neq j\in s} q_{ij}s_x^2(x_{(i)},x_{(j)})\right\}^2}$$

(4.23)

Under the chi-square (cs) type distance function, the adjusted newly tuned jackknife estimator (4.2) of the finite population variance becomes

$$\hat{\sigma}_{T(cs)}^2 = \sum\sum_{i\neq j\in s}\left[(n(n-1)-1)^2\bar{w}(i,j) - (n(n-1)-2)\right]s_y^2(y_{(i)},y_{(j)})$$

$$=\sum\sum_{i\neq j\in s}\left[\dfrac{(n(n-1)-1)^2}{n(n-1)} + \Delta_{ij}\left(S_x^2 - s_x^2\right) - (n(n-1)-2)\right]s_y^2(y_{(i)},y_{(j)})$$

$$=s_y^2 + \hat{\beta}_T\left(S_x^2 - s_x^2\right)$$

(4.24)

where

$$\hat{\beta}_T = \sum\sum_{i\neq j\in s}\Delta_{ij}s_y^2(y_{(i)},y_{(j)})$$

$$=\frac{\sum\sum_{i\neq j\in s}q_{ij}s_x^2(x_{(i)},x_{(j)})s_y^2(y_{(i)},y_{(j)}) - \dfrac{\sum\sum_{i\neq j\in s}q_{ij}s_x^2(x_{(i)},x_{(j)})\sum\sum_{i\neq j\in s}q_{ij}s_y^2(y_{(i)},y_{(j)})}{\sum\sum_{i\neq j\in s}q_{ij}}}{\sum\sum_{i\neq j\in s}q_{ij}s_x^4(x_{(i)},x_{(j)}) - \dfrac{\left\{\sum\sum_{i\neq j\in s}q_{ij}s_x^2(x_{(i)},x_{(j)})\right\}^2}{\sum\sum_{i\neq j\in s}q_{ij}}}$$

(4.25)

denotes an estimator of the regression coefficient.

If $q_{ij}=1$ for all $i,j\in s$ then the estimator $\hat{\sigma}^2_{T(cs)}$ in Equation (4.24) becomes

$$\hat{\sigma}^2_{T(cs)} = s_y^2 + \hat{\beta}_T^*(S_x^2 - s_x^2) \qquad (4.26)$$

where

$$\hat{\beta}_T^* = \frac{\hat{\mu}_{22}^* - Ks_x^2 s_y^2}{\hat{\mu}_{40}^* - Ks_x^4} \qquad (4.27)$$

with

$$\hat{\mu}_{ab}^* = \frac{1}{2n(n-1)}\sum\sum_{i\neq j\in s}(x_i - x_j)^a(y_i - y_j)^b \qquad (4.28)$$

where a and b are nonnegative integers, and

$$K = 2\left[(n(n-1)-1)^2 + n(n-1)(2-n(n-1))\right] \qquad (4.29)$$

Note that Das and Tripathi (1978), Srivastava and Jhajj (1980), and Isaki (1983) have also studied estimators similar to the newly tuned estimator in Equation (4.26).

4.3.1 Estimation of variance of the estimator of variance and coverage

An estimator of the variance of the estimator $\hat{\sigma}^2_{T(cs)}$ is

$$\hat{v}\left(\hat{\sigma}^2_{T(cs)}\right) = n(n-1)\{n(n-1)-1\}^3 \sum\sum_{i\neq j\in s}\{\bar{w}(i,j)\}^2\left\{\hat{\sigma}^2_{T(cs)((i),(j))} - \hat{\sigma}^2_{T(cs)}\right\}^2$$

(4.30)

Note that each newly tuned doubly jackknifed estimator of the population variance is given by

$$\hat{\sigma}^2_{T(cs)((i),(j))} = \frac{n(n-1)\left[\hat{\sigma}^2_{T(cs)} - \{(n(n-1)-1)^2 \bar{w}(i,j) - \{n(n-1)-2\}\}s^2_y\left(y_{(i)}, y_{(j)}\right)\right]}{n(n-1)-1}$$

(4.31)

for $i \neq j = 1, 2, \ldots, n$.

The coverage by the $(1-\alpha)100\%$ confidence interval estimates obtained from this newly tuned jackknife estimator of population variance is obtained by counting how many times the true population variance S^2_y falls in the interval estimate given by

$$\left[\hat{\sigma}^2_{T(cs)} - F_{c_2}\sqrt{\hat{v}\left(\hat{\sigma}^2_{T(cs)}\right)}, \ \hat{\sigma}^2_{T(cs)} + F_{c_1}\sqrt{\hat{v}\left(\hat{\sigma}^2_{T(cs)}\right)}\right]$$

(4.32)

Note that the two critical values are given by $F_{c_1} = F_{\alpha/2}(n-1, n-1)$ and $F_{c_2} = F_{(1-\alpha/2)}(n-1, n-1)$. Also note that: $F_{c_2} = 1/F_{c_1}$.

We generated a population of $N = 70,000$ pumpkins with finite population variances $S^2_y = 596,582,015$ and $S^2_x = 2137.659$. We studied coverage of the nominally 90%, 95%, and 99% confidence intervals formed using the newly tuned jackknifed estimator of the finite population variance S^2_y assuming the population variance S^2_x of an auxiliary variable is known by selecting 10,000 random samples from the Statistical Jumbo Pumpkin Model (SJPM). The results obtained for various sample sizes are shown in Table 4.2.

Table 4.2 **Performance of the newly tuned jackknife estimator**

Sample size (n)	90% coverage	95% coverage	99% coverage
10	0.7186	0.7483	0.8107
12	0.8061	0.8296	0.8700
14	0.8499	0.8676	0.9009
16	0.8993	0.9111	0.9335
18	0.9335	0.9424	0.9570
20	0.9525	0.9597	0.9699
22	0.9618	0.9667	0.9752
24	0.9719	0.9746	0.9796
26	0.9833	0.9861	0.9895
28	0.9844	0.9861	0.9894
30	0.9903	0.9914	0.9939
32	0.9941	0.9947	0.9962
34	0.9935	0.9948	0.9957
36	0.9938	0.9948	0.9959
38	0.9954	0.9959	0.9967
40	0.9963	0.9967	0.9977

Table 4.2 shows that the coverage by the newly tuned jackknife interval estimators of the population variance is better than that of the interval estimators of the population mean in the previous chapters. In particular, we note that coverage of the 90% interval is approximated by 89.93% for samples of 16 pumpkins, coverage of 95% intervals is approximated by 94.24% for samples of 18 pumpkins, and coverage of the 99% intervals is approximated by 98.94% for samples of 28 pumpkins. Thus, the newly tuned jackknife estimator of the population variance of the weight of the pumpkins performs well in the case of small to moderate sample sizes. The reason for the underestimation in the smaller samples lies in the fact that the relation between the weights and circumferences is not linear.

4.3.2 R code

The following R code, **PUMPKIN41.R**, was used to study the coverage by the newly tuned estimator based on a chi-square type distance function.

```
#PROGRAM PUMPKIN41.R
set.seed(2013)
N<-70000
x<-runif(N, min=30, max=190)
m<-5.5*(exp(0.047*x - 0.0001*x*x))
z<-rnorm(N, 0,2)
y<-m*exp(z)
var(x)->SIGXP; var(y)->SIGYP
nreps<-10000
ESTPP=rep(0,nreps)
ci1.max=ci1.min=ci2.max=ci2.min=ci3.max=ci3.min=ESTPP
vESTP=ESTPP
for (n in seq(10,40,2) )
{
 for (r in 1:nreps)
  {
  us<-sample(N,n)
  xs<-x[us]; ys<-y[us]
  sx2<-matrix(ncol=n,nrow=n);sy2=sx2
  div<-n*(n-1)
  for (i in 1:n) {
   for (j in 1:n) {
    if (i!=j) {
  sx2[i,j]<-(div*var(xs)-0.5*((xs[i]-xs[j])^2) )/(div-1)
  sy2[i,j]<-(div*var(ys)-0.5*((ys[i]-ys[j])^2) )/(div-1)
              }  } }
  delta<-div*sum(sx2^2,na.rm=T)-(sum(sx2,na.rm=T))^2
  deltaij<-(div*sx2-sum(sx2,na.rm=T))/delta
  wbnij<-(1/div)+(deltaij/(div-1)^2)*(SIGXP-var(xs))
  ESTij<- div*((div-1)^2*wbnij-div+2)*sy2
```

```
ESTP<- sum(ESTij,na.rm=T)
EST_IJ<-(ESTP - ESTij)/(div-1)
ESTPP[r]<-ESTP/(div-1)
vj<-(wbnij^2)*((EST_IJ - ESTPP[r])^2)
vESTP[r]<-div*((div-1)^3)*sum(vj,na.rm=T)
ci1.max[r]<- ESTPP[r]+qf(0.95,n-1,n-1)*sqrt(vESTP[r])
ci1.min[r]<- ESTPP[r]-qf(0.05,n-1,n-1)*sqrt(vESTP[r])
ci2.max[r]<- ESTPP[r]+qf(0.975,n-1,n-1)*sqrt(vESTP[r])
ci2.min[r]<- ESTPP[r]-qf(0.025,n-1,n-1)*sqrt(vESTP[r])
ci3.max[r]<- ESTPP[r]+qf(0.995,n-1,n-1)*sqrt(vESTP[r])
ci3.min[r]<- ESTPP[r]-qf(0.005,n-1,n-1)*sqrt(vESTP[r])
}
round(sum(ci1.min<SIGYP & ci1.max>SIGYP)/nreps,4)->cov1
round(sum(ci2.min<SIGYP & ci2.max>SIGYP)/nreps,4)->cov2
round(sum(ci3.min<SIGYP & ci3.max>SIGYP)/nreps,4)->cov3
cat(n,cov1,cov2,cov3,'\n')
}
```

4.3.3 Remark on tuning with a chi-square distance

Let us consider tuning the weights $\bar{w}(i,j)$ in such a way that the chi-square type distance function defined as in Equation (4.18) that is

$$\frac{1}{2}\sum\sum_{i\neq j\in s}\frac{\{1-(n(n-1)-1)\bar{w}(i,j)-1/(n(n-1))\}^2}{q_{ij}/(n(n-1))} \qquad (4.33)$$

is minimum subject only to the single tuning constraint (4.14) where q_{ij} are some given choice of weights.

The Lagrange function can now be taken as

$$L_0 = \frac{1}{2}\sum\sum_{i\neq j\in s}\frac{\{1-(n(n-1)-1)\bar{w}(i,j)-1/(n(n-1))\}^2}{q_{ij}/(n(n-1))}$$

$$+\delta_0\left\{\sum\sum_{i\neq j\in s}\bar{w}(i,j)s_x^2(x_{(i)},x_{(j)}) - \frac{S_x^2 - n(n-1)(2-n(n-1))s_x^2}{(n(n-1)-1)^2}\right\} \qquad (4.34)$$

where δ_0 is the Lagrange multiplier constant.

On setting

$$\frac{\partial L_0}{\partial \bar{w}(i,j)} = 0 \qquad (4.35)$$

we have

$$\bar{w}(i,j) = \frac{1}{n(n-1)} + \frac{\delta_0 q_{ij} s_x^2(x_{(i)}, x_{(j)})}{n(n-1)(n(n-1)-1)} \tag{4.36}$$

Constraint (4.14) yields the value of δ_0 as

$$\delta_0 = \frac{n(n-1)(S_x^2 - s_x^2)}{(n(n-1)-1)\sum\sum_{i \neq j \in s} q_{ij} s_x^4(x_{(i)}, x_{(j)})} \tag{4.37}$$

Thus, the newly tuned jackknifed weights become

$$\bar{w}(i,j) = \frac{1}{n(n-1)} + \frac{\Delta_{ij}^*}{(n(n-1)-1)^2}(S_x^2 - s_x^2) \tag{4.38}$$

where

$$\Delta_{ij}^* = \frac{q_{ij} s_x^2(x_{(i)}, x_{(j)})}{\sum\sum_{i \neq j \in s} q_{ij} s_x^4(x_{(i)}, x_{(j)})} \tag{4.39}$$

So the generalized regression (greg) type estimator of the finite population variance in Equation 4.2 becomes

$$\hat{\sigma}^2_{T(cs^*)} = \sum\sum_{i \neq j \in s} \left[(n(n-1)-1)^2 \bar{w}(i,j) - (n(n-1)-2)\right] s_y^2(y_{(i)}, y_{(j)})$$

$$= \sum\sum_{i \neq j \in s} \left[\frac{(n(n-1)-1)^2}{n(n-1)} + \Delta_{ij}^*(S_x^2 - s_x^2) - (n(n-1)-2)\right] s_y^2(y_{(i)}, y_{(j)})$$

$$= s_y^2 + \hat{\beta}_T^*(S_x^2 - s_x^2) \tag{4.40}$$

where

$$\hat{\beta}_T^* = \sum\sum_{i \neq j \in s} \Delta_{ij}^* s_y^2(y_{(i)}, y_{(j)}) = \frac{\sum\sum_{i \neq j \in s} q_{ij} s_x^2(x_{(i)}, x_{(j)}) s_y^2(y_{(i)}, y_{(j)})}{\sum\sum_{i \neq j \in s} q_{ij} s_x^4(x_{(i)}, x_{(j)})} \tag{4.41}$$

denotes another estimator of the regression coefficient.

Note that if $q_{ij} = 1/s_x^2(x_{(i)}, x_{(j)})$, then the estimator (4.40) becomes a ratio type estimator, namely

$$\hat{\sigma}^2_{T(cs^*)} = s_y^2 \frac{S_x^2}{s_x^2} \tag{4.42}$$

Similarly, if

$$q_{ij} = 1 - \frac{\sum\sum_{i \neq j \in s} s_x^2(x_{(i)}, x_{(j)})}{n(n-1)s_x^2(x_{(i)}, x_{(j)})} \tag{4.43}$$

then the modified greg type estimator becomes the linear regression type estimator, namely

$$\hat{\sigma}^2_{T(cs^*)} = s_y^2 + \hat{\beta}_T \left(S_x^2 - s_x^2 \right) \tag{4.44}$$

where $\hat{\beta}_T$ is the same as in Equation (4.25) for $q_{ij} = 1$. However, it is difficult to find a choice of the weights q_{ij} that makes the proposed newly tuned estimators equivalent to product estimator or the traditional greg type estimators.

4.3.4 Numerical illustration

In the following example, we explain the steps involved in the computation of an estimator with the proposed method.

Example 4.1 Consider the following sample of $n = 7$ pumpkins, where x and y are the circumference (in.) and weight (lbs) of the pumpkins.

x	122	67.0	106.5	98.0	115.2	132	101.1
y	6400	800	3084	1042	4500	6700	2397

Construct a 99% confidence interval estimate of the variance of the weight of pumpkins by assuming the population variance of circumference, $S_x^2 = 440$ is known.
Solution. One can easily compute

$$s_y^2 = \frac{1}{(n-1)} \sum_{i=1}^{n} (y_i - \bar{y})^2 = 5724311.29$$

and

$$s_x^2 = \frac{1}{(n-1)} \sum_{i=1}^{n} (x_i - \bar{x})^2 = 437.65$$

The values of various steps involved are given here.

Drop (i,j)		$s_x^2(i,j)$	$s_y^2(i,j)$	$\overline{w}(i,j)$	$\hat{\sigma}^2_{T(cs)((i),(j))}$
1	2	411.4	5481489.6	0.023804	5794457.3
1	3	445.4	5729832.8	0.023811	5726249.0
1	4	441.3	5513829.1	0.023810	5739530.7
1	5	447.8	5819904.2	0.023811	5719271.3
1	6	447.1	5862831.1	0.023811	5719351.1
1	7	443.0	5668513.9	0.023811	5732391.7
2	1	411.4	5481489.6	0.023804	5794457.3
2	3	429.3	5800310.9	0.023808	5755204.6
2	4	436.6	5863214.4	0.023809	5739684.7
2	5	420.0	5696977.4	0.023806	5774956.9
2	6	396.8	5439416.4	0.023802	5821423.7
2	7	434.1	5832826.1	0.023809	5745159.1
3	1	445.4	5729832.8	0.023811	5726249.0
3	2	429.3	5800310.9	0.023808	5755204.6
3	4	447.4	5813077.9	0.023811	5720069.9
3	5	447.4	5839476.7	0.023811	5719421.6
3	6	440.4	5704471.9	0.023810	5736357.8
3	7	448.0	5858172.9	0.023812	5717806.8
4	1	441.3	5513829.1	0.023810	5739530.7
4	2	436.6	5863214.4	0.023809	5739684.7
4	3	447.4	5813077.9	0.023811	5720069.9
4	5	444.7	5718102.2	0.023811	5727846.1
4	6	434.2	5473526.6	0.023809	5753356.0
4	7	448.2	5841538.1	0.023812	5717809.0
5	1	447.8	5819904.2	0.023811	5719271.3
5	2	420.0	5696977.4	0.023806	5774956.9
5	3	447.4	5839476.7	0.023811	5719421.6
5	4	444.7	5718102.2	0.023811	5727846.1
5	6	444.9	5804904.2	0.023811	5725208.1
5	7	445.9	5809994.4	0.023811	5723118.4
6	1	447.1	5862831.1	0.023811	5719351.1
6	2	396.8	5439416.4	0.023802	5821423.7
6	3	440.4	5704471.9	0.023810	5736357.8
6	4	434.2	5473526.6	0.023809	5753356.0
6	5	444.9	5804904.2	0.023811	5725208.1
6	7	436.7	5638126.1	0.023809	5744956.2
7	1	443.0	5668513.9	0.023811	5732391.7
7	2	434.1	5832826.1	0.023809	5745159.1
7	3	448.0	5858172.9	0.023812	5717806.8
7	4	448.2	5841538.1	0.023812	5717809.0
7	5	445.9	5809994.4	0.023811	5723118.4
7	6	436.7	5638126.1	0.023809	5744956.2

where

$$\hat{\sigma}^2_{T(cs)((i),(j))} = \frac{n(n-1)\left[\hat{\sigma}^2_{T(cs)} - \left\{(n(n-1)-1)^2\bar{w}(i,j) - \{n(n-1)-2\}\right\}s_y^2\left(y_{(i)}, y_{(j)}\right)\right]}{n(n-1)-1}$$

From the preceding table, we obtain

$$\hat{\sigma}^2_{T(cs)} = 5{,}880{,}665$$

and

$$SE\left(\hat{\sigma}^2_{T(cs)}\right) = \sqrt{\hat{v}\left(\hat{\sigma}^2_{T(cs)}\right)} = 37{,}416{,}072$$

Hence the 99% confidence interval estimate for the true variance of the weights of the pumpkins is given by 234,678,113–458,859,689. An F-value of 11.073 was used.

4.3.5 R code used for illustration

We used the following R code, **PUMPKIN41EX.R**, to solve the preceding illustration.

```
# PROGRAM PUMPKIN41EX.R
n<-7; SIGXP<-440
xs<-c(122,67,106.5,98,115.2,132,101.1)
ys<-c(6400,800,3084,1042,4500,6700,2397)
sx2<-matrix(ncol=n,nrow=n)
sy2=sx2
div<-n*(n-1)
for (i in 1:n) {
 for (j in 1:n) {
  if (i!=j) {
sx2[i,j]<-(div*var(xs) - 0.5*((xs[i]-xs[j])^2) )/(div-1)
sy2[i,j]<-(div*var(ys) - 0.5*((ys[i]-ys[j])^2) )/(div-1)
          } } }
delta<-div*sum(sx2^2,na.rm=T)-(sum(sx2,na.rm=T))^2
deltaij<-(div*sx2-sum(sx2,na.rm=T))/delta
wbnij<-(1/div)+(deltaij/(div-1)^2)*(SIGXP-var(xs))
ESTij<- div*((div-1)^2*wbnij-div+2)*sy2
ESTP<- sum(ESTij,na.rm=T)
EST_IJ<-(ESTP - ESTij)/(div-1)
ESTPP<-ESTP/(div-1)
vj<-(wbnij^2)*((EST_IJ - ESTPP)^2)
vESTP<-div*((div-1)^3)*sum(vj,na.rm=T)
```

```
L<-ESTP-qf(.025,n-1,n-1)*sqrt(vESTP)
U<-ESTP+qf(.975,n-1,n-1)*sqrt(vESTP)
cat("Tuned estimate:", ESTPP, "SE: ",vESTP^.5 ,'\n')
cat("Confidence Interval:"," ", L,"; ", U,'\n')
for (i in 1:n) {
  for (j in 1:n) {
    if (i!=j)   {
      cat(cbind(i,j,round(sx2[i,j],1), round(sy2[i,j],1),
      round(wbnij[i,j],6), round(EST_IJ[i,j],1),'\n'))
              } } }
```

4.3.6 F-distribution

Note that for a given α level of significance, we have F-distribution as shown in Figure 4.1, where c_1 and c_2 denote critical values.

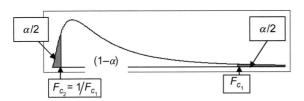

Figure 4.1 F-distribution.

4.4 Tuning of estimator of finite population variance with a dual-to-empirical log-likelihood (dell) function

Next, we consider the following newly tuned dell jackknifed estimator of the finite population variance S_y^2 as

$$\hat{\sigma}^2_{\text{T(dell)}} = \sum\sum_{i \neq j \in s} \left[(n(n-1) - 1)^2 \bar{w}^*(i,j) - (n(n-1) - 2) \right] s_y^2 \big(y_{(i)}, y_{(j)} \big) \quad (4.45)$$

where

$$0 < \bar{w}^*(i,j) = \frac{1 - w^*_{ij}}{n(n-1) - 1} < \frac{1}{n(n-1) - 1} \quad (4.46)$$

are the positive tuned weights constructed so that the following two constraints are satisfied:

$$\sum\sum_{i \neq j \in s} \bar{w}^*(i,j) = 1 \quad (4.47)$$

and

$$\sum\sum_{i \neq j \in s} \bar{w}^*(i,j) \Psi_{ij} = 0 \qquad (4.48)$$

where

$$\Psi_{ij} = s_x^2(x_{(i)}, x_{(j)}) - \frac{S_x^2 - n(n-1)(2-n(n-1))s_x^2}{(n(n-1)-1)^2} \qquad (4.49)$$

is a *pivot*.

Note that the weights w_{ij}^* are similar to the weights w_{ij} in Section 4.2. Also note that for $0 < w_{ij}^* < 1$, we have

$$\sum\sum_{i \neq j \in s} w_{ij}^* = 1 \qquad (4.50)$$

and

$$\sum\sum_{i \neq j \in s} w_{ij}^* \Phi_{ij} = 0 \qquad (4.51)$$

where

$$\Phi_{ij} = (x_i - x_j)^2 - S_x^2 \qquad (4.52)$$

is a *pivot*.

Here we suggest tuning the jackknife weights $\bar{w}^*(i,j)$ in such a way that the dual-to-log-likelihood distance function, defined as

$$\frac{1}{n(n-1)} \sum\sum_{i \neq j \in s} \log(\bar{w}^*(i,j)) \qquad (4.53)$$

is optimum subject to the tuning constraints (4.47) and (4.48).

Obviously the Lagrange function is given by

$$L_2 = \frac{1}{n(n-1)} \sum\sum_{i \neq j \in s} \log(\bar{w}^*(i,j)) - \lambda_0^* \left\{ \sum\sum_{i \neq j \in s} \bar{w}^*(i,j) - 1 \right\} - \lambda_1^* \left\{ \sum\sum_{i \neq j \in s} \bar{w}^*(i,j) \Psi_{ij} \right\} \qquad (4.54)$$

where λ_0^* and λ_1^* are the Lagrange multipliers.

On setting

$$\frac{\partial L_2}{\partial \bar{w}^*(i,j)} = 0 \tag{4.55}$$

we have

$$\bar{w}^*(i,j) = \frac{1/(n(n-1))}{1+\lambda_1^* \Psi_{ij}} \tag{4.56}$$

Constraints (4.47) and (4.48) yield $\lambda_0^* = 1$, and λ_1^* is a solution to the nonlinear equation:

$$\sum\sum_{i \neq j \in s} \frac{\Psi_{ij}}{1+\lambda_1^* \Psi_{ij}} = 0 \tag{4.57}$$

Note that to have $\bar{w}_n(i,j) > 0$, we used $\lambda_1^* > 1/|\max(\Psi_{ij})|$ if $\max(\Psi_{ij}) > 0$ and $\lambda_1^* < 1/|\min(\Psi_{ij})|$ if $\max(\Psi_{ij}) < 0$. Thus, using the dell distance function, the newly tuned jackknife estimator (4.45) of the finite population variance becomes

$$\hat{\sigma}^2_{T(\text{dell})} = \sum\sum_{i \neq j \in s} \left[(n(n-1)-1)^2 \bar{w}^*(i,j) - (n(n-1)-2) \right] s_y^2(y_{(i)}, y_{(j)})$$

$$= \sum\sum_{i \neq j \in s} \left[\frac{(n(n-1)-1)^2/(n(n-1))}{1+\lambda_1^* \Psi_{ij}} - (n(n-1)-2) \right] s_y^2(y_{(i)}, y_{(j)})$$

$$\tag{4.58}$$

4.4.1 Estimation of variance and coverage

An estimator of the variance of the dell-based estimator of the finite population variance $\hat{\sigma}^2_{T(\text{dell})}$ is

$$\hat{v}\left(\hat{\sigma}^2_{T(\text{dell})}\right) = n(n-1)(n(n-1)-1)^3 \sum\sum_{i \neq j \in s} \{\bar{w}^*(i,j)\}^2 \left\{ \hat{\sigma}^2_{T(\text{dell})((i),(j))} - \hat{\sigma}^2_{T(\text{dell})} \right\}^2 \tag{4.59}$$

Note that each newly tuned jackknifed dell estimator of the population variance is given by

$$\hat{\sigma}^2_{T(\text{dell})((i),(j))} = \frac{n(n-1)\left[\hat{\sigma}^2_{T(\text{dell})} - \left\{(n(n-1)-1)^2 \bar{w}^*(i,j) - \{n(n-1)-2\}\right\} s_y^2\left(y_{(i)}, y_{(j)}\right)\right]}{n(n-1)-1}$$

(4.60)

for $i \neq j = 1, 2, \ldots, n$.

The coverage by the $(1-\alpha)100\%$ confidence interval estimates, obtained from this newly tuned jackknife dell estimator of finite population variance is estimated by counting the number of times the true finite population variance S_y^2 falls within the interval estimates given by

$$\left[\hat{\sigma}^2_{T(\text{dell})} - F_{c_2}\sqrt{\hat{v}\left(\hat{\sigma}^2_{T(\text{dell})}\right)}, \; \hat{\sigma}^2_{T(\text{dell})} + F_{c_1}\sqrt{\hat{v}\left(\hat{\sigma}^2_{T(\text{dell})}\right)}\right]$$

(4.61)

Note that the two critical values are given by $F_{c_1} = F_{\alpha/2}(n-1, n-1)$ and $F_{c_2} = F_{(1-\alpha/2)}(n-1, n-1)$. Also note that: $F_{c_2} = 1/F_{c_1}$.

In the simulation study, we approximated the value of λ_1^* by

$$\lambda_1^* \approx \frac{\sum\sum_{i \neq j \in s} \Psi_{ij}}{\sum\sum_{i \neq j \in s} \Psi_{ij}^2} = \frac{n(n-1)\left(s_x^2 - S_x^2\right)}{(n(n-1)-1)^2 \sum\sum_{i \neq j \in s} \Psi_{ij}^2}$$

(4.62)

by assuming that

$$\left|\lambda_1^* \Psi_{ij}\right| < 1$$

(4.63)

However, a better solution to the nonlinear equation may be used, if available. Note that we have not expanded the denominator $\sum\sum_{i \neq j \in s} \Psi_{ij}^2$ because it becomes a lengthy expression. The condition (4.63) will hold because in Equation (4.62) the numerator converges to zero as the sample size increases.

In the simulation study, we generated a population of $N = 70,000$ pumpkins with finite population variances $S_y^2 = 269309177.2$ and $S_x^2 = 1876.67$. We studied coverage of nominal 90%, 95%, and 99% intervals based on the newly tuned jackknifed estimator of population variance by selecting 10,000 random samples from the SJPM. The results obtained for various sample sizes are shown in Table 4.3.

Table 4.3 shows that the coverage by the newly tuned dell estimator of the population variance provides low coverage for moderate sample sizes as compared to the estimator based on chi-square type distance function. In particular, the nominal 90% coverage is shown as 90.95% for a sample of 50 pumpkins, the nominal 95% coverage

Table 4.3 **Performance of the newly tuned jackknifed dual-to-empirical log-likelihood (dell) estimator**

Sample size (n)	90% coverage	95% coverage	99% coverage
10	0.3798	0.4143	0.4861
15	0.4849	0.5163	0.5823
20	0.5759	0.6074	0.6620
25	0.6643	0.6913	0.7362
30	0.7398	0.7622	0.8009
35	0.7895	0.8083	0.8450
40	0.8349	0.8504	0.8805
45	0.8747	0.8889	0.9110
50	0.9095	0.9202	0.9380
55	0.9282	0.9368	0.9519
60	0.9463	0.9534	0.9649
65	0.9586	0.9657	0.9747
70	0.9678	0.9709	0.9791
75	0.9758	0.9795	0.9846
80	0.9862	0.9879	0.9914
85	0.9864	0.9882	0.9917
90	0.9920	0.9938	0.9953
95	0.9951	0.9958	0.9968
100	0.9972	0.9978	0.9988

shown as 95.34% for a sample of 60 pumpkins, and the nominal 99% coverage is shown as 99.14% for a sample of 80 pumpkins. Thus, the newly tuned dell estimator of the population variance of the weight of the pumpkins provides nominal coverage for 90%, 95%, and 99% for large sample sizes as compared to the estimator based on the chi-square distance function so long as estimation of variance of weight of pumpkins is concerned.

4.4.2 R code

We used the following R code, **PUMPKIN42.R**, to study the coverage of the newly tuned estimator of the finite population variance based on the dell distance function.

```
# PROGRAM PUMPKIN42.R
set.seed(2013)
N<-70000
x<-runif(N, min=30, max=190)
m<-5.5*(exp(0.047*x - 0.0001*x*x))
z<-rnorm(N, 0, 2)
y<-m*exp(z)
```

```
var(x)->SIGXP; var(y)->SIGYP
nreps<-10000
ESTPP=rep(0,nreps)
ci1.max=ci1.min=ci2.max=ci2.min=ci3.max=ci3.min= ESTPP
vESTP=ESTPP
for (n in seq(10,50,2) )
 {
 for (r in 1:nreps)
  {
  us<-sample(N,n)
  xs<-x[us]; ys<-y[us]
  sx2<-matrix(ncol=n,nrow=n)
  EST_IJ=ESTij=wbnij=sy2=sx2
  div<-n*(n-1)
  for (i in 1:n) {
    for (j in 1:n) {
      if (i!=j) {
  sx2[i,j]<-(div*var(xs)-0.5*((xs[i]-xs[j])^2) )/(div-1)
  sy2[i,j]<-(div*var(ys)-0.5*((ys[i]-ys[j])^2) )/(div-1)
                }
       }
   }
  shij<-sx2 - (SIGXP - div*(2-div)*var(xs))/(div-1)
  ll<-sum(shij,na.rm=T)/sum(shij^2,na.rm=T)
  wbnij<-(1/div)*(1/(ll*shij))
  ESTij<- div*((div-1)^2*wbnij-div+2)*sy2
  ESTP<- sum(ESTij,na.rm=T)
  EST_IJ<-(ESTP - ESTij)/(div-1)
  ESTPP[r]<-ESTP/(div-1)
  vj<-(wbnij^2)*((EST_IJ - ESTPP[r])^2)
  vESTP[r]< div*((div 1)^3)*sum(vj,na.rm=T)
  ci1.max[r]<- ESTPP[r]+qf(0.95,n-1,n-1)*sqrt(vESTP[r])
  ci1.min[r]<- ESTPP[r]-qf(0.05,n-1,n-1)*sqrt(vESTP[r])
  ci2.max[r]<- ESTPP[r]+qf(0.975,n-1,n-1)*sqrt(vESTP[r])
  ci2.min[r]<- ESTPP[r]-qf(0.025,n-1,n-1)*sqrt(vESTP[r])
  ci3.max[r]<- ESTPP[r]+qf(0.995,n-1,n-1)*sqrt(vESTP[r])
  ci3.min[r]<- ESTPP[r]-qf(0.005,n-1,n-1)*sqrt(vESTP[r])
}
round(sum(ci1.min<SIGYP & ci1.max>SIGYP)/nreps,4)->cov1
round(sum(ci2.min<SIGYP & ci2.max>SIGYP)/nreps,4)->cov2
round(sum(ci3.min<SIGYP & ci3.max>SIGYP)/nreps,4)->cov3
cat(n, cov1,cov2,cov3,'\n')
}
```

4.4.3 Numerical illustration

The following example explains the steps involved in the computation of the use of dell estimator of variance.

Example 4.2 Consider the following sample of $n=7$ pumpkins where x is the circumference (in.) and y is the weight (lbs).

x	122	67.0	106.5	98.0	115.2	132	101.1
y	6400	800	3084	1042	4500	6700	2397

Construct a 99% confidence interval estimate of the variance of the weight of the pumpkins by assuming that the population variance of circumference $S_x^2 = 440$ is known.

Solution. One can easily compute

$$s_y^2 = \frac{1}{(n-1)} \sum_{i=1}^{n} (y_i - \bar{y})^2 = 5724311.29$$

and

$$s_x^2 = \frac{1}{(n-1)} \sum_{i=1}^{n} (x_i - \bar{x})^2 = 437.65$$

The values of various steps involved are given here.

Drop (i,j)		$s_x^2(i,j)$	$s_y^2(i,j)$	$\bar{w}(i,j)$	$\hat{\sigma}^2_{T(\text{dell})((i),(j))}$
1	2	411.4	5481489.6	0.023774	6195778.5
1	3	445.4	5729832.8	0.02382	5749687.3
1	4	441.3	5513829.1	0.023815	5811789.9
1	5	447.8	5819904.2	0.023823	5713571.5
1	6	447.1	5862831.1	0.023822	5720518.7
1	7	443.0	5668513.9	0.023817	5784144.2
2	1	411.4	5481489.6	0.023774	6195778.5
2	3	429.3	5800310.9	0.023798	5965359.7
2	4	436.6	5863214.4	0.023808	5864755.7
2	5	420.0	5696977.4	0.023786	6089827.4
2	6	396.8	5439416.4	0.023754	6379944.3
2	7	434.1	5832826.1	0.023805	5899035
3	1	445.4	5729832.8	0.02382	5749687.3
3	2	429.3	5800310.9	0.023798	5965359.7
3	4	447.4	5813077.9	0.023823	5718223.2
3	5	447.4	5839476.7	0.023823	5717547.3
3	6	440.4	5704471.9	0.023813	5817607.5
3	7	448.0	5858172.9	0.023824	5708869.4

4	1	441.3	5513829.1	0.023815	5811789.9
4	2	436.6	5863214.4	0.023808	5864755.7
4	3	447.4	5813077.9	0.023823	5718223.2
4	5	444.7	5718102.2	0.023819	5759273.1
4	6	434.2	5473526.6	0.023805	5903792.6
4	7	448.2	5841538.1	0.023824	5706412
5	1	447.8	5819904.2	0.023823	5713571.5
5	2	420.0	5696977.4	0.023786	6089827.4
5	3	447.4	5839476.7	0.023823	5717547.3
5	4	444.7	5718102.2	0.023819	5759273.1
5	6	444.9	5804904.2	0.023819	5753460.1
5	7	445.9	5809994.4	0.023821	5739392.7
6	1	447.1	5862831.1	0.023822	5720518.7
6	2	396.8	5439416.4	0.023754	6379944.3
6	3	440.4	5704471.9	0.023813	5817607.5
6	4	434.2	5473526.6	0.023805	5903792.6
6	5	444.9	5804904.2	0.023819	5753460.1
6	7	436.7	5638126.1	0.023808	5868703.3
7	1	443.0	5668513.9	0.023817	5784144.2
7	2	434.1	5832826.1	0.023805	5899035
7	3	448.0	5858172.9	0.023824	5708869.4
7	4	448.2	5841538.1	0.023824	5706412
7	5	445.9	5809994.4	0.023821	5739392.7
7	6	436.7	5638126.1	0.023808	5868703.3

where

$$\hat{\sigma}^2_{T(\text{dell})((i),(j))} = \frac{n(n-1)\left[\hat{\sigma}^2_{T(\text{dell})} - \{(n(n-1)-1)^2 \bar{w}^*(i,j) - \{n(n-1)-2\}\} s_y^2\left(y_{(i)}, y_{(j)}\right)\right]}{n(n-1)-1}$$

The value of λ_1^* is approximated as

$$\lambda_1^* \approx \frac{\sum\sum_{i \neq j \in s} \Psi_{ij}}{\sum\sum_{i \neq j \in s} \Psi_{ij}^2} \approx -8.06 \times 10^{-8} \tag{4.64}$$

although a better approximation may be used if available. In this particular example, note that due to a very small value of λ_1^* there is essentially not much impact on the values of the weights $\bar{w}^*(i,j)$.

We present this example to make the reader aware that such a result is possible. From the preceding table, we obviously have

$$\hat{\sigma}^2_{T(\text{dell})} = 5,993,546$$

and

$$SE\left(\hat{\sigma}^2_{T(dell)}\right) = \sqrt{\hat{v}\left(\hat{\sigma}^2_{T(dell)}\right)} = 59,042,919$$

The 99% confidence interval estimate for the true variance of the weights of the pumpkins is given by 235,590,131–589,350,803. Note that the estimator may perform better with an increase in the sample size. Therefore, in the case of small samples, the use of chi-square distance is suggested.

4.4.4 R code used for illustration

We used the following R code, **PUMPKIN42EX.R**, to solve the preceding illustration.

```
#PROGRAM PUMPKIN42EX.R
n<-7; SIGXP<-440
xs<-c(122,67,106.5,98,115.2,132,101.1)
ys<-c(6400,800,3084,1042,4500,6700,2397)
sx2<-matrix(ncol=n,nrow=n)
sy2=sx2
div<-n*(n-1)
for (i in 1:n) {
  for (j in 1:n) {
    if (i!=j) {
    sx2[i,j]<-(div*var(xs)-0.5*((xs[i]-xs[j])^2) )/(div-1)
    sy2[i,j]<-(div*var(ys -0.5*((ys[i]-ys[j])^2) )/(div-1)
              } } }
shij<-sx2 - (SIGXP - div*(2-div)*var(xs))/(div-1)
ll<-sum(shij,na.rm=T)/sum(shij^2,na.rm=T)
wbnij<-(1/div)*(1/(ll*shij))
ESTij<- div*((div-1)^2*wbnij-div+2)*sy2
ESTP<- sum(ESTij,na.rm=T)
EST_IJ<-(ESTP - ESTij)/(div-1)
ESTPP<-ESTP/(div-1)
vj<-(wbnij^2)*((EST_IJ - ESTPP)^2)
vESTP<-div*((div-1)^3)*sum(vj,na.rm=T)
L<-ESTP-qf(.025,n-1,n-1)*sqrt(vESTP)
U<-ESTP+qf(.975,n-1,n-1)*sqrt(vESTP)
cat("Tuned estimate:", ESTPP, "SE: ",vESTP^.5 ,'\n')
cat("Confidence Interval:"," ", L,"; ", U,'\n')
for (i in 1:n) {
  for (j in 1:n) {
    if (i!=j) {
      cat(cbind(i,j,round(sx2[i,j],1),round(sy2[i,j],1),
      round(wbnij[i,j],6),round(EST_IJ[i,j],1),'\n'))
          }
      } }
```

4.5 Alternative tuning with a chi-square distance

Let us define

$$\sigma_y^2 = \frac{N-1}{N}S_y^2, \quad \sigma_x^2 = \frac{N-1}{N}S_x^2, \quad \hat{\sigma}_y^2 = \frac{n-1}{n}s_y^2, \quad \text{and} \quad \hat{\sigma}_x^2 = \frac{n-1}{n}s_x^2$$

Now we consider an alternative tuned estimator of the finite population variance σ_y^2:

$$\hat{\sigma}_{T(a_1)}^2 = \sum_{j \in s} \left[(n-1)^2 \bar{w}_v(j) - (n-2)\right] \hat{\sigma}_y^2(j) \tag{4.65}$$

where

$$\hat{\sigma}_y^2(j) = \frac{n\hat{\sigma}_y^2 - (y_j - \bar{y}_n)^2}{(n-1)} \quad \text{for} \quad j \in s \tag{4.66}$$

and $\bar{w}_v(j)$ for $j \in s$ is a set of tuned jackknifed weights such that the following two tuning constraints are satisfied:

$$\sum_{j \in s} \bar{w}_v(j) = 1 \tag{4.67}$$

and

$$\sum_{j \in s} \bar{w}_v(j) \hat{\sigma}_x^2(j) = \frac{\sigma_x^2 - n(2-n)\hat{\sigma}_x^2}{(n-1)^2} \tag{4.68}$$

where $\hat{\sigma}_x^2(j) = \left[n\hat{\sigma}_x^2 - (x_j - \bar{x}_n)^2\right]/(n-1)$ for $j \in s$ and any not explicitly defined symbol has its usual meaning. The suffix v in $\bar{w}_v(j)$ indicates that these weights are for variance estimation. Note that here $\hat{\sigma}_x^2(j)$ and $\hat{\sigma}_y^2(j)$ are not exactly the same as the jackknife estimators of variance that were defined in the previous section. We consider optimizing the chi-square distance:

$$\frac{1}{2} \sum_{j \in s} \frac{(1-(n-1)\bar{w}_v(j) - n^{-1})^2}{q_j n^{-1}} \tag{4.69}$$

where q_j is a given set of weights, subject to both tuning constraints (4.67) and (4.68). The resultant tuned weights are given by

$$\bar{w}_v(j) = \frac{1}{n} + \frac{\Delta_j}{(n-1)^2}(\sigma_x^2 - \hat{\sigma}_x^2) \tag{4.70}$$

where

$$\Delta_j = \frac{q_j\hat{\sigma}_x^2(j)\left(\sum_{j\in s}q_j\right) - q_j\left(\sum_{j\in s}q_j\hat{\sigma}_x^2(j)\right)}{\left(\sum_{j\in s}q_j\right)\left(\sum_{j\in s}q_j\hat{\sigma}_x^4(j)\right) - \left(\sum_{j\in s}q_j\hat{\sigma}_x^2(j)\right)^2} \qquad (4.71)$$

Then the alternative newly tuned estimator of the finite population variance σ_y^2 becomes

$$\begin{aligned}\hat{\sigma}_{T(a_1)}^2 &= \sum_{j\in s}\left[(n-1)^2\bar{w}_v(j)-(n-2)\right]\hat{\sigma}_y^2(j) \\ &= \hat{\sigma}_y^2 + \hat{\beta}_v\left(\sigma_x^2 - \hat{\sigma}_x^2\right)\end{aligned} \qquad (4.72)$$

where

$$\hat{\beta}_v = \frac{\left(\sum_{j\in s}q_j\right)\left(\sum_{j\in s}q_j\hat{\sigma}_x^2(j)\hat{\sigma}_y^2(j)\right) - \left(\sum_{j\in s}q_j\hat{\sigma}_y^2(j)\right)\left(\sum_{j\in s}q_j\hat{\sigma}_x^2(j)\right)}{\left(\sum_{j\in s}q_j\right)\left(\sum_{j\in s}q_j\hat{\sigma}_x^4(j)\right) - \left(\sum_{j\in s}q_j\hat{\sigma}_x^2(j)\right)^2} \qquad (4.73)$$

4.5.1 Estimation of variance and coverage

An estimator of variance of the estimator of finite population variance $\hat{\sigma}_{T(a_1)}^2$ is

$$\hat{v}\left(\hat{\sigma}_{T(a_1)}^2\right) = n(n-1)^3\sum_{j\in s}\{\bar{w}_v(j)\}^2\left\{\hat{\sigma}_{T(a_1(j))}^2 - \hat{\sigma}_{T(a_1)}^2\right\}^2 \qquad (4.74)$$

Note that each alternative newly tuned doubly jackknifed estimator of population variance is given by

$$\hat{\sigma}_{T(a_1(j))}^2 = \frac{n\hat{\sigma}_{T(a_1)}^2 - n\left((n-1)^2\bar{w}_v(j) - (n-2)\right)\hat{\sigma}_y^2(j)}{n-1} \qquad (4.75)$$

for $j = 1, 2, \ldots, n$.

The coverage by the $(1-\alpha)100\%$ confidence interval estimates obtained by the newly tuned jackknifed estimator of the finite population variance is obtained by counting the number of times the true finite population variance S_y^2 falls within the interval estimate given by

$$\left[\hat{\sigma}^2_{T(a_1)} - F_{c_2} \sqrt{\hat{v}\left(\hat{\sigma}^2_{T(a_1)}\right)}, \ \hat{\sigma}^2_{T(a_1)} + F_{c_1} \sqrt{\hat{v}\left(\hat{\sigma}^2_{T(a_1)}\right)} \right] \tag{4.76}$$

where $F_{c_1} = F_{\alpha/2}(n-1, n-1)$, $F_{c_2} = F_{(1-\alpha/2)}(n-1, n-1)$, and $F_{c_2} = 1/F_{c_1}$.

We generated a population of 70,000 pumpkins with true variance of weight $\sigma^2_y = 808{,}648{,}830$ and variance of circumference $\sigma^2_x = 1882.56$. Note that these parameters would change slightly if we were to run the program again because these values depend on the automatically generated seed values that are set up with the time clock of the computer. We studied the coverage of nominal 90%, 95%, and 99% confidence intervals using the newly tuned jackknife estimator of population variance by selecting 10,000 random samples from the SJPM. The results obtained for various sample sizes are shown in Table 4.4.

As examples, we note that the attained coverage of the 90% interval for 120 pumpkins was 92.13%, that of the 95% interval for 160 pumpkins was 95.19%, and that of the 99% interval for 380 pumpkins was 99.03%.

Table 4.4 **Alternatively tuned jackknife estimator of variance**

Sample size (n)	90% coverage	95% coverage	99% coverage
100	0.8854	0.8907	0.9032
120	0.9213	0.9265	0.9333
140	0.9385	0.9423	0.9480
160	0.9490	0.9519	0.9567
180	0.9612	0.9626	0.9652
200	0.9685	0.9702	0.9719
220	0.9723	0.9732	0.9756
240	0.9763	0.9774	0.9791
260	0.9801	0.9809	0.9830
280	0.9826	0.9835	0.9849
300	0.9859	0.9865	0.9869
320	0.9867	0.9868	0.9875
340	0.9876	0.9881	0.9888
360	0.9881	0.9887	0.9893
380	0.9890	0.9895	0.9903
400	0.9896	0.9897	0.9902
420	0.9915	0.9915	0.9924
440	0.9921	0.9925	0.9928
460	0.9918	0.9918	0.9921
480	0.9935	0.9938	0.9944
500	0.9922	0.9923	0.9927
520	0.9935	0.9934	0.9937
540	0.9935	0.9936	0.9936
560	0.9954	0.9956	0.9959
580	0.9934	0.9937	0.9937

Continued

Table 4.4 **Continued**

Sample size (n)	90% coverage	95% coverage	99% coverage
600	0.9949	0.9950	0.9953
620	0.9953	0.9955	0.9957
640	0.9956	0.9957	0.9960
660	0.9959	0.9959	0.9961
680	0.9954	0.9956	0.9956
700	0.9962	0.9963	0.9969

4.5.2 R code

The following R code, **PUMPKIN43.R**, was used to study the coverage of the confidence intervals constructed using the alternative newly tuned estimator of variance based on the chi-square distance function.

```
# PROGRAM PUMPKIN43.R
set.seed(2013)
N<-70000
x<-runif(N, min=30, max=190)
m<-5.5*(exp(0.047*x - 0.0001*x*x))
z<-rnorm(N, 0, 2)
y<-m*exp(z)
var(x)->SIGXP; var(y)->SIGYP
nreps<-10000
ESTPP=rep(0,nreps)
ci1.max=ci1.min=ci2.max=ci2.min=ci3.max=ci3.min=ESTPP
vESTP=ESTPP
for (n in seq(100,700,20) )
  {
  for (r in 1:nreps)
    {
    us<-sample(N,n)
    xs<-x[us]; ys<-y[us]
    sigxs<-var(xs)*(n-1)/n;sigys<-var(ys)*(n-1)/n
    sigx2<-rep(0,n)
    deltai=ESTi=wbni=sigy2=sigx2
    sigx2<-(n*sigxs - (xs-mean(xs))^2)/(n-1)
    sigy2<-(n*sigys - (ys-mean(ys))^2)/(n-1)
    delta<-n*sum(sigx2^2)-(sum(sigx2))^2
    deltai<-(n*sigx2-sum(sigx2))/delta
    wbni<-(1/n)+deltai/((n-1)^2)*(SIGXP-sigxs)
    ESTi<- n*((n-1)^2*wbni-n+2)*sigy2
```

```
            ESTP<- sum(ESTi)
            EST_I<-(ESTP - ESTi)/(n-1)
            ESTP[r]<-ESTP/n
            vj<-(wbni^2)*((EST_I - ESTP[r])^2)
            vESTP[r]<-n*((n-1)^3)*sum(vj)
            ci1.max[r]<- ESTP[r]+qf(0.95,n-1,n-1)*sqrt(vESTP[r])
            ci1.min[r]<- ESTP[r]-qf(0.05,n-1,n-1)*sqrt(vESTP[r])
            ci2.max[r]<- ESTP[r]+qf(0.975,n-1,n-1)*sqrt(vESTP[r])
            ci2.min[r]<- ESTP[r]-qf(0.025,n-1,n-1)*sqrt(vESTP[r])
            ci3.max[r]<- ESTP[r]+qf(0.995,n-1,n-1)*sqrt(vESTP[r])
            ci3.min[r]<- ESTP[r]-qf(0.005,n-1,n-1)*sqrt(vESTP[r])
         }
   round(sum(ci1.min<SIGYP & ci1.max>SIGYP)/nreps,4)->cov1
   round(sum(ci2.min<SIGYP & ci2.max>SIGYP)/nreps,4)->cov2
   round(sum(ci3.min<SIGYP & ci3.max>SIGYP)/nreps,4)->cov3
   cat(n, cov1,cov2,cov3,'\n')
      }
```

4.5.3 Numerical illustration

The following example is used to explain the computational work involved in the estimation of variance by this method.

Example 4.3 Consider the following sample of $n=7$ pumpkins where x is the circumference (in.) and y is the weight (lbs) of the pumpkins.

x	122	67.0	106.5	98.0	115.2	132	101.1
y	6400	800	3084	1042	4500	6700	2397

Construct a 99% confidence interval estimate of the variance of the pumpkin weights by assuming the population variance of circumference $S_x^2 = 440$ is known.

Solution. One can easily compute

$\hat{\sigma}_x^2(j)$	$\hat{\sigma}_y^2(j)$	$\bar{w}_v(j)$	$\hat{\sigma}_{T(a_1(j))}^2$
394.8299	4,380,450	0.1435449	5,191,414
184.5203	4,454,317	0.1362038	6,550,355
437.6025	5,686,481	0.1450379	4,579,435
427.0584	4,667,231	0.1446698	4,914,812
423.4546	5,577,179	0.1445441	4,723,343
324.7346	4,081,493	0.1410981	5,669,306
433.6939	5,498,717	0.1449015	4,659,435

where

$$\hat{\sigma}^2_{T(a_1(j))} = \frac{n\hat{\sigma}^2_{T(a_1)} - n\left((n-1)^2 \bar{w}_v(j) - (n-2)\right)\hat{\sigma}^2_y(j)}{n-1}$$

Thus,

$$\hat{\sigma}^2_{T(a_1)} = \frac{1}{n}\sum_{j \in s} \hat{\sigma}^2_{T(a_1(j))} = 5,184,014$$

and

$$SE\left(\hat{\sigma}^2_{T(a_1)}\right) = \sqrt{\hat{v}\left(\hat{\sigma}^2_{T(a_1)}\right)} = 9,424,228$$

Thus, the required 99% (F-value = 11.07) confidence interval estimate of the population variance is 4,332,917–109,538,854.

4.5.4 R code used for illustration

We used the following R code, **PUMPKIN43EX.R**, to solve the preceding illustration.

#PROGRAM PUMPKIN43EX.R
```
n<-7; SIGXP<-440
xs<-c(122,67,106.5,98,115.2,132,101.1)
ys<-c(6400,800,3084,1042,4500,6700,2397)
sigxs<-var(xs)*(n-1)/n;sigys<-var(ys)*(n-1)/n
sigx2<-rep(0,n)
sigy2=sigx2
sigx2<-(n*sigxs - (xs-mean(xs))^2)/(n-1)
sigy2<-(n*sigys - (ys-mean(ys))^2)/(n-1)
delta<-n*sum(sigx2^2)-(sum(sigx2))^2
deltai<-(n*sigx2-sum(sigx2))/delta
wbni<-(1/n)+deltai/((n-1)^2)*(SIGXP-sigxs)
ESTi<- n*((n-1)^2*wbni-n+2)*sigy2
ESTP<- sum(ESTi)
EST_I<-(ESTP - ESTi)/(n-1)
ESTP<-ESTP/n
vj<-(wbni^2)*((EST_I - ESTP)^2)
vESTP<-n*((n-1)^3)*sum(vj)
L<-ESTP-qf(.005,n-1,n-1)*sqrt(vESTP)
U<-ESTP+qf(.995,n-1,n-1)*sqrt(vESTP)
cbind(sigx2,sigy2,wbni,EST_I)
cat("Tuned estimate:", ESTP, "SE: ",vESTP^.5 ,'\n')
cat("Confidence Interval:"," ", L,"; ", U,'\n')
```

4.6 Alternative tuning with a dell function

We again consider an alternative tuned estimator of the finite population variance σ_y^2 defined by

$$\hat{\sigma}_{T(a_2)}^2 = \sum_{j \in s} \left[(n-1)^2 \bar{w}_v^*(j) - (n-2) \right] \hat{\sigma}_y^2(j) \tag{4.77}$$

We consider optimizing the dell function defined by

$$\frac{1}{n} \sum_{j \in s} \ln\left(\bar{w}_v^*(j)\right), \quad \text{where } 0 < \bar{w}_v^*(j) = \frac{1 - w_j^*}{n-1} < \frac{1}{(n-1)} \tag{4.78}$$

for certain weights $0 < w_j^* < 1$ with unit total subject to the tuning constraints given by

$$\sum_{j \in s} \bar{w}_v^*(j) = 1 \tag{4.79}$$

and

$$\sum_{j \in s} \bar{w}_v^*(j) \Psi_j^* = 0 \tag{4.80}$$

where

$$\Psi_j^* = \hat{\sigma}_x^2(j) - \frac{\sigma_x^2 - n(2-n)\hat{\sigma}_x^2}{(n-1)^2} \tag{4.81}$$

The resulting tuned empirical log-likelihood weights are given by

$$\bar{w}_v^*(j) = \frac{1/n}{1 + \lambda_1 \Psi_j^*} \tag{4.82}$$

The alternative newly tuned dell estimator of the finite population variance is

$$\begin{aligned}
\hat{\sigma}_{T(a_2)}^2 &= \sum_{j \in s} \left[(n-1)^2 \bar{w}_v^*(j) - (n-2) \right] \hat{\sigma}_y^2(j) \\
&= \sum_{j \in s} \left\{ \frac{(n-1)^2}{n\{1 + \lambda_1 \Psi_j^*\}} - (n-2) \right\} \hat{\sigma}_y^2(j)
\end{aligned} \tag{4.83}$$

where the value of λ_1 is obtained by solving the nonlinear equation

$$\sum_{j\in s}\frac{\Psi_j^*}{1+\lambda_1\Psi_j^*}=0 \qquad (4.84)$$

In the simulation study, we approximated the value of λ_1 as

$$\lambda_1 \approx \frac{\sum_{i\in s}\Psi_j^*}{\sum_{i\in s}\Psi_j^{*2}} \qquad (4.85)$$

By assuming that $\left|\lambda_1 \Psi_j^*\right| < 1$, a binomial expansion of $\left(1+\lambda_1 \Psi_j^*\right)^{-1}$ in Equation (4.84) leads to the approximation in Equation (4.85). It is easy to see that $\sum_{j\in s}\Psi_j^*$ approaches zero as the sample size increases. Note that a better approximation to λ_1 may also be used if available.

4.6.1 Estimation of variance and coverage

An estimator of variance of the estimator of the finite population variance $\hat{\sigma}_{T(a_2)}^2$ is

$$\hat{v}\left(\hat{\sigma}_{T(a_2)}^2\right) = n(n-1)^3 \sum_{j\in s}\{\bar{w}_v^*(j)\}^2 \left\{\hat{\sigma}_{T(a_2(j))}^2 - \hat{\sigma}_{T(a_2)}^2\right\}^2 \qquad (4.86)$$

Note that each alternative newly tuned doubly jackknifed estimator of population variance is given by

$$\hat{\sigma}_{T(a_2(j))}^2 = \frac{n\hat{\sigma}_{T(a_2)}^2 - n\left((n-1)^2 \bar{w}_v^*(j) - (n-2)\right)\hat{\sigma}_y^2(j)}{n-1} \qquad (4.87)$$

for $j=1,2,\ldots,n$.

The coverage by the $(1-\alpha)100\%$ confidence interval estimates obtained by the alternatively tuned jackknife empirical log-likelihood estimator of the finite population variance is obtained by counting the number of times the true finite population variance S_y^2 falls in the interval estimate given by

$$\left[\hat{\sigma}_{T(a_2)}^2 - F_{c_2}\sqrt{\hat{v}\left(\hat{\sigma}_{T(a_2)}^2\right)},\ \hat{\sigma}_{T(a_2)}^2 + F_{c_1}\sqrt{\hat{v}\left(\hat{\sigma}_{T(a_2)}^2\right)}\right] \qquad (4.88)$$

where $F_{c_1} = F_{\alpha/2}(n-1, n-1)$, $F_{c_2} = F_{(1-\alpha/2)}(n-1, n-1)$, and $F_{c_2} = 1/F_{c_1}$.

We studied coverage by the nominal 90%, 95%, and 99% confidence intervals based on the alternatively tuned jackknife estimator of the population variance by selecting 10,000 random samples from the SJPM. The results obtained for various sample sizes are shown in Table 4.5.

Table 4.5 shows that the coverage by the newly tuned jackknifed empirical log-likelihood estimator of finite population variance is quite good for large sample sizes. Coverage by the nominal 90%, 95%, and 99% intervals are, respectively, 92.13% for 120 pumpkins, 95.19% for 160 pumpkins, and 99.03% for 380 pumpkins. Large samples may be necessary to achieve the required level of confidence. The advantage here is that the weights are always positive, although in this simulation, alternative methods estimate nominal coverages equally effectively for equal sample size.

Table 4.5 **Performance of the alternatively tuned dell jackknife estimator of variance**

Sample size (n)	90% coverage	95% coverage	99% coverage
100	0.8854	0.8906	0.9031
120	0.9213	0.9265	0.9333
140	0.9385	0.9423	0.9480
160	0.9490	0.9519	0.9566
180	0.9612	0.9626	0.9652
200	0.9685	0.9702	0.9719
220	0.9723	0.9732	0.9756
240	0.9763	0.9774	0.9791
260	0.9801	0.9809	0.9830
280	0.9826	0.9835	0.9849
300	0.9859	0.9865	0.9869
320	0.9867	0.9868	0.9875
340	0.9876	0.9881	0.9888
360	0.9881	0.9887	0.9893
380	0.9890	0.9895	0.9903
400	0.9896	0.9897	0.9902
420	0.9915	0.9915	0.9924
440	0.9921	0.9925	0.9928
460	0.9918	0.9918	0.9921
480	0.9935	0.9938	0.9944
500	0.9922	0.9923	0.9927
520	0.9935	0.9934	0.9937
540	0.9935	0.9936	0.9936
560	0.9954	0.9956	0.9959
580	0.9934	0.9937	0.9937

4.6.2 R code

The following R code, **PUMPKIN44.R**, was used to study the coverage of intervals with the alternative tuned estimator of variance.

```
# PROGRAM PUMPKIN44.R
set.seed(2013)
N<-70000
x<-runif(N, min=30, max=190)
m<-5.5*(exp(0.047*x - 0.0001*x*x))
z<-rnorm(N, 0, 2)
y<-m*exp(z)
var(x)->SIGXP; var(y)->SIGYP
nreps<-10000
ESTPP=rep(0,nreps)
ci1.max=ci1.min=ci2.max=ci2.min=ci3.max=ci3.min=vESTPP
vESTP=ESTPP
for (n in seq(100,600,20) )
  {
  for (r in 1:nreps)
   {
   us<-sample(N,n)
   xs<-x[us]; ys<-y[us]
   sigxs<-var(xs)*(n-1)/n;sigys<-var(ys)*(n-1)/n
   sigx2<-rep(0,n)
   shi=ESTi=wbni=sigy2=sigx2
   sigx2<-(n*sigxs - (xs-mean(xs))^2)/(n-1)
   sigy2<-(n*sigys - (ys-mean(ys))^2)/(n-1)
   shi<- sigx2 - (SIGXP-n*(2-n)*sigxs)/((n-1)^2)
   wbni<-1/(n*(1 + sum(shi)/sum(shi^2)*shi))
   ESTi<- n*((n-1)^2*wbni-n+2)*sigy2
   ESTP<- sum(ESTi)
   EST_I<-(ESTP - ESTi)/(n-1)
   ESTP[r]<-ESTP/n
   vj<-(wbni^2)*((EST_I - ESTP[r])^2)
   vESTP[r]<-n*((n-1)^3)*sum(vj)
   ci1.max[r]<- ESTP[r]+qf(0.95,n-1,n-1)*sqrt(vESTP[r])
   ci1.min[r]<- ESTP[r]-qf(0.05,n-1,n-1)*sqrt(vESTP[r])
   ci2.max[r]<- ESTP[r]+qf(0.975,n-1,n-1)*sqrt(vESTP[r])
   ci2.min[r]<- ESTP[r]-qf(0.025,n-1,n-1)*sqrt(vESTP[r])
   ci3.max[r]<- ESTP[r]+qf(0.995,n-1,n-1)*sqrt(vESTP[r])
   ci3.min[r]<- ESTP[r]-qf(0.005,n-1,n-1)*sqrt(vESTP[r])
   }
  round(sum(ci1.min<SIGYP & ci1.max>SIGYP)/nreps,4)->cov1
  round(sum(ci2.min<SIGYP & ci2.max>SIGYP)/nreps,4)->cov2
  round(sum(ci3.min<SIGYP & ci3.max>SIGYP)/nreps,4)->cov3
  cat(n, cov1,cov2,cov3,'\n')
  }
```

4.6.3 Numerical illustration

We explain the steps involved in computation of estimator of variance using the dell method with the following example.

Example 4.4 Consider a sample of $n=7$ pumpkins where x is the circumference (in.) and y is the weight (lbs) of the pumpkins.

x	122	67.0	106.5	98.0	115.2	132	101.1
y	6400	800	3084	1042	4500	6700	2397

Construct a 99% confidence interval estimate of the population variance of the pumpkin weights by assuming that the population variance of circumference $S_x^2 = 440$ is known.

Solution. One can easily compute

$\hat{\sigma}_x^2(j)$	$\hat{\sigma}_y^2(j)$	$\bar{w}_y^*(j)$	$\hat{\sigma}_{T(a_2(j))}^2$
394.8299	4,380,450	0.1434844	5,195,155
184.5203	4,454,317	0.1364451	6,497,820
437.6025	5,686,481	0.1450059	4,579,697
427.0584	4,667,231	0.1446279	4,915,665
423.4546	5,577,179	0.1444991	4,726,496
324.7300	4,081,493	0.14110000	5668642.4
433.6900	5,498,717	0.14490000	4660356.4

where

$$\hat{\sigma}_{T(a_2(j))}^2 = \frac{n\hat{\sigma}_{T(a_2)}^2 - n\left((n-1)^2 \bar{w}_v^*(j) - (n-2)\right)\hat{\sigma}_y^2(j)}{n-1}$$

From the tabulated values, we obtain

$$\hat{\sigma}_{T(a_2)}^2 = \frac{1}{n}\sum_{j\in s}\hat{\sigma}_{T(a_2(j))}^2 = 5,177,690$$

and

$$SE\left(\hat{\sigma}_{T(a_2)}^2\right) = \sqrt{\hat{v}\left(\hat{\sigma}_{T(a_2)}^2\right)} = 9,207,481$$

Thus, the required 99% (F-value $= 11.07$) confidence interval estimate of the finite population variance is 4,346,168–107,132,480.

4.6.4 R code used for illustration

The following R code, **PUMPKIN44EX.R**, was used to solve the preceding illustration.

#PROGRAM PUMPKIN44EX.F95

```
n<-7; SIGXP<-440
xs<-c(122,67,106.5,98,115.2,132,101.1)
ys<-c(6400,800,3084,1042,4500,6700,2397)
sigxs<-var(xs)*(n-1)/n;sigys<-var(ys)*(n-1)/n
sigx2<-rep(0,n)
sigy2=sigx2
sigx2<-(n*sigxs-(xs-mean(xs))^2)/(n-1)
sigy2<-(n*sigys-(ys-mean(ys))^2)/(n-1)
shi<-sigx2 - (SIGXP-n*(2-n)*sigxs)/((n-1)^2)
wbni<-1/(n*(1+sum(shi)/sum(shi^2)*shi))
ESTi<- n*((n-1)^2*wbni-n+2)*sigy2
ESTP<- sum(ESTi)
EST_I<-(ESTP - ESTi)/(n-1)
ESTP<-ESTP/n
vj<-(wbni^2)*((EST_I - ESTP)^2)
vESTP<-n*((n-1)^3)*sum(vj)
L<-ESTP-qf(.005,n-1,n-1)*sqrt(vESTP)
U<-ESTP+qf(.995,n-1,n-1)*sqrt(vESTP)
cbind(sigx2,sigy2,wbni,EST_I)
cat("Tuned estimate:", ESTP, "SE: ",vESTP^.5 ,'\n')
cat("Confidence Interval:"," ", L,"; ", U,'\n')
```

4.7 Exercises

Exercise 4.1 Consider an estimator of the finite population variance $S_y^2 = \{2N(N-1)\}^{-1} \sum\sum_{i\neq j\in\Omega}(y_i-y_j)^2$ given by

$$\hat{\sigma}^2_{\text{Tuned}} = \sum\sum_{i\neq j\in s}\left[(n(n-1)-1)^2\bar{w}(i,j) - (n(n-1)-2)\right]s_y^2\left(y_{(i)},y_{(j)}\right) \quad (4.89)$$

where

$$s_y^2\left(y_{(i)},y_{(j)}\right) = \frac{n(n-1)s_y^2 - 0.5(y_i-y_j)^2}{n(n-1)-1} \quad (4.90)$$

and where

$$s_y^2 = \{2n(n-1)\}^{-1}\sum\sum_{i\neq j\in s}(y_i-y_j)^2 \qquad (4.91)$$

is the sample variance of the study variable obtained by removing the ith and jth units from the sample s. The tuned weights $\bar{w}(i,j)$, are constructed so that the following two constraints are satisfied:

$$\sum\sum_{i\neq j\in s}\bar{w}(i,j) = 1 \qquad (4.92)$$

and

$$\sum\sum_{i\neq j\in s}\bar{w}(i,j)s_x^2\left(x_{(i)},x_{(j)}\right) = \frac{S_x^2 - n(n-1)(2-n(n-1))s_x^2}{(n(n-1)-1)^2} \qquad (4.93)$$

where

$$\bar{w}(i,j) = \frac{1-w_{ij}}{n(n-1)-1} \quad \text{for } 0 < w_{ij} < 1 \qquad (4.94)$$

$$s_x^2\left(x_{(i)},x_{(j)}\right) = \frac{n(n-1)s_x^2 - 0.5(x_i-x_j)^2}{n(n-1)-1} \qquad (4.95)$$

$$S_x^2 = \{2N(N-1)\}^{-1}\sum\sum_{i\neq j\in\Omega}(x_i-x_j)^2 \qquad (4.96)$$

and

$$s_x^2 = \{2n(n-1)\}^{-1}\sum\sum_{i\neq j\in s}(x_i-x_j)^2 \qquad (4.97)$$

Now optimize each of the following distance functions:

$$D_{11} = \frac{1}{2}\sum\sum_{i\neq j\in s}\frac{\{1-(n(n-1)-1)\bar{w}(i,j)-1/(n(n-1))\}^2}{q_{ij}/(n(n-1))} \qquad (4.98)$$

$$D_{22} = \frac{1}{2}\sum\sum_{i\neq j\in s}\left(\sqrt{\bar{w}(i,j)}-1/\sqrt{n(n-1)}\right)^2, \quad \bar{w}(i,j) > 0 \qquad (4.99)$$

and

$$D_{33} = \sum\sum_{i \neq j \in s} \frac{\{1 - (n(n-1)-1)\bar{w}(i,j) - 1/(n(n-1))\}^2}{2q_{ij}/(n(n-1))}$$
$$+ \sum\sum_{i \neq j \in s} \frac{\varphi_{ij}\{\bar{w}(i,j)\}^2 n(n-1)}{2q_{ij}\{n(n-1)-1\}^{-2}} \quad (4.100)$$

where q_{ij} are suitably chosen weights that form various estimators, and φ_{ij} is a penalty, subject to either both tuning constraints (4.92) and (4.93) or only tuning constraint (4.93). Write code in any scientific language, such as FORTRAN, C++, or R, to study these distance functions. Discuss the nature of the tuned weights in each situation. Construct the 90%, 95%, and 99% confidence interval estimates of pumpkin weights by estimating variance using the method of double jackknifing discussed earlier in the chapter.

Exercise 4.2 Consider an estimator of the finite population variance $S_y^2 = \{2N(N-1)\}^{-1} \sum\sum_{i \neq j \in \Omega} (y_i - y_j)^2$ as

$$\hat{\sigma}^2_{\text{Tuned}} = \sum\sum_{i \neq j \in s} \left[(n(n-1)-1)^2 \bar{w}(i,j) - (n(n-1)-2) \right] s_y^2\left(y_{(i)}, y_{(j)}\right) \quad (4.101)$$

where $s_y^2(y_{(i)}, y_{(j)})$ is as defined in Exercise 4.1, and $\bar{w}(i,j)$ are the jackknifed tuned weights such that the following two constraints are satisfied:

$$\sum\sum_{i \neq j \in s} \bar{w}(i,j) = 1 \quad (4.102)$$

and

$$\sum\sum_{i \neq j \in s} \bar{w}(i,j) \left\{ s_x^2(x_{(i)}, x_{(j)}) - \frac{S_x^2 - n(n-1)(2-n(n-1))s_x^2}{(n(n-1)-1)^2} \right\} = 0 \quad (4.103)$$

where

$$s_x^2(x_{(i)}, x_{(j)}) = \frac{n(n-1)s_x^2 - 0.5(x_i - x_j)^2}{n(n-1)-1} \quad (4.104)$$

$$S_x^2 = \{2N(N-1)\}^{-1} \sum\sum_{i \neq j \in \Omega} (x_i - x_j)^2 \quad (4.105)$$

and

$$s_x^2 = \{2n(n-1)\}^{-1}\sum\sum_{i\neq j\in s}(x_i-x_j)^2 \qquad (4.106)$$

Optimize each of the following distance functions:

$$D_{44} = \frac{1}{n(n-1)}\sum\sum_{i\neq j\in s}[\bar{w}(i,j)\,\ln(\bar{w}(i,j))], \quad 0<\bar{w}(i,j)<1/(n(n-1)-1) \qquad (4.107)$$

$$D_{55} = \frac{1}{n(n-1)}\sum\sum_{i\neq j\in s}\ln(\bar{w}(i,j)), \quad 0<\bar{w}(i,j)<1/(n(n-1)-1) \qquad (4.108)$$

and

$$D_{66} = \frac{1}{n(n-1)}\sum\sum_{i\neq j\in s}\tanh^{-1}\left(\frac{\{\bar{w}(i,j)\}^2-1}{\{\bar{w}(i,j)\}^2+1}\right) \qquad (4.109)$$

where $\tanh^{-1}()$ is the hyperbolic tangent function, subject to the two tuning constraints (4.102) and (4.103). Write code in any scientific language, such as FORTRAN, C++, or R, to study these distance functions. Discuss the nature of the tuned weights in each situation. Construct the 90%, 95%, and 99% confidence interval estimates by estimating variance using the method of double jackknifing discussed earlier in the chapter. Comment on the distribution of $-2D_{55}$.

Exercise 4.3 Consider an estimator of the finite population variance $S_y^2 = \{2N(N-1)\}^{-1}\sum\sum_{i\neq j\in\Omega}(y_i-y_j)^2$ defined by

$$\hat{\sigma}_{\text{Tuned}}^2 = \sum\sum_{i\neq j\in s}\left[(n(n-1)-1)^2\bar{w}(i,j)-(n(n-1)-2)\right]s_y^2(y_{(i)},y_{(j)}) \qquad (4.110)$$

where

$$s_y^2(y_{(i)},y_{(j)}) = \frac{n(n-1)s_y^2 - 0.5(y_i-y_j)^2}{n(n-1)-1} \qquad (4.111)$$

and where

$$s_y^2 = \{2n(n-1)\}^{-1}\sum\sum_{i\neq j\in s}(y_i-y_j)^2 \qquad (4.112)$$

is the sample variance of the study variable obtained by removing the ith and jth units from the sample s. Let w_{ij} be any calibrated or design weights such that the following two constraints are satisfied:

$$\sum\sum_{i \neq j \in s} w_{ij} = 1 \qquad (4.113)$$

and

$$E_m \sum\sum_{i \neq j \in s} w_{ij}(y_i - y_j)^2 = E_m\left(S_y^2\right) \qquad (4.114)$$

where E_m denotes the expected value over the linear heteroscedastic model

$$y_i = \beta x_i + e_i \qquad (4.115)$$

such that

$$E_m(e_i|x_i) = 0, \quad E_m(e_i^2|x_i) = \sigma^2 v(x_i), \quad v(x_i) > 0, \quad \text{and} \quad E_m(e_i e_i|x_i x_j) = 0$$

Let

$$\bar{w}(i,j) = \frac{1 - w_{ij}}{n(n-1) - 1} \qquad (4.116)$$

be the jackknifed weights so that

$$\sum\sum_{i \neq j \in s} \bar{w}(i,j) = 1 \qquad (4.117)$$

Note that the new model assisted tuning constraint (4.114) is equivalent to two new tuning constraints given by

$$\sum\sum_{i \neq j \in s} \bar{w}(i,j) s_x^2\left(x_{(i)}, x_{(j)}\right) = \frac{S_x^2 - n(n-1)\{2 - n(n-1)\}s_x^2}{(n(n-1) - 1)^2} \qquad (4.118)$$

and

$$\sum\sum_{i \neq j \in s} \bar{w}(i,j) v_{s(x)}\left(x_{(i)}, x_{(j)}\right) = \frac{v_{\Omega(x)} - n(n-1)\{2 - n(n-1)\}v_{s(x)}}{(n(n-1) - 1)^2} \qquad (4.119)$$

where

$$s_x^2(x_{(i)}, x_{(j)}) = \frac{n(n-1)s_x^2 - 0.5(x_i - x_j)^2}{n(n-1) - 1} \quad (4.120)$$

and

$$v_{s(x)}(x_{(i)}, x_{(j)}) = \frac{n(n-1)v_{s(x)} - 0.5(v(x_i) + v(x_j))}{n(n-1) - 1} \quad (4.121)$$

with

$$s_x^2 = \{2n(n-1)\}^{-1} \sum\sum_{i \neq j \in s} (x_i - x_j)^2 \quad (4.122)$$

$$S_x^2 = \{2N(N-1)\}^{-1} \sum\sum_{i \neq j \in \Omega} (x_i - x_j)^2 \quad (4.123)$$

$$v_{s(x)} = \{2n(n-1)\}^{-1} \sum\sum_{i \neq j \in s} (v(x_i) + v(x_j)) \quad (4.124)$$

and

$$v_{\Omega(x)} = \{2N(N-1)\}^{-1} \sum\sum_{i \neq j \in \Omega} (v(x_i) + v(x_j)) \quad (4.125)$$

Optimize each of the following distance functions:

$$D_{11} = \frac{1}{2} \sum\sum_{i \neq j \in s} \frac{\{1 - (n(n-1) - 1)\bar{w}(i,j) - 1/(n(n-1))\}^2}{q_{ij}/(n(n-1))} \quad (4.126)$$

$$D_{22} = \frac{1}{2} \sum\sum_{i \neq j \in s} \left(\sqrt{\bar{w}(i,j)} - 1/\sqrt{n(n-1)}\right)^2, \quad \bar{w}(i,j) > 0 \quad (4.127)$$

and

$$D_{33} = \sum\sum_{i \neq j \in s} \frac{\{1 - (n(n-1) - 1)\bar{w}(i,j) - 1/(n(n-1))\}^2}{2q_{ij}/(n(n-1))}$$
$$+ \sum\sum_{i \neq j \in s} \frac{\Phi_{ij}\{\bar{w}(i,j)\}^2 n(n-1)}{2q_{ij}\{n(n-1) - 1\}^{-2}} \quad (4.128)$$

where q_{ij} are suitably chosen weights that form various estimators, and Φ_{ij} is a penalty, subject to tuning constraints (4.117)–(4.119). Write code in any scientific language, such as FORTRAN, C++, or R, to study these distance functions. Discuss the nature of the tuned weights in each situation. Construct the 90%, 95%, and 99% confidence interval estimates of the weight of pumpkins. Estimate the variance using the method of double jackknifing discussed earlier in the chapter. Discuss the special cases with $v(x_i) = x_i^g$, where g is any real value.

Exercise 4.4 Consider an estimator of the finite population variance $S_y^2 = \{2N(N-1)\}^{-1} \sum\sum_{i \neq j \in \Omega} (y_i - y_j)^2$ defined by

$$\hat{\sigma}_{\text{Tuned}}^2 = \sum\sum_{i \neq j \in s} \left[(n(n-1)-1)^2 \bar{w}(i,j) - (n(n-1)-2) \right] s_y^2(y_{(i)}, y_{(j)}) \quad (4.129)$$

where

$$s_y^2(y_{(i)}, y_{(j)}) = \frac{n(n-1)s_y^2 - 0.5(y_i - y_j)^2}{n(n-1) - 1} \quad (4.130)$$

and where

$$s_y^2 = \{2n(n-1)\}^{-1} \sum\sum_{i \neq j \in s} (y_i - y_j)^2 \quad (4.131)$$

is the sample variance of the study variable obtained by removing the ith and jth units from the sample s. Let w_{ij} be any calibrated or design weights such that the following two constraints are satisfied:

$$\sum\sum_{i \neq j \in s} w_{ij} = 1 \quad (4.132)$$

and

$$E_m \sum\sum_{i \neq j \in s} w_{ij} (y_i - y_j)^2 = E_m \left(S_y^2 \right) \quad (4.133)$$

where E_m denotes the expected value over the linear heteroscedastic model

$$y_i = \beta x_i + e_i \quad (4.134)$$

such that

$$E_m(e_i | x_i) = 0, \quad E_m(e_i^2 | x_i) = \sigma^2 v(x_i), \quad v(x_i) > 0, \quad \text{and} \quad E_m(e_i e_j | x_i x_j) = 0$$

Let

$$\bar{w}(i,j) = \frac{1-w_{ij}}{n(n-1)-1}, \quad 0 < \bar{w}(i,j) < 1/(n(n-1)-1) \qquad (4.135)$$

be the jackknifed weights, so that

$$\sum\sum_{i\neq j \in s} \bar{w}(i,j) = 1 \qquad (4.136)$$

Note that the new model assisted tuning constraint (4.133) is equivalent to two new tuning constraints given by

$$\sum\sum_{i\neq j \in s} \bar{w}(i,j)\left[s_x^2(x_{(i)},x_{(j)}) - \frac{S_x^2 - n(n-1)\{2-n(n-1)\}s_x^2}{(n(n-1)-1)^2}\right] = 0 \qquad (4.137)$$

and

$$\sum\sum_{i\neq j \in s} \bar{w}(i,j)\left\{v_{s(x)}(x_{(i)},x_{(j)}) - \frac{v_{\Omega(x)} - n(n-1)\{n(n-1)-1\}v_{s(x)}}{(n(n-1)-1)^2}\right\} = 0 \qquad (4.138)$$

where

$$s_x^2(x_{(i)},x_{(j)}) = \frac{n(n-1)s_x^2 - 0.5(x_i - x_j)^2}{n(n-1)-1} \qquad (4.139)$$

and

$$v_{s(x)}(x_{(i)},x_{(j)}) = \frac{n(n-1)v_{s(x)} - 0.5(v(x_i) + v(x_j))}{n(n-1)-1} \qquad (4.140)$$

with

$$s_x^2 = \{2n(n-1)\}^{-1}\sum\sum_{i\neq j \in s}(x_i - x_j)^2 \qquad (4.141)$$

$$S_x^2 = \{2N(N-1)\}^{-1}\sum\sum_{i\neq j \in \Omega}(x_i - x_j)^2 \qquad (4.142)$$

$$v_{s(x)} = \{2n(n-1)\}^{-1}\sum\sum_{i\neq j \in s}(v(x_i) + v(x_j)) \qquad (4.143)$$

and

$$v_{\Omega(x)} = \{2N(N-1)\}^{-1} \sum\sum_{i \neq j \in \Omega} (v(x_i) + v(x_j)) \tag{4.144}$$

Optimize each of the following distance functions:

$$D_{44} = \frac{1}{n(n-1)} \sum\sum_{i \neq j \in s} [\bar{w}(i,j) \ln(\bar{w}(i,j))], \quad \bar{w}(i,j) > 0 \tag{4.145}$$

$$D_{55} = \frac{1}{n(n-1)} \sum\sum_{i \neq j \in s} \ln(\bar{w}(i,j)), \quad \bar{w}(i,j) > 0 \tag{4.146}$$

and

$$D_{66} = \frac{1}{n(n-1)} \sum\sum_{i \neq j \in s} \tanh^{-1}\left(\frac{\{\bar{w}(i,j)\}^2 - 1}{\{\bar{w}(i,j)\}^2 + 1}\right) \tag{4.147}$$

where $\tanh^{-1}()$ is the hyperbolic tangent function, subject to three tuning constraints (4.136)–(4.138). Write code in any scientific language, such as FORTRAN, C++, or R, to study these distance functions. Discuss the nature of the tuned weights in each situation. Construct the 90%, 95%, and 99% confidence interval estimates by estimating variance using the method of double jackknifing discussed in the chapter. Discuss the distribution of $-2D_{55}$ and comment on it. Discuss the special cases with $v(x_i) = x_i^g$, where g is any real value.

Exercise 4.5 Consider an estimator of the finite population variance $S_y^2 = \{2N(N-1)\}^{-1} \sum\sum_{i \neq j \in \Omega} (y_i - y_j)^2$, defined by

$$\hat{\sigma}_{Tuned}^2 = \sum\sum_{i \neq j \in s} \left[(n(n-1)-1)^2 \bar{w}(i,j) - (n(n-1)-2)\right] s_y^2(y_{(i)}, y_{(j)}) \tag{4.148}$$

where

$$s_y^2(y_{(i)}, y_{(j)}) = \frac{n(n-1)s_y^2 - 0.5(y_i - y_j)^2}{n(n-1) - 1} \tag{4.149}$$

Let $\bar{w}(i,j)$ be a set of jackknife tuned weights such that the following two tuning constraints are satisfied:

$$\sum\sum_{i \neq j \in s} \bar{w}(i,j) = 1 \tag{4.150}$$

and

$$\sum\sum_{i\neq j\in s} \bar{w}(i,j)\left\{s_{\hat{y}}^2\left(\hat{y}_{(i)},\hat{y}_{(j)}\right) - \frac{S_{\hat{y}}^2 - n(n-1)\{2-n(n-1)\}}{\{n(n-1)-1\}^2}\right\} = 0 \qquad (4.151)$$

where

$$\hat{y}_i = h(x_i, \hat{\beta}) \qquad (4.152)$$

denotes the predicted value of the study variable y_i based on any linear or nonlinear model using the known information on the auxiliary variable x_i for $i \in \Omega$,

$$S_{\hat{y}}^2 = \frac{1}{2N(N-1)} \sum\sum_{i\neq j\in\Omega} \left(\hat{y}_i - \hat{y}_j\right)^2 \qquad (4.153)$$

$$s_{\hat{y}}^2 = \{2n(n-1)\}^{-1} \sum\sum_{i\neq j\in s} \left(\hat{y}_i - \hat{y}_j\right)^2 \qquad (4.154)$$

and

$$s_{\hat{y}}^2\left(\hat{y}_{(i)},\hat{y}_{(j)}\right) = \frac{n(n-1)s_{\hat{y}}^2 - 0.5\left(\hat{y}_i - \hat{y}_j\right)^2}{n(n-1)-1} \qquad (4.155)$$

Optimize each of the following distance functions:

$$D_{44} = \frac{1}{n(n-1)} \sum\sum_{i\neq j\in s} [\bar{w}(i,j) \ln(\bar{w}(i,j))], \quad \bar{w}(i,j) > 0 \qquad (4.156)$$

$$D_{55} = \frac{1}{n(n-1)} \sum\sum_{i\neq j\in s} \ln(\bar{w}(i,j)), \quad \bar{w}(i,j) > 0 \qquad (4.157)$$

and

$$D_{66} = \frac{1}{n(n-1)} \sum\sum_{i\neq j\in s} \tanh^{-1}\left(\frac{\{\bar{w}(i,j)\}^2 - 1}{\{\bar{w}(i,j)\}^2 + 1}\right) \qquad (4.158)$$

where $\tanh^{-1}()$ is the hyperbolic tangent function, subject to tuning constraints (4.150) and (4.151). Write code in any scientific language, such as FORTRAN, SAS, or R, to study these distance functions. Discuss the nature of the tuned weights in each situation. Construct the 90%, 95%, and 99% confidence interval estimates by estimating variance using the method of double jackknifing discussed in the chapter. Discuss the distribution of $-2D_{55}$ and comment on it.

Exercise 4.6 Consider an estimator of the finite population variance $S_y^2 = \{2N(N-1)\}^{-1} \sum\sum_{i \neq j \in \Omega} (y_i - y_j)^2$, defined by

$$\hat{\sigma}^2_{\text{Tuned}} = \sum\sum_{i \neq j \in s} \left[(n(n-1)-1)^2 \bar{w}(i,j) - (n(n-1)-2) \right] s_y^2 \left(y_{(i)}, y_{(j)} \right) \quad (4.159)$$

where

$$s_y^2 \left(y_{(i)}, y_{(j)} \right) = \frac{n(n-1)s_y^2 - 0.5(y_i - y_j)^2}{n(n-1) - 1} \quad (4.160)$$

and $\bar{w}(i,j)$ are the jackknife tuned weights such that the following two tuning constraints are satisfied:

$$\sum\sum_{i \neq j \in s} \bar{w}(i,j) = 1 \quad (4.161)$$

and

$$\sum\sum_{i \neq j \in s} \bar{w}(i,j) s_{\hat{y}}^2 \left(\hat{y}_{(i)}, \hat{y}_{(j)} \right) = \frac{S_{\hat{y}}^2 - n(n-1)(2 - n(n-1))s_{\hat{y}}^2}{(n(n-1)-1)^2} \quad (4.162)$$

where

$$\hat{y}_i = h(x_i, \hat{\beta}) \quad (4.163)$$

denotes the predicted value of the study variable y_i based on any linear or nonlinear model using the known information on the auxiliary variable x_i for $i \in \Omega$,

$$S_{\hat{y}}^2 = \frac{1}{2N(N-1)} \sum\sum_{i \neq j \in \Omega} (\hat{y}_i - \hat{y}_j)^2 \quad (4.164)$$

$$s_{\hat{y}}^2 = \{2n(n-1)\}^{-1} \sum\sum_{i \neq j \in s} (\hat{y}_i - \hat{y}_j)^2 \quad (4.165)$$

and

$$s_{\hat{y}}^2 \left(\hat{y}_{(i)}, \hat{y}_{(j)} \right) = \frac{n(n-1)s_{\hat{y}}^2 - 0.5(\hat{y}_i - \hat{y}_j)^2}{n(n-1) - 1} \quad (4.166)$$

Optimize each of the following distance functions:

$$D_{11} = \frac{1}{2}\sum\sum_{i\neq j\in s}\frac{\{1-(n(n-1)-1)\bar{w}(i,j)-1/(n(n-1))\}^2}{q_{ij}/(n(n-1))} \quad (4.167)$$

$$D_{22} = \frac{1}{2}\sum\sum_{i\neq j\in s}\left(\sqrt{\bar{w}(i,j)}-1/\sqrt{n(n-1)}\right)^2, \quad \bar{w}(i,j)>0 \quad (4.168)$$

and

$$D_{33} = \sum\sum_{i\neq j\in s}\frac{\{1-(n(n-1)-1)\bar{w}(i,j)-1/(n(n-1))\}^2}{2q_{ij}/(n(n-1))}$$
$$+ \sum\sum_{i\neq j\in s}\frac{\Phi_{ij}\{\bar{w}(i,j)\}^2 n(n-1)}{2q_{ij}\{n(n-1)-1\}^{-2}} \quad (4.169)$$

where q_{ij} are suitably chosen weights that form various estimators and Φ_{ij} is a penalty, subject to the two tuning constraints (4.161) and (4.162). Write code in any scientific language, such as FORTRAN, C++, SAS, or R, to study these distance functions. Discuss the nature of tuned weights in each situation. Construct the 90%, 95%, and 99% confidence interval estimates of the weight of pumpkins by estimating variance using the method of double jackknifing discussed earlier in the chapter.

Exercise 4.7 Consider an alternative tuned estimator of the finite population variance $\sigma_y^2 = N^{-1}\sum_{j\in\Omega}(y_i-\bar{Y})^2$, defined by

$$\hat{\sigma}_{T(a_1)}^2 = \sum_{j\in s}\left[(n-1)^2\bar{w}_y(j)-(n-2)\right]\hat{\sigma}_y^2(j) \quad (4.170)$$

where

$$\hat{\sigma}_y^2(j) = \frac{n\hat{\sigma}_y^2 - (y_j-\bar{y}_n)^2}{(n-1)} \quad \text{for } j\in s \quad (4.171)$$

and $\bar{w}_y(j)$ for $j\in s$ is a set of jackknifed tuned weights such that the following three tuning constraints are satisfied:

$$\sum_{j\in s}\bar{w}_y(j) = 1 \quad (4.172)$$

$$\sum_{j\in s}\bar{w}_y(j)\bar{x}_n(j) = \frac{\bar{X}-n(2-n)\bar{x}_n}{(n-1)^2} \quad (4.173)$$

and

$$\sum_{j\in s}\bar{w}_v(j)\hat{\sigma}_x^2(j) = \frac{\sigma_x^2 - n(2-n)\hat{\sigma}_x^2}{(n-1)^2} \tag{4.174}$$

where any other symbols have their usual meanings. Optimize each of the following distance functions:

$$D_1 = (2^{-1}n)\sum_{j\in s} q_j^{-1}\left(1 - (n-1)\bar{w}_v(j) - n^{-1}\right)^2 \tag{4.175}$$

$$D_2 = \sum_{j\in s}[\bar{w}_v(j)\ln(\bar{w}_v(j))], \quad 0 < \bar{w}_v(j) < 1/(n-1) \tag{4.176}$$

$$D_3 = 2\sum_{j\in s}\left(\sqrt{1-(n-1)\bar{w}_v(j)} - \sqrt{n^{-1}}\right)^2, \quad 0 < \bar{w}_n(j) < 1/(n-1) \tag{4.177}$$

$$D_4 = \sum_{j\in s}[-n^{-1}\ln(\bar{w}_v(j))], \quad 0 < \bar{w}_n(j) < 1/(n-1) \tag{4.178}$$

$$D_5 = \sum_{j\in s}\frac{(1-(n-1)\bar{w}_v(j) - n^{-1})^2}{2(1-(n-1)\bar{w}_v(j))}, \quad 0 < \bar{w}_n(j) < 1/(n-1) \tag{4.179}$$

$$D_6 = \frac{1}{n}\sum_{j\in s}\tanh^{-1}\left(\frac{\{\bar{w}_v(j)\}^2 - 1}{\{\bar{w}_v(j)\}^2 + 1}\right) \tag{4.180}$$

and

$$D_7 = \frac{1}{2}\sum_{j\in s}\frac{\left(1-(n-1)\bar{w}_v(j) - n^{-1}\right)^2}{q_j n^{-1}} + \frac{1}{2}\sum_{j\in s}\frac{\varphi_j\{\bar{w}_v(j)\}^2}{q_j n^{-1}(n-1)^{-2}} \tag{4.181}$$

where q_j are suitably chosen weights that form various estimators, φ_j is a penalty, and $\tanh^{-1}()$ is the hyperbolic tangent function, subject to tuning constraints (4.172)–(4.174). Write code in any scientific language, such as FORTRAN, C++, or R, to study these distance functions. Discuss the nature of the tuned weights in each situation. Construct the 90%, 95%, and 99% confidence interval estimates by estimating variance using the method of double jackknifing discussed in the chapter. Simulate and discuss the distributions of $-2D_4$ and other distance functions.

Exercise 4.8 Consider an alternative tuned estimator of the finite population variance $\sigma_y^2 = N^{-1}\sum_{i=1}^{N}(y_i - \bar{Y})^2$, defined by

$$\hat{\sigma}^2_{T(a_1)} = \sum_{j \in s} \left[(n-1)^2 \bar{w}_v(j) - (n-2)\right] \hat{\sigma}^2_y(j) \tag{4.182}$$

where

$$\bar{w}_v(j) = \frac{1 - w_j}{n - 1} \tag{4.183}$$

for $j \in s$ is a set of alternative tuned weights. Obtain the weights $\bar{w}_v(j)$ such that the chi-square distance:

$$D_1 = (2^{-1}n) \sum_{j \in s} q_j^{-1} \left(1 - (n-1)w_v(j) - n^{-1}\right)^2 \tag{4.184}$$

where q_j is a set of suitably chosen weights, is optimum subject to the three constraints

$$\sum_{j \in s} \bar{w}_v(j) = 1 \tag{4.185}$$

$$\sum_{j \in s} \bar{w}_v(j) \bar{x}_n(j) = \frac{\bar{X} - n(2-n)\bar{x}_n}{(n-1)^2} \tag{4.186}$$

and

$$E_m \sum_{j \in s} w_j (y_j - \bar{y})^2 = E_m \left(\sigma_y^2\right) \tag{4.187}$$

where E_m denotes the expected value under the model:

$$y_i = \beta x_i + e_i \tag{4.188}$$

such that $E_m(e_i|x_i) = 0$, $E_m(e_i^2|x_i) = \sigma^2 v(x_i)$, $v(x_i) > 0$, and $E_m(e_i e_i|x_i x_j) = 0$.
Note that the model assisted constraint $E_m \sum_{j \in s} w_j (y_j - \bar{y})^2 = E_m\left(\sigma_y^2\right)$ is equivalent to the following two tuning constraints:

$$\sum_{j \in s} \bar{w}_v(j) \hat{\sigma}_x^2(j) = \frac{\sigma_x^2 - n(2-n)\hat{\sigma}_x^2}{(n-1)^2} \tag{4.189}$$

and

$$\sum_{j \in s} \bar{w}_v(j) v^*_{s(x)}(j) = \frac{v^*_{\Omega(x)} - n(2-n)v^*_{s(x)}}{(n-1)^2} \tag{4.190}$$

where

$$v^*_{\Omega(x)} = \frac{1}{N}\sum_{i\in\Omega} v(x_i) \qquad (4.191)$$

$$v^*_{s(x)} = \frac{1}{n}\sum_{i\in s} v(x_i) \qquad (4.192)$$

and

$$v^*_{s(x)}(j) = \frac{nv^*_{s(x)} - v(x_j)}{n-1} \qquad (4.193)$$

for any function $v(x_i) > 0$. Write code in any scientific programming language, such as SAS, FORTRAN, C, or R, to study the chi-square distance function. Discuss the nature of the tuned weights. Construct the 90%, 95%, and 99% confidence interval estimates by estimating the variance using the method of double jackknifing discussed earlier in the chapter. Also study distance functions of your own choice.

Exercise 4.9 Consider taking a first-phase sample s_1 of m units using a simple random sampling (SRS) scheme from a population Ω consisting of N units. Only collect information on the auxiliary variable x_i, for $i = 1, 2, \ldots, m$, in the sample s_1. Let $s^{*2}_x = (m-1)^{-1}\sum_{i=1}^{m}(x_i - \bar{x}_m)^2$ with $\bar{x}_m = m^{-1}\sum_{i=1}^{m} x_i$ be an estimator of the finite population variance of the auxiliary variable in the first-phase sample. Assume a sample s_2 of n units is taken using an SRS scheme from the given first-phase sample s_1. Assume that both the study variable y_i and the auxiliary variable x_i are measured for $i = 1, 2, \ldots, n$, constituting the second-phase sample s_2. Consider an alternative tuned estimator of the finite population variance given by

$$\hat{\sigma}^2_{\text{TP(cs)}} = \sum\sum_{i\neq j \in s_2}\left[(n(n-1)-1)^2 \bar{w}(i,j) - (n(n-1)-2)\right] s^2_y(y_{(i)}, y_{(j)}) \qquad (4.194)$$

where

$$s^2_y(y_{(i)}, y_{(j)}) = \frac{n(n-1)s^2_y - 0.5(y_i - y_j)^2}{n(n-1)-1} \qquad (4.195)$$

and

$$s^2_y = \{2n(n-1)\}^{-1}\sum\sum_{i\neq j \in s_2}(y_i - y_j)^2 \qquad (4.196)$$

have their usual meanings. Tuned jackknifed weights $\bar{w}(i,j)$ are obtained by minimizing the distance function

$$D = \frac{1}{2}\sum\sum_{i\neq j \in s_2}\frac{\{1-(n(n-1)-1)\bar{w}(i,j)-1/(n(n-1))\}^2}{q_{ij}/(n(n-1))} \quad (4.197)$$

subject to the following two tuning constraints:

$$\sum\sum_{i\neq j \in s_2}\bar{w}(i,j) = 1 \quad (4.198)$$

and

$$\sum\sum_{i\neq j \in s_2}\bar{w}(i,j)s_x^2(x_{(i)},x_{(j)}) = \frac{s_x^{*2} - n(n-1)(2-n(n-1))s_x^2}{(n(n-1)-1)^2} \quad (4.199)$$

where

$$s_x^2(x_{(i)},x_{(j)}) = \frac{n(n-1)s_x^2 - 0.5(x_i - x_j)^2}{n(n-1)-1} \quad (4.200)$$

and

$$s_x^2 = \{2n(n-1)\}^{-1}\sum\sum_{i\neq j \in s_2}(x_i - x_j)^2 \quad (4.201)$$

Write code in any scientific programming language, such as C, R, or SAS, to investigate the nominal 90%, 95%, and 99% confidence interval estimates using double jackknifing method for different sizes of the first-phase and second-phase samples.

Exercise 4.10 Consider an estimator of the finite population variance $S_y^2 = \{2N(N-1)\}^{-1}\sum\sum_{i\neq j \in \Omega}(y_i - y_j)^2$ given by

$$\hat{\sigma}_{Tuned}^2 = \sum\sum_{i\neq j\in s}\left[(n(n-1)-1)^2\bar{w}(i,j) - (n(n-1)-2)\right]s_y^2(y_{(i)},y_{(j)}) \quad (4.202)$$

where

$$s_y^2(y_{(i)},y_{(j)}) = \frac{n(n-1)s_y^2 - 0.5(y_i - y_j)^2}{n(n-1)-1} \quad (4.203)$$

$$s_y^2 = \{2n(n-1)\}^{-1} \sum\sum_{i\neq j\in s} (y_i - y_j)^2 \qquad (4.204)$$

is the sample variance of the study variable obtained by removing the ith and jth units from the sample s, and $\bar{w}(i,j)$ is the tuned weight, constructed so that the following two constraints are satisfied:

$$\sum\sum_{i\neq j\in s} \bar{w}(i,j) = 1 \qquad (4.205)$$

and

$$\sum\sum_{i\neq j\in s} \bar{w}(i,j) \left[\hat{G}_x^{1-n}(i) - \hat{G}_x^{1-n}(j)\right]^2 = \frac{\sum\sum_{i\neq j\in s} \left[\hat{G}_x^{1-n}(i) - \hat{G}_x^{1-n}(j)\right]^2 - 2\left(S_x^2(\hat{G}_x)^{-2n}\right)}{n(n-1)-1} \qquad (4.206)$$

where

$$S_x^2 = \{2N(N-1)\}^{-1} \sum\sum_{i\neq j\in \Omega} (x_i - x_j)^2 \qquad (4.207)$$

Above $\hat{G}_x(i)$ and $\hat{G}_x(j)$ are the jackknifed sample geometric means after removing the ith and jth units, respectively, from the sample geometric mean, which is given by

$$\hat{G}_x = \left(\prod_{i=1}^{n} x_i\right)^{1/n} \qquad (4.208)$$

Now optimize the following distance function

$$D_{11} = \frac{1}{2} \sum\sum_{i\neq j\in s} \frac{\{1 - (n(n-1)-1)\bar{w}(i,j) - 1/(n(n-1))\}^2}{q_{ij}/(n(n-1))} \qquad (4.209)$$

subject to the preceding two tuning constraints.

Generate a population from your own simulating model, such as the SJPM. Write R code to study the behavior of nominally 90%, 95%, and 99% confidence interval estimates based on the resultant confidence interval estimators for various sample sizes. Discuss your findings.

Exercise 4.11 Repeat Exercise 4.10, replacing the second constraint by

$$\sum\sum_{i\neq j\in s}\frac{\bar{w}(i,j)\left(\hat{H}_x(i)-\hat{H}_x(j)\right)^2}{\left[\hat{H}_x+n\left(\hat{H}_x(i)-\hat{H}_x\right)\right]^2\left[\hat{H}_x+n\left(\hat{H}_x(j)-\hat{H}_x\right)\right]^2}$$
$$=\sum\sum_{i\neq j\in s}\frac{(n(n-1)-1)^{-1}\left(\hat{H}_x(i)-\hat{H}_x(j)\right)^2}{\left[\hat{H}_x+n\left(\hat{H}_x(i)-\hat{H}_x\right)\right]^2\left[\hat{H}_x+n\left(\hat{H}_x(j)-\hat{H}_x\right)\right]^2}-\frac{2(n(n-1)-1)^{-1}S_x^2}{(n-1)^2\hat{H}_x^4}$$
(4.210)

where $\hat{H}_x(i)$ and $\hat{H}_x(j)$ are the jackknifed sample harmonic means after dropping the ith and jth units, respectively, from the sample harmonic mean.

Exercise 4.12 Repeat Exercise 4.10, replacing the second constraint by the following:

$$\sum\sum_{i\neq j\in s}\bar{w}(i,j)[\bar{x}_n(i)-\bar{x}_n(j)]^2 = \frac{1}{(n(n-1)-1)}\left[\sum\sum_{i\neq j\in s}(\bar{x}_n(i)-\bar{x}_n(j))^2 - \frac{2S_x^2}{(n-1)^2}\right]$$
(4.211)

where $\bar{x}_n(i)=\dfrac{n\bar{x}_n-x_i}{n-1}$ and $\bar{x}_n(j)=\dfrac{n\bar{x}_n-x_j}{n-1}$ are the jackknifed sample means after dropping the ith and jth units, respectively, from the sample mean $\bar{x}_n$.

Exercise 4.13 Consider an alternative tuned estimator of the finite population variance $\sigma_y^2 = N^{-1}\sum_{i=1}^{N}(y_i-\bar{Y})^2$, defined by

$$\hat{\sigma}_{T(a_1)}^2 = \sum_{j\in s}\left[(n-1)^2\bar{w}_y(j)-(n-2)\right]\hat{\sigma}_y^2(j) \quad (4.212)$$

Obtain weights $\bar{w}_y(j)$ such that the chi-squared distance:

$$D_1 = (2^{-1}n)\sum_{j\in s}q_j^{-1}\left(1-(n-1)\bar{w}_y(j)-n^{-1}\right)^2 \quad (4.213)$$

where q_j is a set of suitably chosen weights, is optimum subject to the two constraints:

$$\sum_{j\in s}\bar{w}_y(j)=1 \quad (4.214)$$

and

$$\sum_{j\in s}\bar{w}_y(j)\left[\hat{G}_x(j)\right]^{(1-n)} = \frac{1}{(n-1)}\left[\sum_{j\in s}(\hat{G}_x(j))^{(1-n)}-\bar{X}(\hat{G}_x)^{-n}\right] \quad (4.215)$$

where $\bar{X} = N^{-1}\sum_{i=1}^{N} x_i$ denotes the known population arithmetic mean of the auxiliary variable and $\hat{G}_x(j)$ is the jth jackknifed sample geometric mean $\hat{G}_x$ of the auxiliary variable.

Exercise 4.14 Consider an alternative tuned estimator of the finite population variance $\sigma_y^2 = N^{-1}\sum_{i=1}^{N}(y_i - \bar{Y})^2$, defined by

$$\hat{\sigma}_{T(a_1)}^2 = \sum_{j \in s}\left[(n-1)^2 \bar{w}_y(j) - (n-2)\right]\hat{\sigma}_y^2(j) \tag{4.216}$$

Obtain weights $\bar{w}_y(j)$ such that the chi-squared distance

$$D_1 = (2^{-1}n)\sum_{j \in s} q_j^{-1}\left(1 - (n-1)\bar{w}_y(j) - n^{-1}\right)^2 \tag{4.217}$$

where q_j is a set of suitably chosen weights, is optimum subject to the two constraints:

$$\sum_{j \in s} \bar{w}_y(j) = 1 \tag{4.218}$$

and

$$\sum_{j \in s}\frac{\bar{w}_y(j)\hat{H}_x(j)}{n\hat{H}_x(j) - (n-1)\hat{H}_x} = \frac{1}{(n-1)}\left[\sum_{j \in s}\frac{\hat{H}_x(j)}{n\hat{H}_x(j) - (n-1)\hat{H}_x} - \frac{\bar{X}}{\hat{H}_x}\right] \tag{4.219}$$

where $\bar{X} = N^{-1}\sum_{i=1}^{N} x_i$ denotes the known population arithmetic mean of the auxiliary variable, and

$$\hat{H}_x(j) = (n-1)\left(\sum_{i \neq j=1}^{n} x_i^{-1}\right)^{-1}, \quad j = 1, 2, \ldots, n \tag{4.220}$$

is the jth jackknifed estimator of the sample harmonic mean estimator

$$\hat{H}_x = n\left(\sum_{i=1}^{n} x_i^{-1}\right)^{-1}$$

Write code in any scientific language, such as SAS, FORTRAN, C, or R, to generate a population from the SJPM. Discuss the nature of the tuned weights. Suggest an estimator of variance and a confidence interval estimator. Investigate the nominal 90%, 95%, and 99% coverage by the confidence interval estimates by the method of double jackknifing discussed earlier in the chapter. In addition, study distance functions of your own choice, and make comments.

Exercise 4.15 Assume a sample s of n units is taken using SRSWOR scheme from a population Ω of N units. Consider a new estimator of the finite population variance as

$$\hat{\sigma}^2_{\text{new}} = \sum\sum_{i \neq j \in s} w^*_{ij} [\bar{y}_n(i) - \bar{y}_n(j)]^2 \tag{4.221}$$

where $\bar{y}_n(i)$ and $\bar{y}_n(j)$ are the jackknifed sample means obtained after removing ith and jth units from the sample mean $\bar{y}_n$. The weights w^*_{ij} are obtained by optimizing the chi-squared type distance function

$$D = \frac{n}{(n-1)} \sum\sum_{i \neq j \in s} \frac{\left(w^*_{ij} - 0.5 n^{-1}(n-1)\right)^2}{q_{ij}} \tag{4.222}$$

subject to the two constraints given by

$$\sum\sum_{i \neq j \in s} w^*_{ij} = \frac{1}{2}(n-1)^2 \tag{4.223}$$

and

$$\sum\sum_{i \neq j \in s} w^*_{ij} [x_n(i) - x_n(j)]^2 = S^2_x \tag{4.224}$$

where $\bar{x}_n(i)$ and $\bar{x}_n(j)$ are the jackknifed sample means obtained after removing ith and jth units from the sample mean $\bar{x}_n$, and

$$S^2_x = \frac{1}{2N(N-1)} \sum\sum_{i \neq j \in \Omega} (x_i - x_j)^2 \tag{4.225}$$

is the known population mean squared error of the auxiliary variable.

Write code in any scientific language, such as SAS, FORTRAN, C, or R, to generate a suitable population of your choice and to study the nature of the calibrated weights w^*_{ij}. Suggest a confidence interval estimator. Study the nominal 90%, 95%, and 99% coverage by the confidence interval estimates for various sample sizes. In addition, suggest changes and study other distance functions of your own choice and make comments.

Tuned estimators of correlation coefficient

5.1 Introduction

In this chapter, we consider the problem of estimating the population correlation coefficient. New estimators for the Pearson's correlation coefficient are suggested and are investigated through empirical studies. At the end of the chapter, ideas for estimating the regression coefficient, the ratio of two population means, and estimators of finite population variance are suggested in the form of unsolved exercises.

5.2 Correlation coefficient

The problem of estimation of the finite population correlation coefficient

$$\rho_{xy} = \frac{S_{xy}}{S_x S_y} \qquad (5.1)$$

is well known in the survey sampling literature. The symbols S_x^2, S_y^2, and S_{xy} have their usual meaning. For the reader's convenience, these are defined again as $(N-1)S_x^2 = \sum_{i \in \Omega}(x_i - \bar{X})^2$, $(N-1)S_y^2 = \sum_{i \in \Omega}(y_i - \bar{Y})^2$, and $(N-1)S_{xy} = \sum_{i \in \Omega}(x_i - \bar{X})(y_i - \bar{Y})$, with $\bar{X} = N^{-1}\sum_{i \in \Omega} x_i$ and $\bar{Y} = N^{-1}\sum_{i \in \Omega} y_i$.

For a bivariate normal population, an estimator of the finite population correlation coefficient ρ_{xy}, due to Pearson (1896), has been defined as:

$$r = \frac{s_{xy}}{s_x s_y} \qquad (5.2)$$

where $(n-1)s_{xy} = \sum_{i \in s}(x_i - \bar{x}_n)(y_i - \bar{y}_n)$, $(n-1)s_x^2 = \sum_{i \in s}(x_i - \bar{x}_n)^2$, $(n-1)s_y^2 = \sum_{i \in s}(y_i - \bar{y}_n)^2$, $n\bar{x}_n = \sum_{i \in s} x_i$, and $n\bar{y}_n = \sum_{i \in s} y_i$.

Wakimoto (1971) and Gupta, Singh, and Kashani (1993) studied the behavior of the estimator r under different sampling schemes. Srivastava and Jhajj (1986) have proposed a class of estimators of the population correlation coefficient as

$$r_{class} = rH(u,v) \qquad (5.3)$$

where $u = \bar{x}/\bar{X}$, $v = s_x^2/S_x^2$ for $N\bar{X} = \sum_{i \in \Omega} x_i$ and $(N-1)S_x^2 = \sum_{i \in \Omega} (x_i - \bar{X})^2$, the function $H(.,.)$ is parametric with $H(1,1) = 1$ and satisfies certain regularity conditions.

Singh, Mangat, and Gupta (1996) have reviewed literature on the problem of estimation of the correlation coefficient and have shown that the class of estimators due to Srivastava and Jhajj (1986) can take an inadmissible value, that is, outside the range $[-1.0, +1.0]$, when applied to a sample also reported by Singh (2003). Singh, Sedory, and Kim (2014) introduced a new empirical likelihood estimator of the correlation coefficient.

It is interesting to note that the tuned dual-to-empirical log-likelihood (dell) estimate of the correlation coefficient provides only admissible values, like the Singh, Sedory, and Kim (2014) estimate, and makes use of auxiliary information at the estimation stage.

5.3 Tuned estimator of correlation coefficient

The newly tuned estimator of the finite population correlation coefficient ρ_{xy} is defined as

$$r_{\text{Tuned}} = \sum_{j \in s} \bar{w}_n(j) r(j) \tag{5.4}$$

where

$$r(j) = \frac{s_{xy}(j)}{\sqrt{s_x^2(j)} \sqrt{s_y^2(j)}} \tag{5.5}$$

with

$$s_{xy}(j) = \frac{(n-1)s_{xy}}{(n-2)} - \frac{n}{(n-1)(n-2)} (x_j - \bar{x}_n)(y_j - \bar{y}_n) \tag{5.6}$$

$$s_x^2(j) = \frac{(n-1)s_x^2}{(n-2)} - \frac{n}{(n-1)(n-2)} (x_j - \bar{x}_n)^2 \tag{5.7}$$

and

$$s_y^2(j) = \frac{(n-1)s_y^2}{(n-2)} - \frac{n}{(n-1)(n-2)} (y_j - \bar{y}_n)^2 \tag{5.8}$$

The tuned weights $\bar{w}_n(j)$ are obtained such that the dell function:

$$\frac{1}{n}\sum_{j\in s}\ln(\bar{w}_n(j)) \text{ with } 0<\bar{w}_n(j)=\frac{1-w_j}{n-1}<\frac{1}{n-1} \tag{5.9}$$

for some unit length weights $w_j>0$, is optimized subject to the three constraints:

$$\sum_{j\in s}\bar{w}_n(j)=1 \tag{5.10}$$

$$\sum_{j\in s}\bar{w}_n(j)\left\{\bar{x}_n(j)-\frac{(\bar{X}-n(2-n)\bar{x}_n)}{(n-1)^2}\right\}=0 \tag{5.11}$$

and

$$\sum_{j\in s}\bar{w}_n(j)\left\{\bar{\eta}(j)-\frac{n\bar{\eta}-1}{n-1}\right\}=0 \tag{5.12}$$

where

$$\bar{\eta}(j)=\frac{n\bar{\eta}-\eta_j}{n-1} \tag{5.13}$$

$$\bar{\eta}=\frac{1}{n}\sum_{j\in s}\eta_j \tag{5.14}$$

and

$$\eta_j=\frac{\frac{(n-1)}{(n-2)}-\frac{n(x_j-\bar{x}_n)(y_j-\bar{y}_n)}{(n-1)(n-2)s_{xy}}}{\sqrt{\frac{(n-1)}{(n-2)}-\frac{n(x_j-\bar{x}_n)^2}{(n-1)(n-2)s_x^2}}\sqrt{\frac{(n-1)}{(n-2)}-\frac{n(y_j-\bar{y}_n)^2}{(n-1)(n-2)s_y^2}}} \tag{5.15}$$

Note that

$$r(j)=r\eta_j \text{ for all } j\in s \tag{5.16}$$

The Lagrange function is given by

$$L=\frac{1}{n}\sum_{j\in s}\ln(\bar{w}_n(j))-\lambda_0\left\{\sum_{j\in s}\bar{w}_n(j)-1\right\}-\lambda_1\left\{\sum_{j\in s}\bar{w}_n(j)\Psi_{1j}\right\}-\lambda_2\left\{\sum_{j\in s}\bar{w}_n(j)\Psi_{2j}\right\} \tag{5.17}$$

where λ_0, λ_1, and λ_2 are Lagrange multipliers,

$$\Psi_{1j} = \bar{x}_n(j) - \frac{\bar{X} - n(2-n)\bar{x}_n}{(n-1)^2} \quad (5.18)$$

and

$$\Psi_{2j} = \left(\bar{\eta}(j) - \frac{n\bar{\bar{\eta}} - 1}{n-1}\right) \quad (5.19)$$

On setting

$$\frac{\partial L}{\partial \bar{w}_n(j)} = 0 \quad (5.20)$$

the set of tuned positive weights is given by

$$\bar{w}_n(j) = \frac{1}{n}\left[\frac{1}{1 + \lambda_1 \Psi_{1j} + \lambda_2 \Psi_{2j}}\right] \quad (5.21)$$

Constraints (5.10), (5.11) and (5.12) yield $\lambda_0 = 1$, and λ_1 and λ_2 are given by a solution to two nonlinear equations:

$$\sum_{j \in s} \frac{\Psi_{1j}}{1 + \lambda_1 \Psi_{1j} + \lambda_2 \Psi_{2j}} = 0 \quad (5.22)$$

and

$$\sum_{j \in s} \frac{\Psi_{2j}}{1 + \lambda_1 \Psi_{1j} + \lambda_2 \Psi_{2j}} = 0 \quad (5.23)$$

Thus, the newly tuned estimator of the finite population correlation coefficient ρ_{xy} in Equation (5.4) becomes

$$r_{\text{Tuned}} = \frac{1}{n}\sum_{j \in s} \frac{r(j)}{1 + \lambda_1 \Psi_{1j} + \lambda_2 \Psi_{2j}} = r \quad (5.24)$$

The most important feature of the newly tuned estimator r_{Tuned} is the same as that of the usual sample correlation coefficient, that is, it always lies between -1 and $+1$. This fact is due to the new calibration constraint in Equation (5.12).

5.3.1 Estimation of variance of the estimator of correlation coefficient and coverage

Based on some simulation trials, we consider a tuned estimator of the variance of the estimator of the finite population correlation coefficient r_{Tuned}, defined as

$$\hat{v}(r_{\text{Tuned}}) = n^2 \left(\frac{C_1}{C_2}\right)^2 \sum_{j \in s} (\bar{w}_n(j))^2 \{r_{\text{Tuned}(j)} - r_{\text{Tuned}}\}^2 \tag{5.25}$$

where $C_1 = (n-1)/(n-2)$ and $C_2 = n/((n-1)(n-2))$. Note that each newly tuned doubly jackknifed estimator of the finite population correlation coefficient ρ_{xy} is given by

$$r_{\text{Tuned}(j)} = \frac{n \sum_{j \in s} r(j) \bar{w}_n(j) - nr(j) \bar{w}_n(j)}{n-1} \quad \text{for } j = 1, 2, \ldots, n \tag{5.26}$$

The coverage of the $(1-\alpha)100\%$ confidence interval estimates obtained by the newly tuned estimator of the finite population correlation coefficient is obtained by counting the number of times the true value of ρ_{xy} falls in the interval estimate given by

$$r_{\text{Tuned}} \mp t_{\alpha/2}(\text{df} = n-2)\sqrt{\hat{v}(r_{\text{Tuned}})} \tag{5.27}$$

Note that in the simulation study we approximated the values of λ_1 and λ_2, under some assumptions, as

$$\lambda_1 \approx \frac{\left(\sum_{j \in s} \Psi_{1j}\right)\left(\sum_{j \in s} \Psi_{2j}^2\right) - \left(\sum_{j \in s} \Psi_{2j}\right)\left(\sum_{j \in s} \Psi_{1j}\Psi_{2j}\right)}{\left(\sum_{j \in s} \Psi_{1j}^2\right)\left(\sum_{j \in s} \Psi_{2j}^2\right) - \left(\sum_{j \in s} \Psi_{1j}\Psi_{2j}\right)^2} \tag{5.28}$$

and

$$\lambda_2 \approx \frac{\left(\sum_{j \in s} \Psi_{2j}\right)\left(\sum_{j \in s} \Psi_{1j}^2\right) - \left(\sum_{j \in s} \Psi_{1j}\right)\left(\sum_{j \in s} \Psi_{1j}\Psi_{2j}\right)}{\left(\sum_{j \in s} \Psi_{1j}^2\right)\left(\sum_{j \in s} \Psi_{2j}^2\right) - \left(\sum_{j \in s} \Psi_{1j}\Psi_{2j}\right)^2} \tag{5.29}$$

It may be worth noting that due to these approximations the numerical values of r and r_{Tuned} may differ, but theoretically r and r_{Tuned} are equal. A better approximation to the nonlinear equation could be used, if available.

We have

$$\sum_{j \in s} \{r_{\text{Tuned}(j)} - r_{\text{Tuned}}\}^2 = \sum_{j \in s} \left[\frac{nr - nr(j)\bar{w}_n(j)}{n-1} - r\right]^2$$

$$= \sum_{j\in s} \left[\frac{nr - nr(j)\bar{w}_n(j) - nr + r}{n-1} \right]^2$$

$$= \frac{1}{(n-1)^2} \sum_{j\in s} [r - nr(j)\bar{w}_n(j)]^2$$

$$= \frac{1}{(n-1)^2} \sum_{j\in s} [r - n(r\eta_j)\bar{w}_n(j)]^2$$

$$= \frac{1}{(n-1)^2} \sum_{j\in s} \left[r - \frac{r\eta_j}{1 + \lambda_1 \Psi_{1j} + \lambda_2 \Psi_{2j}} \right]^2$$

$$= \frac{r^2}{(n-1)^2} \sum_{j\in s} \left[\frac{(1-\eta_j) + \lambda_1 \Psi_{1j} + \lambda_2 \Psi_{2j}}{1 + \lambda_1 \Psi_{1j} + \lambda_2 \Psi_{2j}} \right]^2$$

$$= \frac{r^2}{(n-1)^2} \sum_{j\in s} \left[\frac{(1-\eta_j)^2 + (\lambda_1 \Psi_{1j} + \lambda_2 \Psi_{2j})^2 + 2(1-\eta_j)(\lambda_1 \Psi_{1j} + \lambda_2 \Psi_{2j})}{(1 + \lambda_1 \Psi_{1j}, + \lambda_2 \Psi_{2j})^2} \right]$$

$$= \frac{r^2}{(n-1)^2} \sum_{j\in s} \left[\frac{(1-\eta_j)^2 + (\lambda_1 \Psi_{1j} + \lambda_2 \Psi_{2j})^2 + 2(1-\eta_j)(\lambda_1 \Psi_{1j} + \lambda_2 \Psi_{2j})}{(1 + \lambda_1 \Psi_{1j}, + \lambda_2 \Psi_{2j})^2} \right]$$

(5.30)

Assuming $|\lambda_1 \Psi_{1j} + \lambda_2 \Psi_{2j}| < 1$, applying binomial expansion, then under certain regularity conditions, we have

$$\sum_{j\in s} \{r_{\text{Tuned}(j)} - r_{\text{Tuned}}\}^2 = \frac{r^2}{(n-1)^2} \sum_{j\in s} (1-\eta_j)^2 + f(\Psi_{1j}, \Psi_{2j}, \eta_j) \quad (5.31)$$

where $f(\Psi_{1j}, \Psi_{2j}, \eta_j)$ is some function of order $O(n^{-2})$ conversing to zero as the sample size increases.

Again, assuming $\left| \left(\frac{C_2}{C_1}\right) \frac{(x_j - \bar{x}_n)^2}{s_x^2} \right| < 1$ and $\left| \left(\frac{C_2}{C_1}\right) \frac{(y_j - \bar{y}_n)^2}{s_y^2} \right| < 1$, we have

$$\eta_j = \frac{\frac{(n-1)}{(n-2)} - \frac{n(x_j - \bar{x}_n)(y_j - \bar{y}_n)}{(n-1)(n-2)s_{xy}}}{\sqrt{\frac{(n-1)}{(n-2)} - \frac{n(x_j - \bar{x}_n)^2}{(n-1)(n-2)s_x^2}} \sqrt{\frac{(n-1)}{(n-2)} - \frac{n(y_j - \bar{y}_n)^2}{(n-1)(n-2)s_y^2}}}$$

$$= \left[1 - \left(\frac{C_2}{C_1}\right) \frac{(x_j - \bar{x}_n)(y_j - \bar{y}_n)}{s_{xy}} \right] \left[1 - \left(\frac{C_2}{C_1}\right) \frac{(x_j - \bar{x}_n)^2}{s_x^2} \right]^{-\frac{1}{2}} \left[1 - \left(\frac{C_2}{C_1}\right) \frac{(y_j - \bar{y}_n)^2}{s_y^2} \right]^{-\frac{1}{2}}$$

$$\approx 1 - \left(\frac{C_2}{C_1}\right) \frac{(x_j - \bar{x}_n)(y_j - \bar{y}_n)}{s_{xy}} + \frac{1}{2}\left(\frac{C_2}{C_1}\right) \frac{(x_j - \bar{x}_n)^2}{s_x^2} + \frac{1}{2}\left(\frac{C_2}{C_1}\right) \frac{(y_j - y_n)^2}{s_y^2} + \ldots$$

(5.32)

Thus, we have

$$(1-\eta_j)^2 \approx \left(\frac{C_2}{C_1}\right)^2 \left[\frac{(x_j-\bar{x}_n)(y_j-\bar{y}_n)}{s_{xy}} - \frac{1}{2}\frac{(x_j-\bar{x}_n)^2}{s_x^2} - \frac{1}{2}\frac{(y_j-y_n)^2}{s_y^2} + \ldots\right]^2$$

and

$$\sum_{j=1}^{n}(1-\eta_j)^2 \approx (n-1)\left(\frac{C_2}{C_1}\right)^2 \left[\frac{\hat{\mu}_{22}}{s_{xy}^2} + \frac{\hat{\mu}_{40}}{4s_x^4} + \frac{\hat{\mu}_{04}}{4s_y^4} - \frac{\hat{\mu}_{31}}{s_x^2 s_{xy}} + \frac{\hat{\mu}_{13}}{s_{xy} s_y^2} + \frac{\hat{\mu}_{22}}{2s_x^2 s_y^2}\right] \quad (5.33)$$

where $\hat{\mu}_{ab} = \frac{1}{(n-1)}\sum_{i=1}^{n}(x_i-\bar{x}_n)^a(y_i-\bar{y}_n)^b$, $a,b=1,2,3,4$, have their usual meanings.

The approximation in Equation (5.33) allows us to estimate the variance of the estimator of the correlation estimator with the proposed newly tuned methodology and also justifies the multiplier in front of the developed estimator of variance in Equation (5.25).

We generated two random variables, $y_i^* \sim N(0,1)$ and $x_i^* \sim N(0,1)$, for $i=1,2,\ldots,N$ (with $N=70{,}000$) from two independent standard normal variables. For each of the values 0.1, 0.3, 0.5, 0.7, and 0.9, we generated a population of x_i, y_i pairs having those values as correlation coefficient (ρ_{xy}) by means of the formula:

$$x_i = \bar{X} + \sigma_x x_i^* \quad (5.34)$$

and

$$y_i = \bar{Y} + \sigma_y \sqrt{\left(1-\rho_{xy}^2\right)} y_i^* + \rho_{xy}\sigma_y x_i^* \quad (5.35)$$

with $\bar{X}=30.0$, $\bar{Y}=45.5$, $\sigma_x=20.1$, and $\sigma_y=23.5$. Note the use of very small average values of both variables; here we are considering a different kind of Statistical Jumbo Pumpkin Model (SJPM). As said earlier, the choice of a particular SJPM depends on the problem being considered by an investigator. Here we are interested in various amounts of correlation between the weight y_i (lbs) and circumference x_i (in.) on a particular farm where pumpkin weights and circumferences are not too scattered. Then we select 10,000 random samples, in the range 5–100, from each one of these populations as shown in Table 5.1, and found the proportion of times the true value of the correlation coefficient falls in the preceding interval estimate. In Table 5.1, ρ_{xy} indicates the correlation coefficient we were trying to simulate, and Rhoxy indicates the actual correlation coefficient from our population of 70,000 paired values. This latter value is the one used to estimate coverage. For $\rho_{xy}=0.1$ (Rhoxy$=0.1019759$) the estimated coverages by nominally 90%, 95%, and 99% intervals were 89.24%, 94.32%, and 98.63% for samples of sizes 65, 95, and 95, respectively. Now for $\rho_{xy}=0.3$ (or true Rhoxy$=0.2981710$) the estimated coverages 90%, 95%, and 99% were 89.21%,

Table 5.1 **Performance of the newly tuned jackknife estimator of the correlation coefficient**

ρ_{xy}	Rhoxy	n	Cov1	Cov2	Cov3
0.1	0.1019759	5	0.6312	0.6840	0.7549
0.1	0.1019759	10	0.8182	0.8731	0.9323
0.1	0.1019759	15	0.8454	0.9012	0.9542
0.1	0.1019759	20	0.8646	0.9186	0.9645
0.1	0.1019759	25	0.8681	0.9187	0.9684
0.1	0.1019759	30	0.8784	0.9253	0.9732
0.1	0.1019759	35	0.8792	0.9306	0.9760
0.1	0.1019759	40	0.8838	0.9346	0.9826
0.1	0.1019759	45	0.8874	0.9379	0.9786
0.1	0.1019759	50	0.8881	0.9403	0.9835
0.1	0.1019759	55	0.8867	0.9376	0.9832
0.1	0.1019759	60	0.8880	0.9371	0.9813
0.1	0.1019759	65	0.8924	0.9396	0.9829
0.1	0.1019759	70	0.8895	0.9386	0.9817
0.1	0.1019759	75	0.8904	0.9412	0.9830
0.1	0.1019759	80	0.8905	0.9398	0.9834
0.1	0.1019759	85	0.8964	0.9430	0.9850
0.1	0.1019759	90	0.8891	0.9377	0.9840
0.1	0.1019759	95	0.8908	0.9432	0.9863
0.1	0.1019759	100	0.8920	0.9420	0.9854
0.3	0.2981710	5	0.6392	0.6804	0.7404
0.3	0.2981710	10	0.8342	0.8814	0.9367
0.3	0.2981710	15	0.8564	0.9072	0.9574
0.3	0.2981710	20	0.8686	0.9155	0.9691
0.3	0.2981710	25	0.8733	0.9214	0.9711
0.3	0.2981710	30	0.8773	0.9271	0.9732
0.3	0.2981710	35	0.8834	0.9311	0.9767
0.3	0.2981710	40	0.8853	0.9347	0.9796
0.3	0.2981710	45	0.8841	0.9348	0.9801
0.3	0.2981710	50	0.8910	0.9389	0.9826
0.3	0.2981710	55	0.8847	0.9383	0.9807
0.3	0.2981710	60	0.8854	0.9353	0.9819
0.3	0.2981710	65	0.8886	0.9368	0.9801
0.3	0.2981710	70	0.8919	0.9397	0.9844
0.3	0.2981710	75	0.8910	0.9430	0.9856
0.3	0.2981710	80	0.8903	0.9411	0.9835
0.3	0.2981710	85	0.8921	0.9414	0.9842
0.3	0.2981710	90	0.8898	0.9393	0.9851
0.3	0.2981710	95	0.8973	0.9432	0.9841
0.3	0.2981710	100	0.8905	0.9401	0.9849
0.5	0.495181	5	0.6741	0.7084	0.7559
0.5	0.495181	10	0.8717	0.9131	0.9559
0.5	0.495181	15	0.8803	0.9235	0.9687
0.5	0.495181	20	0.8769	0.9211	0.9702

Continued

Table 5.1 Continued

ρ_{xy}	Rhoxy	n	Cov1	Cov2	Cov3
0.5	0.495181	25	0.8790	0.9245	0.9702
0.5	0.495181	30	0.8810	0.9271	0.9736
0.5	0.495181	35	0.8824	0.9305	0.9753
0.5	0.495181	40	0.8840	0.9297	0.9765
0.5	0.495181	45	0.8815	0.9295	0.9775
0.5	0.495181	50	0.8950	0.9371	0.9806
0.5	0.495181	55	0.8833	0.9347	0.9778
0.5	0.495181	60	0.8829	0.9348	0.9785
0.5	0.495181	65	0.8870	0.9377	0.9820
0.5	0.495181	70	0.8946	0.9389	0.9800
0.5	0.495181	75	0.8905	0.9372	0.9803
0.5	0.495181	80	0.8963	0.9433	0.9826
0.5	0.495181	85	0.8964	0.9446	0.9822
0.5	0.495181	90	0.8916	0.9410	0.9824
0.5	0.495181	95	0.8914	0.9417	0.9833
0.5	0.495181	100	0.8932	0.9415	0.9815
0.7	0.7008931	5	0.7389	0.7590	0.7869
0.7	0.7008931	10	0.9046	0.9348	0.9673
0.7	0.7008931	15	0.9066	0.9424	0.9784
0.7	0.7008931	20	0.9025	0.9408	0.9773
0.7	0.7008931	25	0.8909	0.9341	0.9754
0.7	0.7008931	30	0.8873	0.9332	0.9763
0.7	0.7008931	35	0.8909	0.9351	0.9771
0.7	0.7008931	40	0.8930	0.9384	0.9805
0.7	0.7008931	45	0.8924	0.9400	0.9796
0.7	0.7008931	50	0.8873	0.9362	0.9800
0.7	0.7008931	55	0.8931	0.9366	0.9793
0.7	0.7008931	60	0.8963	0.9376	0.9767
0.7	0.7008931	65	0.8947	0.9400	0.9808
0.7	0.7008931	70	0.8959	0.9423	0.9833
0.7	0.7008931	75	0.8965	0.9419	0.9828
0.7	0.7008931	80	0.8987	0.9446	0.9815
0.7	0.7008931	85	0.8913	0.9412	0.9806
0.7	0.7008931	90	0.8980	0.9416	0.9826
0.7	0.7008931	95	0.8973	0.9437	0.9849
0.7	0.7008931	100	0.8953	0.9423	0.9814
0.9	0.8997493	5	0.7231	0.7296	0.7364
0.9	0.8997493	10	0.9215	0.9345	0.9461
0.9	0.8997493	15	0.9574	0.9727	0.9845
0.9	0.8997493	20	0.9566	0.9742	0.9888
0.9	0.8997493	25	0.9525	0.9754	0.9904
0.9	0.8997493	30	0.9477	0.9690	0.9892
0.9	0.8997493	35	0.9457	0.9703	0.9901
0.9	0.8997493	40	0.9453	0.9721	0.9905
0.9	0.8997493	45	0.9439	0.9696	0.9903

Continued

Table 5.1 **Continued**

ρ_{xy}	Rhoxy	n	Cov1	Cov2	Cov3
0.9	0.8997493	50	0.9432	0.9684	0.9893
0.9	0.8997493	55	0.9395	0.9659	0.9886
0.9	0.8997493	60	0.9365	0.9674	0.9917
0.9	0.8997493	65	0.9344	0.9648	0.9903
0.9	0.8997493	70	0.9309	0.9643	0.9903
0.9	0.8997493	75	0.9303	0.9614	0.9878
0.9	0.8997493	80	0.9316	0.9630	0.9888
0.9	0.8997493	85	0.9253	0.9603	0.9883
0.9	0.8997493	90	0.9224	0.9588	0.9895
0.9	0.8997493	95	0.9258	0.9618	0.9891
0.9	0.8997493	100	0.9275	0.9631	0.9898

94.32%, and 98.56% for samples of sizes 85, 95, and 75, respectively. For $\rho_{xy} = 0.9$ (or true Rhoxy = 0.8997493) the estimated 90%, 95%, and 99% coverages were 92.15%, 95.88%, and 99.01% for samples of sizes 10, 90, and 35, respectively.

Therefore, for a moderate sample size, the suggested confidence interval estimator based on the proposed tuned jackknife technique gives approximately the nominal coverage.

5.3.2 R code

The following R code, **PUMPKIN51.R**, was used to study the 90%, 95%, and 99% coverage by newly the tuned estimator of the finite population correlation coefficient based on dell distance function.

```
#PROGRAM PUMPKIN51.R
set.seed(2013)
N<-70000
for (rho in seq(.1,.9,0.2)) {
xe<-rnorm(N,0,1); ye<-rnorm(N,0,1)
x<-30 + 20.1*xe
y<-45.5 + 23.5*sqrt((1-rho^2))*ye + rho*23.5*xe
XPMEAN<-mean(x);rhoxy<-cor(x,y)
nreps<-10000
ESTP=rep(0,nreps)
ci1.max=ci1.min=ci2.max=ci2.min=ci3.max=ci3.min= ESTP
vESTP=ESTP
for (n in seq(5,100,5))
  {
  for (r in 1:nreps)
```

```
{
  us<-sample(N,n)
  xs<-x[us]; ys<-y[us]
  RXY<-cor(xs,ys)
  SERXY<-(1-RXY^2)/sqrt(n)
  c1<-(n-1)/(n-2);c2<-n/((n-1)*(n-2))
  sxy<-RXY*sqrt(var(xs)*var(ys))
  etan<- c1-(c2/sxy)*(xs - mean(xs))*(ys - mean(ys))
  etad1<-c1-(c2/var(xs))*(xs-mean(xs))*(xs - mean(xs))
  etad2<-c1-(c2/var(ys))*(ys-mean(ys))*(ys - mean(ys))
  etaj<-etan/(sqrt(etad1 * etad2))
  rr<-RXY*etaj
  xmj<-(sum(xs)-xs)/(n-1)
  AETA<-sum(etaj)/n
  eta_j<-(sum(etaj) - etaj)/(n-1)
  shi1<- xmj - (XPMEAN-n*(2-n)*mean(xs))/((n-1)^2)
  shi2<- eta_j - (sum(etaj)-1)/(n-1)
  del<-sum(shi1^2)*sum(shi2^2) - (sum(shi1*shi2))^2
  al1<-sum(shi1)*sum(shi2^2) - sum(shi2)*sum(shi1*shi2)
  al2<-sum(shi2)*sum(shi1^2) - sum(shi1)*sum(shi1*shi2)
  al1<-al1/del;al12<-al2/del
  wbni<-1/(n*(1 + al1*shi1 + al2*shi2))
  ESTi<- n*rr*wbni
  ESTP[r]<- sum(ESTi)
  EST_I<- (ESTP[r] - ESTi)/(n-1)
  ESTP[r]<-ESTP[r]/n
  vj<- (wbni)^2*(EST_I - ESTP[r])^2
  vESTP[r]<-n^2*(c1/c2)^2*sum(vj)
  ci1.max[r]<- RXY -qt(.05,n-2)*sqrt(vESTP[r])
  ci1.min[r]<- RXY +qt(.05,n-2)*sqrt(vESTP[r])
  ci2.max[r]<- RXY -qt(.025,n-2)*sqrt(vESTP[r])
  ci2.min[r]<- RXY +qt(.025,n-2)*sqrt(vESTP[r])
  ci3.max[r]<- RXY -qt(.005,n-2)*sqrt(vESTP[r])
  ci3.min[r]<- RXY +qt(.005,n-2)*sqrt(vESTP[r])
}
sum(abs(ESTP) >1)->out
for (r in 1:nreps) if (abs(ESTP[r])>1) {
      ci1.max[r]<-NaN;ci1.min[r]<-NaN
      ci2.max[r]<-NaN;ci2.min[r]<-NaN
      ci3.max[r]<-NaN;ci3.min[r]<-NaN}
round(sum(ci1.min<rhoxy & ci1.max>rhoxy,na.rm=T)/nreps,4)->cov1
round(sum(ci2.min<rhoxy & ci2.max>rhoxy,na.rm=T)/nreps,4)->cov2
round(sum(ci3.min<rhoxy & ci3.max>rhoxy,na.rm=T)/nreps,4)->cov3
cat(rho,rhoxy,n, cov1,cov2,cov3,'\n')
   }
}
```

5.3.3 Numerical illustration

We explain the main steps in constructing the confidence interval estimate using the proposed method with the following example.

Example 5.1 Consider a sample of $n = 10$ pumpkins with x and y as circumference (in.) and weight (lbs) as follows:

x	405	177	196	279	303	345	415	269	253	286
y	7801	710	296	3946	5072	1950	5392	983	4163	2180

Construct the 70% confidence interval estimate of the finite population correlation coefficient between the weight and circumference of pumpkins by assuming that the population mean circumference of 295 in. is known.

Solution. One can easily compute the jackknife tuned estimates of the correlation coefficient and the jackknife weights as follows.

$\bar{w}_n(j)$	$r_{\text{Tuned}(j)}$
0.09507095	0.7786556
0.09823759	0.7703889
0.09765982	0.7720279
0.10063455	0.7624706
0.10081483	0.7617713
0.10418488	0.7512382
0.09884569	0.7670773
0.10070172	0.7623289
0.10237428	0.7573421
0.10027052	0.7635347

Thus, the tuned estimator of the correlation coefficient is computed as

$$r_{\text{Tuned}} = 0.7646836 \quad \text{and} \quad SE(r_{\text{Tuned}}) = 0.1879029$$

Hence using $|t_{0.15}(\text{df} = 8)| = 1.1081$, the required 70% confidence interval estimate of the correlation coefficient is computed as $0.5564598 - 0.9729073$.

5.3.4 R code used for illustration

The following R code, **PUMPKIN51.EX.R**, was used to derive the results in the preceding illustration.

```
#PROGRAM PUMPKIN51EX.R
n<-10; XPMEAN<-295
xs<-c(405,177,196,279,303,345,415,269,253,286)
ys<-c(7801,710,296,3946,5072,1950,5392,983,4163,2180)
```

```
RXY<-cor(xs,ys)
SERXY<-(1-RXY^2)/sqrt(n)
c1<-(n-1)/(n-2);c2<-n/((n-1)*(n-2))
sxy<-RXY*sqrt(var(xs)*var(ys))
etan<- c1-(c2/sxy)     *(xs - mean(xs))*(ys - mean(ys))
etad1<-c1-(c2/var(xs)) *(xs - mean(xs))*(xs - mean(xs))
etad2<-c1-(c2/var(ys)) *(ys - mean(ys))*(ys - mean(ys))
etaj<-etan/(sqrt(etad1 * etad2))
r<-RXY*etaj
xmj<-(sum(xs)-xs)/(n-1)
AETA<-sum(etaj)/n
eta_j<-(sum(etaj) - etaj)/(n-1)
shi1<- xmj - (XPMEAN-n*(2-n)*mean(xs))/((n-1)^2)
shi2<- eta_j - (sum(etaj)-1)/(n-1)
del<-sum(shi1^2)*sum(shi2^2) - (sum(shi1*shi2))^2
al1<-sum(shi1)*sum(shi2^2) - sum(shi2)*sum(shi1*shi2)
al2<-sum(shi2)*sum(shi1^2) - sum(shi1)*sum(shi1*shi2)
al1<-al1/del;al12<-al2/del
wbni<-1/(n*(1 + al1*shi1 + al2*shi2))
ESTi<- n*r*wbni
ESTP<- sum(ESTi)
EST_I<-(ESTP - ESTi)/(n-1)
ESTP<-ESTP/n
vj<- (wbni^2)*(EST_I - ESTP)^2
vESTP<-n^2*(c1/c2)^2*sum(vj)
L<-ESTP+qt(.15,n-2)*sqrt(vESTP)
U<-ESTP-qt(.15,n-2)*sqrt(vESTP)
cbind(wbni,EST_I)
cat("Tuned estimate:", ESTP, "SE: ",vESTP^.5,'\n')
cat("Confidence Interval:"," ", L,"; ", U,'\n')
```

5.4 Exercises

Exercise 5.1 Consider a tuned estimator of the finite population correlation coefficient ρ_{xy} as

$$r_{\text{Tuned}(1)} = \sum_{j \in s} \bar{w}_n(j) r(j) \qquad (5.36)$$

where $r(j)$ denotes the previously defined estimator of the correlation coefficient, and

$$\bar{w}_n(j) = \frac{1 - w_j}{n - 1} \quad \text{for } j \in s \qquad (5.37)$$

is a set of tuned jackknife weights for the unit length weights w_j, such that the following five constraints are satisfied:

$$\sum_{j \in s} \bar{w}_n(j) = 1 \tag{5.38}$$

$$\sum_{j \in s} \bar{w}_n(j) \left\{ \bar{\eta}(j) - \frac{n\bar{\eta} - 1}{n - 1} \right\} = 0 \tag{5.39}$$

$$\sum_{j \in s} \bar{w}_n(j) \left\{ \bar{x}_n(j) - \frac{(\bar{X} - n(2-n)\bar{x}_n)}{(n-1)^2} \right\} = 0 \tag{5.40}$$

$$\sum_{j \in s} \bar{w}_n(j) \left\{ \hat{\sigma}_x^2(j) - \frac{\sigma_x^2 - n(2-n)\hat{\sigma}_x^2}{(n-1)^2} \right\} = 0 \tag{5.41}$$

and

$$\sum_{j \in s} \bar{w}_n(j) \left\{ v_{s(x)}^*(j) - \frac{v_{\Omega(x)}^* - n(2-n)v_{s(x)}^*}{(n-1)^2} \right\} = 0 \tag{5.42}$$

where

$$\bar{\eta} = \frac{1}{n} \sum_{j=1}^{n} \eta_j, \quad \bar{\eta}(j) = \frac{n\bar{\eta} - \eta_j}{n - 1}$$

$$\eta_j = \frac{\dfrac{(n-1)}{(n-2)} - \dfrac{n(x_j - \bar{x}_n)(y_j - \bar{y}_n)}{(n-1)(n-2)}}{\sqrt{\dfrac{(n-1)}{(n-2)} - \dfrac{n(x_j - \bar{x}_n)^2}{(n-1)(n-2)}} \sqrt{\dfrac{(n-1)}{(n-2)} - \dfrac{n(y_j - \bar{y}_n)^2}{(n-1)(n-2)}}}$$

$$v_{\Omega(x)}^* = \frac{1}{N} \sum_{i \in \Omega} v(x_i), \quad v_{s(x)}^* = \frac{1}{n} \sum_{i \in s} v(x_i), \quad v_{s(x)}^*(j) = \frac{nv_{s(x)}^* - v(x_j)}{n - 1},$$

$$\sigma_x^2 = N^{-1} \sum_{i \in \Omega} (x_i - \bar{X})^2, \quad \bar{X} = N^{-1} \sum_{i \in \Omega} x_i, \quad \hat{\sigma}_x^2 = n^{-1} \sum_{i \in s} (x_i - \bar{x}_n)^2,$$

$$\bar{x}_n = n^{-1} \sum_{i \in s} x_i, \quad \text{and} \quad \hat{\sigma}_x^2(j) = \frac{n\hat{\sigma}_x^2 - (x_j - \bar{x}_n)^2}{n - 1}$$

Note that we assume that under the known heteroscedastic nature of the linear model,

$$y_i = \beta x_i + e_i \tag{5.43}$$

the assumptions $E(e_i|x_i) = 0$, $E(e_i^2|x_i) = \sigma^2 v(x_i)$ and $E(e_i e_j|x_i x_j) = 0$ with $v(x_i) > 0$ are satisfied.

Optimize each of the following two distance functions:

$$D_1 = \frac{1}{n} \sum_{j \in s} \ln(\bar{w}_n(j)), \quad \text{with } 0 < w_j < 1 \tag{5.44}$$

and

$$D_2 = \frac{1}{n} \sum_{j \in s} \tanh^{-1}\left(\frac{\{\bar{w}_n(j)\}^2 - 1}{\{\bar{w}_n(j)\}^2 + 1} \right) \quad \text{with } -\infty < w_j < +\infty \tag{5.45}$$

where $\tanh^{-1}()$ is the hyperbolic tangent function, subject to the five tuning constraints (5.38)–(5.42). Write code in any scientific language, like R, FORTRAN, or C, to study these distance functions. Discuss the nature of tuned weights in each situation. Construct the 90%, 95%, and 99% confidence interval estimates in each situation by estimating the variance using the method of double jackknifing discussed in the chapter. Investigate the nominal coverage through simulation. Also, simulate and discuss the distribution of $-2D_1$ and D_2. Show, if possible, under which conditions the value of the estimator $r_{\text{Tuned}(1)}$ lies between -1 and $+1$.

Exercise 5.2 Consider a tuned estimator of the finite population correlation coefficient ρ_{xy} defined by

$$r_{\text{Tuned}(2)} = \sum \sum_{i \neq j \in s} \bar{w}(i,j) r(i,j) \tag{5.46}$$

where $r(i,j)$ denotes the value of the correlation coefficient after dropping partially two pairs of values (x_i, y_i) and (x_j, y_j), and $\bar{w}(i,j)$ for $i,j \in s$ is a set of jackknife tuned weights such that the following four constraints are satisfied:

$$\sum \sum_{i \neq j \in s} \bar{w}(i,j) = 1 \tag{5.47}$$

$$\sum \sum_{i \neq j \in s} \bar{w}(i,j) \left\{ \bar{\eta}(i,j) - \frac{n(n-1)\bar{\eta}-1}{n(n-1)-1} \right\} = 0 \tag{5.48}$$

$$\sum \sum_{i \neq j \in s} \bar{w}(i,j) \left\{ s_x^2(x_{(i)}, x_{(j)}) - \frac{S_x^2 - n(n-1)(2-n(n-1))s_x^2}{(n(n-1)-1)^2} \right\} = 0 \tag{5.49}$$

$$\sum \sum_{i \neq j \in s} \bar{w}(i,j) \left\{ v_{s(x)}(x_{(i)}, x_{(j)}) - \frac{v_{\Omega(x)} - n(n-1)\{2-n(n-1)\}v_{s(x)}}{(n(n-1)-1)^2} \right\} = 0 \tag{5.50}$$

where

$$s_x^2(x_{(i)}, x_{(j)}) = \frac{n(n-1)s_x^2 - 0.5(x_i - x_j)^2}{n(n-1) - 1} \quad (5.51)$$

$$S_x^2 = \{2N(N-1)\}^{-1} \sum_{i \neq} \sum_{j \in \Omega} (x_i - x_j)^2 \quad (5.52)$$

$$s_x^2 = \{2n(n-1)\}^{-1} \sum_{i \neq} \sum_{j \in s} (x_i - x_j)^2 \quad (5.53)$$

and

$$\bar{\eta}(i,j) = \frac{n(n-1)\bar{\eta} - \eta_{ij}}{n(n-1) - 1} \quad \text{with } \bar{\eta} = \frac{1}{n(n-1)} \sum_{i \neq} \sum_{j \in s} \eta_{ij} \quad (5.54)$$

$$\bar{w}(i,j) = \frac{1 - w_{ij}}{n(n-1) - 1} \quad \text{with } \sum_{i \neq} \sum_{j \in s} w_{ij} = 1 \quad (5.55)$$

$$\eta_{ij} = \frac{1 - \dfrac{(x_i - x_j)(y_i - y_j)}{2n(n-1)s_{xy}}}{\sqrt{1 - \dfrac{(x_i - x_j)^2}{2n(n-1)s_x^2}} \sqrt{1 - \dfrac{(y_i - y_j)^2}{2n(n-1)s_y^2}}} \quad (5.56)$$

$$v_{s(x)}(x_{(i)}, x_{(j)}) = \frac{n(n-1)v_{s(x)} - 0.5(v(x_i) + v(x_j))}{n(n-1) - 1} \quad (5.57)$$

$$v_{s(x)} = \{2n(n-1)\}^{-1} \sum_{i \neq} \sum_{j \in s} (v(x_i) + v(x_j)) \quad (5.58)$$

and

$$v_{\Omega(x)} = \{2N(N-1)\}^{-1} \sum_{i \neq} \sum_{j \in \Omega} (v(x_i) + v(x_j)) \quad (5.59)$$

Assume that for the linear heteroscedastic model,

$$y_i = \beta x_i + e_i \quad (5.60)$$

the following assumptions are satisfied: $E(e_i|x_i) = 0$, $E(e_i^2|x_i) = \sigma^2 v(x_i)$, $v(x_i) > 0$, and $E(e_i e_i|x_i x_j) = 0$.

Optimize each of the following distance functions:

$$D_{11} = \frac{1}{n(n-1)} \sum_{i \neq j \in s} \sum \ln(\bar{w}(i,j)), \quad 0 < \bar{w}(i,j) < 1 \tag{5.61}$$

and

$$D_{22} = \frac{1}{n(n-1)} \sum_{i \neq j \in s} \sum \tanh^{-1}\left(\frac{\{\bar{w}(i,j)\}^2 - 1}{\{\bar{w}(i,j)\}^2 + 1}\right), \quad -\infty < \bar{w}(i,j) < +\infty \tag{5.62}$$

where $\tanh^{-1}()$ is the hyperbolic tangent function, subject to the four tuning constraints (5.47)–(5.50). Write code in any scientific programming language to study these distance functions. Discuss the nature of tuned weights in each situation. Construct the 90%, 95%, and 99% confidence interval estimates in each situation by estimating the variance using the method discussed in this chapter by considering doubly jackknifed estimates of the correlation coefficient. Compare the resulting coverages with the respective nominal coverages for different sample sizes. Simulate and discuss the distribution of $-2D_{11}$ and D_{22}. Show, if possible, under which conditions the value of the estimator $r_{\text{Tuned}(2)}$ lies between -1 and $+1$.

Exercise 5.3 Consider a tuned ratio estimator of the finite population mean $\bar{Y}$ as

$$\bar{y}_{\text{Rat}(1)} = \sum_{j \in s} \bar{w}_n(j) \bar{y}_R(j) \tag{5.63}$$

where

$$\bar{y}_R(j) = \bar{y}_n(j) \frac{\bar{X}}{\bar{x}_n(j)} \tag{5.64}$$

denotes the jth ratio estimator of the population mean after dropping the jth value, and

$$\bar{w}_n(j) = \frac{1 - w_j}{n - 1} \quad \text{for } j \in s \tag{5.65}$$

is a set of tuned jackknife weights for the unit length weights w_j $\left(\sum_{j \in s} w_j = 1\right)$, such that the following five constraints are satisfied:

$$\sum_{j \in s} \bar{w}_n(j) = 1 \tag{5.66}$$

$$\sum_{j \in s} \bar{w}_n(j) \left\{ \kappa(j) - \frac{n\bar{\kappa} - 1}{n - 1} \right\} = 0 \tag{5.67}$$

$$\sum_{j\in s}\bar{w}_n(j)\left\{\bar{x}_n(j)-\frac{(\bar{X}-n(2-n)\bar{x}_n)}{(n-1)^2}\right\}=0 \qquad (5.68)$$

$$\sum_{j\in s}\bar{w}_n(j)\left\{\hat{\sigma}_x^2(j)-\frac{\sigma_x^2-n(2-n)\hat{\sigma}_x^2}{(n-1)^2}\right\}=0 \qquad (5.69)$$

$$\sum_{j\in s}\bar{w}_n(j)\left\{v_{s(x)}^*(j)-\frac{v_{\Omega(x)}^*-n(2-n)v_{s(x)}^*}{(n-1)^2}\right\}=0 \qquad (5.70)$$

where

$$v_{\Omega(x)}^*=\frac{1}{N}\sum_{i\in\Omega}v(x_i),\ v_{s(x)}^*=\frac{1}{n}\sum_{i\in s}v(x_i),\ v_{s(x)}^*(j)=\frac{nv_{s(x)}^*-v(x_j)}{n-1},$$

$$\sigma_x^2=N^{-1}\sum_{i\in\Omega}(x_i-\bar{X})^2,\ \bar{X}=N^{-1}\sum_{i\in\Omega}x_i,\ \hat{\sigma}_x^2=n^{-1}\sum_{i\in\Omega}(x_i-\bar{x}_n)^2,$$

$$\bar{x}_n=n^{-1}\sum_{i\in s}x_i,\ \hat{\sigma}_x^2(j)=\frac{n\hat{\sigma}_x^2-(x_j-\bar{x}_n)^2}{n-1},\ \bar{\kappa}=\frac{1}{n}\sum_{j\in s}\kappa_j,\ \bar{\kappa}(j)=\frac{n\bar{\kappa}-\kappa_j}{n-1}$$

and

$$\kappa_j=\left(1-\frac{y_j}{n\bar{y}_n}\right)\Big/\left(1-\frac{x_j}{n\bar{x}_n}\right)$$

Note that we assume that due to the heteroscedastic nature of the linear model,

$$y_i=\beta x_i+e_i \qquad (5.71)$$

the assumptions $E(e_i|x_i)=0$, $E(e_i^2|x_i)=\sigma^2 v(x_i)$, and $E(e_i e_j|x_i x_j)=0$ with $v(x_i)>0$ are satisfied.

Subject to the preceding five tuning constraints, optimize each one of the following distance functions:

$$D_1=(2^{-1}n)\sum_{j\in s}q_j^{-1}\left(1-(n-1)\bar{w}_n(j)-n^{-1}\right)^2,\ -\infty<\bar{w}_n(j)<+\infty \qquad (5.72)$$

$$D_2=\sum_{j\in s}[\bar{w}_n(j)\ln(\bar{w}_n(j))],\ 0<\bar{w}_n(j)<1/(n-1) \qquad (5.73)$$

$$D_3=2\sum_{j\in s}\left(\sqrt{1-(n-1)\bar{w}_n(j)}-\sqrt{n^{-1}}\right)^2,\ 0<\bar{w}_n(j)<1/(n-1) \qquad (5.74)$$

$$D_4 = \sum_{j \in s} \left[n^{-1} \ln(\bar{w}_n(j)) \right], \quad 0 < \bar{w}_n(j) < 1/(n-1) \tag{5.75}$$

$$D_5 = \sum_{j \in s} \frac{(1-(n-1)\bar{w}_n(j) - n^{-1})^2}{2(1-(n-1)\bar{w}_n(j))}, \quad 0 < \bar{w}_n(j) < 1/(n-1) \tag{5.76}$$

$$D_6 = \frac{1}{2} \sum_{j \in s} \tanh^{-1}\left(\frac{\{\bar{w}_n(j)\}^2 - 1}{\{\bar{w}_n(j)\}^2 + 1} \right), \quad -\infty < \bar{w}_n(j) < +\infty \tag{5.77}$$

and

$$D_7 = \frac{1}{2} \sum_{j \in s} \frac{(1-(n-1)\bar{w}_n(j) - n^{-1})^2}{q_j n^{-1}} + \frac{1}{2} \sum_{j \in s} \frac{\varphi_j \{\bar{w}_n(j)\}^2}{q_j n^{-1}(n-1)^{-2}}, \quad -\infty < \bar{w}_n(j) < +\infty \tag{5.78}$$

where q_j are suitably chosen weights that form different types of estimators, φ_j is a penalty as in Farrell and Singh (2002a), and $\tanh^{-1}()$ is the hyperbolic tangent function as in Singh (2012). Write code in any scientific programming language to study these distance functions. Discuss the nature of tuned weights in each situation. Construct the 90%, 95%, and 99% confidence interval estimates in each situation by estimating the variance with the method discussed in the chapter, by considering all possible doubly jackknifed estimators. Simulate and discuss the distribution of $-2D_4$ and D_6.

Exercise 5.4 Consider a tuned estimator of the finite population regression coefficient β as

$$\hat{\beta}_{\text{Tuned}(1)} = \sum_{j \in s} \bar{w}_n(j) \hat{\beta}(j) \tag{5.79}$$

where $\hat{\beta}(j)$ denotes the estimator of the regression coefficient after dropping the jth pair of values (x_j, y_j) and

$$\bar{w}_n(j) = \frac{1-w_j}{n-1} \quad \text{for } j \in s \tag{5.80}$$

is a set of tuned jackknife weights for the unit length weights w_j $\left(\sum_{j \in s} w_j = 1 \right)$, such that the following five constraints are satisfied:

$$\sum_{j \in s} \bar{w}_n(j) = 1 \tag{5.81}$$

$$\sum_{j \in s} \bar{w}_n(j) \left\{ \zeta(j) - \frac{n\bar{\zeta} - 1}{n-1} \right\} = 0 \qquad (5.82)$$

$$\sum_{j \in s} \bar{w}_n(j) \left\{ \bar{x}_n(j) - \frac{(\bar{X} - n(2-n)\bar{x}_n)}{(n-1)^2} \right\} = 0 \qquad (5.83)$$

$$\sum_{j \in s} \bar{w}_n(j) \left\{ \hat{\sigma}_x^2(j) - \frac{\sigma_x^2 - n(2-n)\hat{\sigma}_x^2}{(n-1)^2} \right\} = 0 \qquad (5.84)$$

$$\sum_{j \in s} \bar{w}_n(j) \left\{ v^*_{s(x)}(j) - \frac{v^*_{\Omega(x)} - n(2-n)v^*_{s(x)}}{(n-1)^2} \right\} = 0 \qquad (5.85)$$

where

$$v^*_{\Omega(x)} = \frac{1}{N} \sum_{i \in \Omega} v(x_i), \quad v^*_{s(x)} = \frac{1}{n} \sum_{i \in s} v(x_i),$$

$$v^*_{s(x)}(j) = \frac{nv^*_{s(x)} - v(x_j)}{n-1}, \quad \sigma_x^2 = N^{-1} \sum_{i \in \Omega} (x_i - \bar{X})^2, \quad \bar{X} = N^{-1} \sum_{i \in \Omega} x_i,$$

$$\hat{\sigma}_x^2 = n^{-1} \sum_{i \in \Omega} (x_i - \bar{x}_n)^2, \quad \bar{x}_n = n^{-1} \sum_{i \in s} x_i, \quad \hat{\sigma}_x^2(j) = \frac{n\hat{\sigma}_x^2 - (x_j - \bar{x}_n)^2}{n-1},$$

$$\bar{\zeta} = \frac{1}{n} \sum_{j \in s} \zeta_j, \quad \bar{\zeta}_j = \frac{n\bar{\zeta} - \zeta_j}{n-1} \text{ and } \zeta_j = \frac{\frac{(n-1)}{(n-2)} - \frac{n(x_j - \bar{x}_n)(y_j - \bar{y}_n)}{(n-1)(n-2)}}{\frac{(n-1)}{(n-2)} - \frac{n(x_j - \bar{x}_n)^2}{(n-1)(n-2)}}$$

Note that here we assume that we have the known heteroscedastic linear model:

$$y_i = \beta x_i + e_i \qquad (5.86)$$

where $E(e_i|x_i) = 0$, $E(e_i^2|x_i) = \sigma^2 v(x_i)$, and $E(e_i e_j|x_i x_j) = 0$ with $v(x_i) > 0$ are satisfied.

Subject to the preceding five tuning constraints, optimize each of the following distance functions:

$$D_1 = (2^{-1}n) \sum_{j \in s} q_j^{-1} \left(1 - (n-1)\bar{w}_n(j) - n^{-1}\right)^2, \quad -\infty < \bar{w}_n(j) < +\infty \qquad (5.87)$$

$$D_2 = \sum_{j \in s} [\bar{w}_n(j) \ln(\bar{w}_n(j))], \quad 0 < \bar{w}_n(j) < 1/(n-1) \qquad (5.88)$$

$$D_3 = 2\sum_{j\in s}\left(\sqrt{1-(n-1)\bar{w}_n(j)} - \sqrt{n^{-1}}\right)^2, \quad 0 < \bar{w}_n(j) < 1/(n-1) \tag{5.89}$$

$$D_4 = \sum_{j\in s}\left[n^{-1}\ln(\bar{w}_n(j))\right], \quad 0 < \bar{w}_n(j) < 1/(n-1) \tag{5.90}$$

$$D_5 = \sum_{j\in s}\frac{(1-(n-1)\bar{w}_n(j)-n^{-1})^2}{2(1-(n-1)\bar{w}_n(j))}, \quad 0 < \bar{w}_n(j) < 1/(n-1) \tag{5.91}$$

$$D_6 = \frac{1}{n}\sum_{j\in s}\tanh^{-1}\left(\frac{\{\bar{w}_n(j)\}^2 - 1}{\{\bar{w}_n(j)\}^2 + 1}\right), \quad -\infty < \bar{w}_n(j) < +\infty \tag{5.92}$$

and

$$D_7 = \frac{1}{2}\sum_{j\in s}\frac{(1-(n-1)\bar{w}_n(j)-n^{-1})^2}{q_j n^{-1}} + \frac{1}{2}\sum_{j\in s}\frac{\varphi_j\{\bar{w}_n(j)\}^2}{q_j n^{-1}(n-1)^{-2}}, \quad -\infty < \bar{w}_n(j) < +\infty \tag{5.93}$$

where q_j are suitably chosen weights used to form different types of estimators, φ_j is a penalty as in Farrell and Singh (2002a), and $\tanh^{-1}()$ is the hyperbolic tangent function as in Singh (2012). Write code in any scientific programming language, such as R, to study these distance functions. Discuss the nature of tuned weights in each situation. Construct the 90%, 95%, and 99% confidence interval estimates in each situation by estimating the variance using the method earlier discussed in the chapter, considering all possible doubly jackknifed estimators. Simulate and discuss the distribution of $-2D_4$ and D_6.

Exercise 5.5 Consider a tuned estimator of the finite population regression coefficient β as

$$\hat{\beta}_{\text{Tuned}(2)} = \sum\sum_{i\neq j \in s}\bar{w}(i,j)\hat{\beta}(i,j) \tag{5.94}$$

where $\hat{\beta}(i,j)$ denotes the value of the regression coefficient estimates after dropping two pairs of values (x_i, y_i) and (x_j, y_j). The weights $\bar{w}(i,j)$ for $i,j \in s$ is a set of jackknife tuned weights such that the following four constraints are satisfied:

$$\sum\sum_{i\neq j\in s}\bar{w}(i,j) = 1 \tag{5.95}$$

$$\sum\sum_{i\neq j\in s}\bar{w}(i,j)\left\{\bar{\eta}^*(i,j) - \frac{n(n-1)\bar{\eta}^* - 1}{n(n-1)-1}\right\} = 0 \tag{5.96}$$

$$\sum\sum_{i\neq j\in s}\bar{w}(i,j)\left\{s_x^2\left(x_{(i)},x_{(j)}\right)-\frac{S_x^2-n(n-1)(2-n(n-1))s_x^2}{(n(n-1)-1)^2}\right\}=0 \qquad (5.97)$$

$$\sum\sum_{i\neq j\in s}\bar{w}(i,j)\left\{v_{s(x)}\left(x_{(i)},x_{(j)}\right)-\frac{v_{\Omega(x)}-n(n-1)\{2-n(n-1)\}v_{s(x)}}{(n(n-1)-1)^2}\right\}=0$$

(5.98)

where

$$\bar{\eta}^* = \frac{1}{n(n-1)}\sum\sum_{i\neq j\in s}\eta_{ij}^*, \quad \bar{\eta}^*(i,j) = \frac{n(n-1)\bar{\eta}^* - \eta_{ij}^*}{n(n-1)-1},$$

$$\eta_{ij}^* = \left\{1 - \frac{(x_i-x_j)(y_i-y_j)}{2n(n-1)s_{xy}}\right\} \Big/ \left\{1 - \frac{(x_i-x_j)^2}{2n(n-1)s_x^2}\right\},$$

$$\bar{w}(i,j) = \frac{1-w_{ij}}{n(n-1)-1} \text{ with } \sum\sum_{i\neq j\in s} w_{ij} = 1,$$

$$v_{s(x)}\left(x_{(i)},x_{(j)}\right) = \frac{n(n-1)v_{s(x)} - 0.5\left(v(x_i)+v(x_j)\right)}{n(n-1)-1},$$

$$v_{s(x)} = \frac{\sum\sum_{i\neq j\in s}\left(v(x_i)+v(x_j)\right)}{2n(n-1)}, \text{ and } v_{\Omega(x)} = \frac{\sum\sum_{i\neq j\in\Omega}\left(v(x_i)+v(x_j)\right)}{2N(N-1)}$$

Assume that for the linear heteroscedastic model

$$y_i = \beta x_i + e_i \qquad (5.99)$$

the following assumptions are satisfied: $E(e_i|x_i)=0$, $E(e_i^2|x_i)=\sigma^2 v(x_i)$, $v(x_i)>0$, and $E(e_i e_i|x_i x_j)=0$

Subject to the preceding four tuning constraints, optimize each of the following distance functions:

$$D_{11} = \frac{1}{2}\sum\sum_{i\neq j\in s}\frac{\{1-(n(n-1)-1)\bar{w}(i,j)-1/n(n-1))\}^2}{q_{ij}/(n(n-1))} \qquad (5.100)$$

$$D_{22} = \frac{1}{2}\sum\sum_{i\neq j\in s}\left(\sqrt{\bar{w}(i,j)}-1/\sqrt{n(n-1)}\right)^2, \quad \bar{w}(i,j)>0 \qquad (5.101)$$

and

$$D_{33} = \sum\sum_{i\neq j\in s}\frac{\{1-(n(n-1)-1)\bar{w}(i,j)-1/(n(n-1))\}^2}{2q_{ij}/(n(n-1))}$$
$$+ \sum\sum_{i\neq j\in s}\frac{\varphi_{ij}\{\bar{w}(i,j)\}^2 n(n-1)}{2q_{ij}\{n(n-1)-1\}^{-2}} \qquad (5.102)$$

where q_{ij} are the suitably chosen weights that form different types of estimators, and φ_{ij} is a penalty,

$$D_{44} = \frac{1}{n(n-1)} \sum\sum_{i \neq j \in s} [\bar{w}(i,j) \ln(\bar{w}(i,j))], \quad 0 < \bar{w}(i,j) < 1/(n(n-1)-1) \tag{5.103}$$

$$D_{55} = \frac{1}{n(n-1)} \sum\sum_{i \neq j \in s} \ln(\bar{w}(i,j)), \quad 0 < \bar{w}(i,j) < 1/(n(n-1)-1) \tag{5.104}$$

and

$$D_{66} = \frac{1}{n(n-1)} \sum\sum_{i \neq j \in s} \tanh^{-1}\left(\frac{\{\bar{w}(i,j)\}^2 - 1}{\{\bar{w}(i,j)\}^2 + 1}\right) \tag{5.105}$$

where $\tanh^{-1}()$ is the hyperbolic tangent function.

Write code in any scientific computer language to study these distance functions. Discuss the nature of tuned weights in each situation. Construct the 90%, 95%, and 99% confidence interval estimates in each situation by estimating the variance using the method discussed in this chapter, considering all possible doubly jackknifed estimates of the regression coefficient. Simulate the distribution of $-2D_{55}$ and D_{66}.

Exercise 5.6 Consider a tuned estimator of the finite population variance S_y^2 as

$$\hat{\sigma}^2_{\text{Tuned}(2)} = \sum\sum_{i \neq j \in s} \bar{w}(i,j) \left[s_y^2(y_i, y_j) \frac{S_x^2}{s_x^2(x_i, x_j)} \right] \tag{5.106}$$

where $s_y^2(y_i, y_j)$ and $s_x^2(x_i, x_j)$ denote the values of the estimators of the finite population variance of the study and auxiliary variables, respectively, after dropping two pairs of values (y_i, y_j) and (x_i, x_j), and $\bar{w}(i,j)$ for $i, j \in s$ is a set of jackknifed tuned weights such that the following four constraints are satisfied:

$$\sum\sum_{i \neq j \in s} \bar{w}(i,j) = 1 \tag{5.107}$$

$$\sum\sum_{i \neq j \in s} \bar{w}(i,j) \left\{ \bar{\delta}(i,j) - \frac{n(n-1)\bar{\delta}-1}{n(n-1)-1} \right\} = 0 \tag{5.108}$$

$$\sum\sum_{i \neq j \in s} \bar{w}(i,j) \left\{ s_x^2(x_{(i)}, x_{(j)}) - \frac{S_x^2 - n(n-1)(2 - n(n-1))s_x^2}{(n(n-1)-1)^2} \right\} = 0 \tag{5.109}$$

$$\sum\sum_{i\neq j\in s}\bar{w}(i,j)\left\{v_{s(x)}\left(x_{(i)},x_{(j)}\right)-\frac{v_{\Omega(x)}-n(n-1)\{2-n(n-1)\}v_{s(x)}}{(n(n-1)-1)^2}\right\}=0$$

(5.110)

where

$$\bar{\delta}=\frac{1}{n(n-1)}\sum\sum_{i\neq j\in s}\delta_{ij}, \;\bar{\delta}(i,j)=\frac{n(n-1)\bar{\delta}-\delta_{ij}}{n(n-1)-1},$$

$$\delta_{ij}=\left\{1-\frac{(y_i-y_j)^2}{2n(n-1)s_y^2}\right\}\bigg/\left\{1-\frac{(x_i-x_j)^2}{2n(n-1)s_x^2}\right\},$$

$$\bar{w}(i,j)=\frac{1-w_{ij}}{n(n-1)-1} \;\text{with}\; \sum\sum_{i\neq j\in s}w_{ij}=1,$$

$$v_{s(x)}\left(x_{(i)},x_{(j)}\right)=\frac{n(n-1)v_{s(x)}-0.5\left(v(x_i)+v(x_j)\right)}{n(n-1)-1},$$

$$v_{s(x)}=\{2n(n-1)\}^{-1}\sum\sum_{i\neq j\in s}\left(v(x_i)+v(x_j)\right)$$

and

$$v_{\Omega(x)}=\{2N(N-1)\}^{-1}\sum\sum_{i\neq j\in s}\left(v(x_i)+v(x_j)\right)$$

Assume that for the linear heteroscedastic model,

$$y_i=\beta x_i+e_i \quad (5.111)$$

the following assumptions are satisfied: $E_m(e_i|x_i)=0$, $E_m\left(e_i^2|x_i\right)=\sigma^2 v(x_i)$, $v(x_i)>0$, and $E_m\left(e_ie_j|x_ix_j\right)=0$.

Subject to the preceding four tuning constraints, optimize each of the following distance functions:

$$D_{11}=\frac{1}{2}\sum\sum_{i\neq j\in s}\frac{\{1-(n(n-1)-1)\bar{w}(i,j)-1/n(n-1))\}^2}{q_{ij}/(n(n-1))} \quad (5.112)$$

$$D_{22}=\frac{1}{2}\sum\sum_{i\neq j\in s}\left(\sqrt{\bar{w}(i,j)}-1/\sqrt{n(n-1)}\right)^2, \;\bar{w}(i,j)>0 \quad (5.113)$$

and

$$D_{33}=\sum\sum_{i\neq j\in s}\frac{\{1-(n(n-1)-1\bar{w}(i,j)-1/(n(n-1)))\}^2}{2q_{ij}/(n(n-1))}$$
$$+\sum\sum_{i\neq j\in s}\frac{\varphi_{ij}\{\bar{w}(i,j)\}^2 n(n-1)}{2q_{ij}\{n(n-1)-1\}^{-2}} \quad (5.114)$$

where q_{ij} are the suitably chosen weights that form different types of estimators, and φ_{ij} is a penalty,

$$D_{44} = \frac{1}{n(n-1)} \sum\sum_{i \neq j \in s} [\bar{w}(i,j) \ln(\bar{w}(i,j))], \quad 0 < \bar{w}(i,j) < 1/(n(n-1)-1)$$

(5.115)

$$D_{55} = \frac{1}{n(n-1)} \sum\sum_{i \neq j \in s} \ln(\bar{w}(i,j)), \quad 0 < \bar{w}(i,j) < 1/(n(n-1)-1) \quad (5.116)$$

and

$$D_{66} = \frac{1}{n(n-1)} \sum\sum_{i \neq j \in s} \tanh^{-1}\left(\frac{\{\bar{w}(i,j)\}^2 - 1}{\{\bar{w}(i,j)\}^2 + 1}\right) \quad (5.117)$$

where $\tanh^{-1}()$ is the hyperbolic tangent function. Write code in any scientific language, such as R, to study the resultant estimators from these distance functions. Discuss the nature of the tuned weights in each situation. Construct the 90%, 95%, and 99% confidence interval estimates in each situation by estimating the variance with the method discussed in the chapter, considering all possible doubly jackknifed estimates of the regression coefficient. Simulate and discuss the distributions of $-2D_{55}$ and D_{66}.

Exercise 5.7 A student of medicine studies the statement made by a team of doctors about the negative relationship between age (years) and duration of sleep (minutes), and from a simple random and with replacement sample six patients obtained the following data:

| Age (x) | 78 | 74 | 87 | 72 | 72 | 66 |
| Duration of sleep (y) | 345 | 381 | 270 | 345 | 364 | 480 |

Apply the newly tuned method of estimating to estimate the correlation coefficient ρ_{xy} between age and duration of sleep. Also apply it to construct the 75% confidence interval for the correlation coefficient by assuming that the average age in the population is 76 years. Comment on your findings.

Tuning of multicharacter survey estimators

6.1 Introduction

In this chapter, we discuss the tuning of multicharacter survey estimators when the sample is selected using a probability proportional to size and with replacement (PPSWR) sampling, where the selection probabilities have low positive correlation with the study variables. The role of two auxiliary variables, one at the selection stage and another at the estimation stage, is discussed. Unsolved exercises are also provided at the end of the chapter.

6.2 Transformation on selection probabilities

Consider a population Ω consisting of N units. Let z_i, $i = 1, 2, ..., N$ be the value of the ith unit of an auxiliary variable associated with a study variable. Consider a sample s of n units that is selected using a PPSWR scheme.

In short, let

$$p_i = \frac{z_i}{Z}, \quad i = 1, 2, ..., N \tag{6.1}$$

where $Z = \sum_{i \in \Omega} z_i$ is known, be the probability of selecting the ith unit in the sample. Note then that

$$\sum_{i \in \Omega} p_i = 1 \tag{6.2}$$

Let ρ_{yz} be the known value of the correlation coefficient between the study variable(s) and the auxiliary variable z_i. Let us make clear that in a multicharacter survey, there are several study variables and one auxiliary variable, which is used at the selection stage of the sample. For those study variables that have high correlation with the auxiliary variable, one can use the well-known Hansen and Hurwitz (1943) estimator of population total for PPSWR sampling. For those study variables that have low correlation with the auxiliary variable used at the selection stage, Bansal and Singh (1985) suggested the following transformation on the selection probabilities p_i as

$$p_i^* = \left(1 + \frac{1}{N}\right)^{(1-\rho_{yz})} (1 + p_i)^{(\rho_{yz})} - 1 \tag{6.3}$$

Note that if the value of the correlation coefficient $\rho_{yz} = 0$ then $p_i^* = 1/N$, which leads to the claim due to Rao (1966), and if $\rho_{yz} = 1$ then $p_i^* = p_i$, which leads to the claim of Hansen and Hurwitz (1943). Thus, the transformation p_i^* in Equation (6.3) is a kind of compromise between the Rao (1966) and Hansen and Hurwitz (1943) methods.

6.3 Tuning with a chi-square distance function

The newly tuned jackknife estimator of the population total Y in multicharacter surveys is defined as

$$\hat{Y}_{\text{MTuned(cs)}} = \sum_{j \in s} \left[\left\{ (n-1)^2 \bar{w}_n(j) - (n-2) \right\} \hat{Y}_M(j) \right] \qquad (6.4)$$

where

$$\hat{Y}_M(j) = \frac{\sum_{i=1}^{n} \frac{y_i}{p_i^*} - \frac{y_j}{p_j^*}}{n-1} \qquad (6.5)$$

is the estimator of the population total obtained by removing the jth unit from the sample s, and $\bar{w}_n(j)$ is the jth jackknife tuned weight constructed so that the following two constraints are satisfied:

$$\sum_{j \in s} \bar{w}_n(j) = 1 \qquad (6.6)$$

$$\sum_{j \in s} \bar{w}_n(j) \hat{X}_M(j) = \frac{(X - n(2-n)\hat{X}_M)}{(n-1)^2} \qquad (6.7)$$

where

$$\hat{X}_M(j) = \frac{n\hat{X}_M - x_j/p_j^*}{n-1} \qquad (6.8)$$

and

$$\hat{X}_M = \frac{1}{n} \sum_{i \in s} \frac{x_i}{p_i^*} \qquad (6.9)$$

is an estimator of the population total X of the second auxiliary variable, which we use at the estimation stage. We suggest tuning the weights $\bar{w}_n(j)$ so that the modified chi-square type distance function, defined as

$$(2^{-1}n) \sum_{j \in s} q_j^{-1} \left(1 - (n-1)\bar{w}_n(j) - n^{-1} \right)^2 \qquad (6.10)$$

is optimum subject to the tuning constraints (6.6) and (6.7), where q_j is a choice of weights.

The Lagrange function becomes

$$L_1 = (2^{-1}n) \sum_{j \in s} q_j^{-1} \left(1 - (n-1)\bar{w}_n(j) - n^{-1}\right)^2 - \lambda_0 \left\{ \sum_{j \in s} \bar{w}_n(j) - 1 \right\}$$
$$- \lambda_1 \left\{ \sum_{j \in s} \bar{w}_n(j) \hat{X}_M(j) - \frac{(X - n(2-n)\hat{X}_M)}{(n-1)^2} \right\} \tag{6.11}$$

where λ_0 and λ_1 are Lagrange multipliers constants.

Note that

$$n(n-1)^2 \left\{ \frac{X - n(2-n)\hat{X}_M}{(n-1)^2} - \frac{1}{n} \sum_{j \in s} \hat{X}_M(j) \right\} = n(X - \hat{X}_M) \tag{6.12}$$

On setting

$$\frac{\partial L_1}{\partial \bar{w}_n(j)} = 0$$

we have

$$\bar{w}_n(j) = \frac{1}{n} \left[1 + \lambda_0 \frac{q_j}{(n-1)^2} + \lambda_1 \frac{q_j}{(n-1)^2} \hat{X}_M(j) \right] \tag{6.13}$$

On using Equation (6.13) in Equations (6.6) and (6.7), one is led to the following set of normal equations that give the optimum values of λ_0 and λ_1:

$$\begin{bmatrix} \sum_{j \in s} q_j, & \sum_{j \in s} q_j \hat{X}_M(j) \\ \sum_{j \in s} q_j \hat{X}_M(j), & \sum_{j \in s} q_j \{\hat{X}_M(j)\}^2 \end{bmatrix} \begin{bmatrix} \lambda_0 \\ \lambda_1 \end{bmatrix} = \begin{bmatrix} 0 \\ n(X - \hat{X}_M) \end{bmatrix} \tag{6.14}$$

The tuned jackknife weights $\bar{w}_n(j)$ are then given by

$$\bar{w}_n(j) = \frac{1}{n} + \frac{\Delta_j}{(n-1)^2} (X - \hat{X}_M) \tag{6.15}$$

where

$$\Delta_j = \frac{\left(\sum_{j\in s} q_j\right) q_j \hat{X}_M(j) - q_j \left(\sum_{j\in s} q_j \hat{X}_M(j)\right)}{\left(\sum_{j\in s} q_j\right) \left\{\sum_{j\in s} q_j (\hat{X}_M(j))^2\right\} - \left\{\sum_{j\in s} q_j \hat{X}_M(j)\right\}^2}. \quad (6.16)$$

Thus, under the chi-square (cs) type distance function, the newly tuned estimator (6.4) of the population total Y for multicharacter surveys becomes

$$\hat{Y}_{\text{MTuned(cs)}} = \hat{Y}_M + \hat{\beta}_{\text{MTuned}} (X - \hat{X}_M) \quad (6.17)$$

where

$$\hat{\beta}_{\text{MTuned}} = \frac{\left(\sum_{j\in s} q_j\right) \left(\sum_{j\in s} q_j \hat{X}_M(j) \hat{Y}_M(j)\right) - \left(\sum_{j\in s} q_j \hat{X}_M(j)\right) \left(\sum_{j\in s} q_j \hat{Y}_M(j)\right)}{\left(\sum_{j\in s} q_j\right) \left(\sum_{j\in s} q_j (\hat{X}_M(j))^2\right) - \left(\sum_{j\in s} q_j \hat{X}_M(j)\right)^2}$$

is the tuned estimator of the regression coefficient.

6.3.1 Estimation of variance and coverage

An *adjusted* estimator, to estimate the variance of the multicharacter survey estimator, $\hat{Y}_{\text{MTuned(cs)}}$, is

$$\hat{v}(\hat{Y}_{\text{MTuned(cs)}}) = n(n-1)^3 \sum_{j\in s} (\bar{w}_n(j))^2 p_j^* \{\hat{Y}_{\text{MTuned(cs)}(j)} - \hat{Y}_{\text{MTuned(cs)}}\}^2 \quad (6.18)$$

Note that in Equation (6.18), instead of using p_i^* one might investigate the possibility of using

$$\bar{p}^*(j) = \frac{n\bar{p}^* - p_j^*}{n-1} \quad (6.19)$$

where

$$\bar{p}^* = \frac{1}{n} \sum_{j=1}^{n} p_j^*$$

This has been left as an exercise.

Note that each newly tuned multicharacter doubly jackknifed estimator of the population total Y is given by

$$\hat{Y}_{\text{MTuned(cs)}(j)} = \frac{n\hat{Y}_{\text{MTuned(cs)}} - n\left\{(n-1)^2 \bar{w}_n(j) - (n-2)\right\}\hat{Y}_M(j)}{n-1} \quad (6.20)$$

for $j = 1, 2, \ldots, n$.

The coverage by the $(1-\alpha)100\%$ confidence interval estimates obtained using this newly tuned multicharacter jackknife estimator of the population total, and an estimate of its variance, is obtained by counting the number of times the true population total Y falls within the interval estimate given by

$$\hat{Y}_{\text{MTuned(cs)}} \mp t_{\alpha/2}(\text{df} = n-1)\sqrt{\hat{v}(\hat{Y}_{\text{MTuned(cs)}})} \quad (6.21)$$

Note the use of degree of freedom df $= n - 1$ by assuming that only one parameter, that is, population total, is being estimated. We studied coverage by the nominally 90%, 95%, and 99% intervals based on the estimator. We investigated the situation with two auxiliary variables X, Z that are independent, but are each correlated with the study variable Y. We generated three random variables $y_i^* \sim N(0,1)$, $x_i^* \sim N(0,1)$, and $z_i^* \sim N(0,1)$ for $i = 1, 2, \ldots, N$ from three independent standard normal variables using the IMSL subroutine RNNOR. For four different values of the correlation coefficient ρ_{xy} and nine different values of the correlation coefficient ρ_{yz}, we generated populations with three variables Y, X, and Z taking values y_i, x_i, and z_i for the ith unit in the population as

$$y_i = \bar{Y} + \sigma_y y_i^* \quad (6.22)$$

$$x_i = \bar{X} + \sigma_x \sqrt{\left(1 - \rho_{xy}^2\right)} x_i^* + \rho_{xy}\sigma_x y_i^* \quad (6.23)$$

and

$$z_i = \bar{Z} + \sigma_z \sqrt{\left(1 - \rho_{yz}^2\right)} z_i^* + \rho_{yz}\sigma_z y_i^* \quad (6.24)$$

In this study, we fixed $\bar{X} = 2325$, $\bar{Y} = 2454$, $\bar{Z} = 322$, $\sigma_x = 232$, $\sigma_y = 245$, and $\sigma_z = 52$. Note that we used the variable X at the estimation stage and assumed it has correlations 0.3, 0.5, 0.7, and 0.9 with the study variable Y to see the effect of low as well as high values of the correlation coefficient. In the same way, we used the variable Z at the selection stage of sampling, and this variable is expected to have low correlation with the study variable, Y (e.g., 0.0, 0.1); however, we considered ten values of ρ_{yz} simply to investigate its effect on the study. The choice of the values of these correlation coefficients has two aims. The first is to save space, and the second is to study the effect of high and low values of the correlation coefficients.

We formed populations of size $N = 2000$ units for each one of the 40 combinations of ρ_{xy} and ρ_{yz}, and for each sample size of 4, 6, 8, 10, 12, and 14, we selected 10,000 samples of that size and formed the three confidence intervals. All samples are selected by using the Lahiri (1951) method of selecting a sample using PPSWR sampling. Coverage was estimated by the proportion of times the 10,000 intervals contained the population total. Results are shown in Table 6.1.

Table 6.1 Performance of the newly tuned multicharacter estimator

ρ_{xy}	RHOXY	ρ_{yz}	RHOYZ	n	90%	95%	99%
0.3	0.2978796	0.0	0.0051677	4	0.8220	0.8669	0.9297
0.3	0.2978796	0.0	0.0051677	6	0.8958	0.9171	0.9496
0.3	0.2978796	0.0	0.0051677	8	0.9295	0.9435	0.9615
0.3	0.2978796	0.0	0.0051677	10	0.9528	0.9623	0.9746
0.3	0.2978796	0.0	0.0051677	12	0.9608	0.9689	0.9774
0.3	0.2978796	0.0	0.0051677	14	0.9658	0.9719	0.9808
0.3	0.2898565	0.1	0.0912027	4	0.8347	0.8769	0.9316
0.3	0.2898565	0.1	0.0912027	6	0.8982	0.9194	0.9487
0.3	0.2898565	0.1	0.0912027	8	0.9294	0.9423	0.9609
0.3	0.2898565	0.1	0.0912027	10	0.9513	0.9601	0.9733
0.3	0.2898565	0.1	0.0912027	12	0.9660	0.9725	0.9809
0.3	0.2898565	0.1	0.0912027	14	0.9720	0.9772	0.9837
0.3	0.2893902	0.2	0.1542838	4	0.8282	0.8692	0.9268
0.3	0.2893902	0.2	0.1542838	6	0.9039	0.9238	0.9536
0.3	0.2893902	0.2	0.1542838	8	0.9337	0.9483	0.9656
0.3	0.2893902	0.2	0.1542838	10	0.9508	0.9606	0.9747
0.3	0.2893902	0.2	0.1542838	12	0.9641	0.9709	0.9792
0.3	0.2893902	0.2	0.1542838	14	0.9698	0.9759	0.9836
0.3	0.2814688	0.3	0.3024993	4	0.8255	0.8705	0.9296
0.3	0.2814688	0.3	0.3024993	6	0.9000	0.9212	0.9497
0.3	0.2814688	0.3	0.3024993	8	0.9328	0.9462	0.9628
0.3	0.2814688	0.3	0.3024993	10	0.9502	0.9582	0.9707
0.3	0.2814688	0.3	0.3024993	12	0.9601	0.9656	0.9756
0.3	0.2814688	0.3	0.3024993	14	0.9701	0.9742	0.9812
0.3	0.2800336	0.4	0.4206022	4	0.8406	0.8816	0.9332
0.3	0.2800336	0.4	0.4206022	6	0.9032	0.9207	0.9514
0.3	0.2800336	0.4	0.4206022	8	0.9327	0.9453	0.9627
0.3	0.2800336	0.4	0.4206022	10	0.9514	0.9620	0.9723
0.3	0.2800336	0.4	0.4206022	12	0.9653	0.9716	0.9786
0.3	0.2800336	0.4	0.4206022	14	0.9740	0.9794	0.9852
0.3	0.2511401	0.5	0.5078852	4	0.8409	0.8833	0.9366
0.3	0.2511401	0.5	0.5078852	6	0.9059	0.9258	0.9529
0.3	0.2511401	0.5	0.5078852	8	0.9380	0.9511	0.9661
0.3	0.2511401	0.5	0.5078852	10	0.9565	0.9657	0.9761

Continued

Table 6.1 Continued

ρ_{xy}	RHOXY	ρ_{yz}	RHOYZ	n	90%	95%	99%
0.3	0.2511401	0.5	0.5078852	12	0.9691	0.9743	0.9822
0.3	0.2511401	0.5	0.5078852	14	0.9774	0.9823	0.9878
0.3	0.3262503	0.6	0.5984581	4	0.8641	0.8993	0.9447
0.3	0.3262503	0.6	0.5984581	6	0.9155	0.9329	0.9575
0.3	0.3262503	0.6	0.5984581	8	0.9416	0.9529	0.9688
0.3	0.3262503	0.6	0.5984581	10	0.9577	0.9662	0.9756
0.3	0.3262503	0.6	0.5984581	12	0.9684	0.9745	0.9819
0.3	0.3262503	0.6	0.5984581	14	0.9736	0.9788	0.9850
0.3	0.3237574	0.7	0.6893118	4	0.8701	0.9031	0.9488
0.3	0.3237574	0.7	0.6893118	6	0.9234	0.9393	0.9607
0.3	0.3237574	0.7	0.6893118	8	0.9504	0.9612	0.9742
0.3	0.3237574	0.7	0.6893118	10	0.9657	0.9720	0.9797
0.3	0.3237574	0.7	0.6893118	12	0.9718	0.9767	0.9832
0.3	0.3237574	0.7	0.6893118	14	0.9772	0.9816	0.9879
0.3	0.3234880	0.8	0.8049675	4	0.8863	0.9155	0.9554
0.3	0.3234880	0.8	0.8049675	6	0.9268	0.9428	0.9637
0.3	0.3234880	0.8	0.8049675	8	0.9536	0.9640	0.9768
0.3	0.3234880	0.8	0.8049675	10	0.9623	0.9693	0.9796
0.3	0.3234880	0.8	0.8049675	12	0.9736	0.9801	0.9865
0.3	0.3234880	0.8	0.8049675	14	0.9772	0.9816	0.9867
0.3	0.3039484	0.9	0.8964122	4	0.9016	0.9261	0.9605
0.3	0.3039484	0.9	0.8964122	6	0.9445	0.9557	0.9715
0.3	0.3039484	0.9	0.8964122	8	0.9604	0.9665	0.9775
0.3	0.3039484	0.9	0.8964122	10	0.9714	0.9759	0.9831
0.3	0.3039484	0.9	0.8964122	12	0.9792	0.9830	0.9879
0.3	0.3039484	0.9	0.8964122	14	0.9843	0.9875	0.9918
0.5	0.4833906	0.0	0.0026272	4	0.8391	0.8817	0.9339
0.5	0.4833906	0.0	0.0026272	6	0.9045	0.9254	0.9540
0.5	0.4833906	0.0	0.0026272	8	0.9354	0.9478	0.9656
0.5	0.4833906	0.0	0.0026272	10	0.9515	0.9592	0.9709
0.5	0.4833906	0.0	0.0026272	12	0.9638	0.9713	0.9782
0.5	0.4833906	0.0	0.0026272	14	0.9734	0.9788	0.9857
0.5	0.5064535	0.1	0.0992117	4	0.8475	0.8890	0.9408
0.5	0.5064535	0.1	0.0992117	6	0.9093	0.9271	0.9514
0.5	0.5064535	0.1	0.0992117	8	0.9361	0.9478	0.9633
0.5	0.5064535	0.1	0.0992117	10	0.9537	0.9621	0.9735
0.5	0.5064535	0.1	0.0992117	12	0.9662	0.9741	0.9813
0.5	0.5064535	0.1	0.0992117	14	0.9724	0.9767	0.9834
0.5	0.5167572	0.2	0.1951615	4	0.8454	0.8869	0.9391
0.5	0.5167572	0.2	0.1951615	6	0.9069	0.9263	0.9529
0.5	0.5167572	0.2	0.1951615	8	0.9381	0.9509	0.9682
0.5	0.5167572	0.2	0.1951615	10	0.9597	0.9682	0.9793
0.5	0.5167572	0.2	0.1951615	12	0.9650	0.9721	0.9802
0.5	0.5167572	0.2	0.1951615	14	0.9730	0.9769	0.9833
0.5	0.5171975	0.3	0.3007205	4	0.8524	0.8899	0.9395

Continued

Table 6.1 **Continued**

ρ_{xy}	RHOXY	ρ_{yz}	RHOYZ	n	90%	95%	99%
0.5	0.5171975	0.3	0.3007205	6	0.9097	0.9284	0.9529
0.5	0.5171975	0.3	0.3007205	8	0.9383	0.9513	0.9685
0.5	0.5171975	0.3	0.3007205	10	0.9574	0.9670	0.9773
0.5	0.5171975	0.3	0.3007205	12	0.9704	0.9753	0.9821
0.5	0.5171975	0.3	0.3007205	14	0.9733	0.9778	0.9847
0.5	0.5010488	0.4	0.4293076	4	0.8489	0.8863	0.9376
0.5	0.5010488	0.4	0.4293076	6	0.9124	0.9319	0.9580
0.5	0.5010488	0.4	0.4293076	8	0.9436	0.9544	0.9694
0.5	0.5010488	0.4	0.4293076	10	0.9565	0.9655	0.9760
0.5	0.5010488	0.4	0.4293076	12	0.9678	0.9737	0.9820
0.5	0.5010488	0.4	0.4293076	14	0.9748	0.9803	0.9858
0.5	0.5044690	0.5	0.5025602	4	0.8611	0.8997	0.9428
0.5	0.5044690	0.5	0.5025602	6	0.9211	0.9385	0.9600
0.5	0.5044690	0.5	0.5025602	8	0.9428	0.9538	0.9692
0.5	0.5044690	0.5	0.5025602	10	0.9599	0.9681	0.9777
0.5	0.5044690	0.5	0.5025602	12	0.9701	0.9759	0.9828
0.5	0.5044690	0.5	0.5025602	14	0.9769	0.9814	0.9860
0.5	0.4796009	0.6	0.6105799	4	0.8714	0.9035	0.9474
0.5	0.4796009	0.6	0.6105799	6	0.9217	0.9382	0.9607
0.5	0.4796009	0.6	0.6105799	8	0.9504	0.9597	0.9732
0.5	0.4796009	0.6	0.6105799	10	0.9654	0.9714	0.9800
0.5	0.4796009	0.6	0.6105799	12	0.9718	0.9768	0.9823
0.5	0.4796009	0.6	0.6105799	14	0.9782	0.9820	0.9870
0.5	0.4774821	0.7	0.6929410	4	0.8706	0.9064	0.9494
0.5	0.4774821	0.7	0.6929410	6	0.9271	0.9436	0.9652
0.5	0.4774821	0.7	0.6929410	8	0.9533	0.9629	0.9742
0.5	0.4774821	0.7	0.6929410	10	0.9648	0.9720	0.9806
0.5	0.4774821	0.7	0.6929410	12	0.9715	0.9774	0.9841
0.5	0.4774821	0.7	0.6929410	14	0.9778	0.9823	0.9865
0.5	0.4933499	0.8	0.8119242	4	0.8846	0.9178	0.9556
0.5	0.4933499	0.8	0.8119242	6	0.9276	0.9438	0.9647
0.5	0.4933499	0.8	0.8119242	8	0.9594	0.9676	0.9778
0.5	0.4933499	0.8	0.8119242	10	0.9706	0.9752	0.9817
0.5	0.4933499	0.8	0.8119242	12	0.9767	0.9811	0.9866
0.5	0.4933499	0.8	0.8119242	14	0.9822	0.9856	0.9893
0.5	0.5033757	0.9	0.9006591	4	0.9092	0.9332	0.9634
0.5	0.5033757	0.9	0.9006591	6	0.9452	0.9588	0.9728
0.5	0.5033757	0.9	0.9006591	8	0.9616	0.9698	0.9792
0.5	0.5033757	0.9	0.9006591	10	0.9694	0.9749	0.9824
0.5	0.5033757	0.9	0.9006591	12	0.9802	0.9841	0.9891
0.5	0.5033757	0.9	0.9006591	14	0.9834	0.9868	0.9899
0.7	0.7068314	0.0	−0.0153304	4	0.8739	0.9088	0.9517
0.7	0.7068314	0.0	−0.0153304	6	0.9224	0.9401	0.9612
0.7	0.7068314	0.0	−0.0153304	8	0.9469	0.9589	0.9733
0.7	0.7068314	0.0	−0.0153304	10	0.9591	0.9674	0.9774

Continued

Table 6.1 Continued

ρ_{xy}	RHOXY	ρ_{yz}	RHOYZ	n	90%	95%	99%
0.7	0.7068314	0.0	−0.0153304	12	0.9744	0.9787	0.9847
0.7	0.7068314	0.0	−0.0153304	14	0.9748	0.9800	0.9863
0.7	0.7037323	0.1	0.0930306	4	0.8697	0.9028	0.9446
0.7	0.7037323	0.1	0.0930306	6	0.9242	0.9412	0.9625
0.7	0.7037323	0.1	0.0930306	8	0.9538	0.9644	0.9746
0.7	0.7037323	0.1	0.0930306	10	0.9619	0.9701	0.9786
0.7	0.7037323	0.1	0.0930306	12	0.9710	0.9772	0.9839
0.7	0.7037323	0.1	0.0930306	14	0.9765	0.9806	0.9869
0.7	0.7091130	0.2	0.2079550	4	0.8678	0.9006	0.9464
0.7	0.7091130	0.2	0.2079550	6	0.9233	0.9385	0.9630
0.7	0.7091130	0.2	0.2079550	8	0.9481	0.9589	0.9732
0.7	0.7091130	0.2	0.2079550	10	0.9619	0.9685	0.9783
0.7	0.7091130	0.2	0.2079550	12	0.9712	0.9761	0.9861
0.7	0.7091130	0.2	0.2079550	14	0.9773	0.9817	0.9868
0.7	0.7228749	0.3	0.2823218	4	0.8776	0.9091	0.9504
0.7	0.7228749	0.3	0.2823218	6	0.9302	0.9460	0.9667
0.7	0.7228749	0.3	0.2823218	8	0.9541	0.9638	0.9754
0.7	0.7228749	0.3	0.2823218	10	0.9644	0.9714	0.9807
0.7	0.7228749	0.3	0.2823218	12	0.9729	0.9785	0.9840
0.7	0.7228749	0.3	0.2823218	14	0.9790	0.9829	0.9873
0.7	0.7043535	0.4	0.4168755	4	0.8753	0.9077	0.9494
0.7	0.7043535	0.4	0.4168755	6	0.9243	0.9402	0.9615
0.7	0.7043535	0.4	0.4168755	8	0.9481	0.9587	0.9717
0.7	0.7043535	0.4	0.4168755	10	0.9649	0.9717	0.9801
0.7	0.7043535	0.4	0.4168755	12	0.9713	0.9769	0.9831
0.7	0.7043535	0.4	0.4168755	14	0.9751	0.9792	0.9843
0.7	0.6969780	0.5	0.5394821	4	0.8877	0.9168	0.9553
0.7	0.6969780	0.5	0.5394821	6	0.9327	0.9460	0.9645
0.7	0.6969780	0.5	0.5394821	8	0.9489	0.9598	0.9733
0.7	0.6969780	0.5	0.5394821	10	0.9623	0.9701	0.9785
0.7	0.6969780	0.5	0.5394821	12	0.9709	0.9766	0.9836
0.7	0.6969780	0.5	0.5394821	14	0.9789	0.9827	0.9875
0.7	0.7091886	0.6	0.5960213	4	0.8856	0.9132	0.9525
0.7	0.7091886	0.6	0.5960213	6	0.9360	0.9495	0.9683
0.7	0.7091886	0.6	0.5960213	8	0.9549	0.9653	0.9774
0.7	0.7091886	0.6	0.5960213	10	0.9701	0.9748	0.9835
0.7	0.7091886	0.6	0.5960213	12	0.9744	0.9796	0.9850
0.7	0.7091886	0.6	0.5960213	14	0.9768	0.9810	0.9867
0.7	0.7066548	0.7	0.7085645	4	0.8921	0.9205	0.9572
0.7	0.7066548	0.7	0.7085645	6	0.9361	0.9513	0.9692
0.7	0.7066548	0.7	0.7085645	8	0.9581	0.9658	0.9770
0.7	0.7066548	0.7	0.7085645	10	0.9712	0.9767	0.9839
0.7	0.7066548	0.7	0.7085645	12	0.9770	0.9822	0.9875
0.7	0.7066548	0.7	0.7085645	14	0.9819	0.9847	0.9893
0.7	0.6880201	0.8	0.8009049	4	0.9008	0.9262	0.9612

Continued

Table 6.1 Continued

ρ_{xy}	RHOXY	ρ_{yz}	RHOYZ	n	90%	95%	99%
0.7	0.6880201	0.8	0.8009049	6	0.9399	0.9524	0.9701
0.7	0.6880201	0.8	0.8009049	8	0.9573	0.9648	0.9767
0.7	0.6880201	0.8	0.8009049	10	0.9686	0.9743	0.9827
0.7	0.6880201	0.8	0.8009049	12	0.9750	0.9804	0.9860
0.7	0.6880201	0.8	0.8009049	14	0.9806	0.9832	0.9871
0.7	0.6878207	0.9	0.8979229	4	0.9060	0.9275	0.9622
0.7	0.6878207	0.9	0.8979229	6	0.9506	0.9612	0.9774
0.7	0.6878207	0.9	0.8979229	8	0.9674	0.9746	0.9823
0.7	0.6878207	0.9	0.8979229	10	0.9743	0.9787	0.9843
0.7	0.6878207	0.9	0.8979229	12	0.9804	0.9829	0.9881
0.7	0.6878207	0.9	0.8979229	14	0.9826	0.9861	0.9900
0.9	0.8976998	0.0	0.0042819	4	0.9196	0.9386	0.9672
0.9	0.8976998	0.0	0.0042819	6	0.9494	0.9611	0.9756
0.9	0.8976998	0.0	0.0042819	8	0.9668	0.9737	0.9820
0.9	0.8976998	0.0	0.0042819	10	0.9784	0.9825	0.9883
0.9	0.8976998	0.0	0.0042819	12	0.9843	0.9868	0.9908
0.9	0.8976998	0.0	0.0042819	14	0.9866	0.9887	0.9921
0.9	0.8999227	0.1	0.1200173	4	0.9163	0.9381	0.9679
0.9	0.8999227	0.1	0.1200173	6	0.9535	0.9633	0.9759
0.9	0.8999227	0.1	0.1200173	8	0.9663	0.9731	0.9818
0.9	0.8999227	0.1	0.1200173	10	0.9790	0.9831	0.9885
0.9	0.8999227	0.1	0.1200173	12	0.9824	0.9865	0.9905
0.9	0.8999227	0.1	0.1200173	14	0.9854	0.9889	0.9915
0.9	0.8966213	0.2	0.1928654	4	0.9247	0.9445	0.9711
0.9	0.8966213	0.2	0.1928654	6	0.9550	0.9643	0.9777
0.9	0.8966213	0.2	0.1928654	8	0.9695	0.9745	0.9835
0.9	0.8966213	0.2	0.1928654	10	0.9757	0.9816	0.9878
0.9	0.8966213	0.2	0.1928654	12	0.9801	0.9834	0.9891
0.9	0.8966213	0.2	0.1928654	14	0.9867	0.9886	0.9917
0.9	0.8977677	0.3	0.2522655	4	0.9213	0.9422	0.9681
0.9	0.8977677	0.3	0.2522655	6	0.9545	0.9638	0.9775
0.9	0.8977677	0.3	0.2522655	8	0.9694	0.9742	0.9820
0.9	0.8977677	0.3	0.2522655	10	0.9773	0.9820	0.9881
0.9	0.8977677	0.3	0.2522655	12	0.9818	0.9856	0.9902
0.9	0.8977677	0.3	0.2522655	14	0.9889	0.9908	0.9934
0.9	0.9045316	0.4	0.3869980	4	0.9206	0.9424	0.9674
0.9	0.9045316	0.4	0.3869980	6	0.9555	0.9652	0.9781
0.9	0.9045316	0.4	0.3869980	8	0.9689	0.9751	0.9837
0.9	0.9045316	0.4	0.3869980	10	0.9795	0.9832	0.9884
0.9	0.9045316	0.4	0.3869980	12	0.9825	0.9856	0.9907
0.9	0.9045316	0.4	0.3869980	14	0.9852	0.9886	0.9918
0.9	0.8984540	0.5	0.4761073	4	0.9178	0.9390	0.9684
0.9	0.8984540	0.5	0.4761073	6	0.9570	0.9662	0.9790
0.9	0.8984540	0.5	0.4761073	8	0.9684	0.9742	0.9823
0.9	0.8984540	0.5	0.4761073	10	0.9779	0.9814	0.9869

Continued

Table 6.1 Continued

ρ_{xy}	RHOXY	ρ_{yz}	RHOYZ	n	90%	95%	99%
0.9	0.8984540	0.5	0.4761073	12	0.9829	0.9861	0.9905
0.9	0.8984540	0.5	0.4761073	14	0.9859	0.9883	0.9914
0.9	0.9003546	0.6	0.5882043	4	0.9209	0.9406	0.9699
0.9	0.9003546	0.6	0.5882043	6	0.9573	0.9668	0.9773
0.9	0.9003546	0.6	0.5882043	8	0.9709	0.9774	0.9858
0.9	0.9003546	0.6	0.5882043	10	0.9779	0.9826	0.9876
0.9	0.9003546	0.6	0.5882043	12	0.9852	0.9880	0.9912
0.9	0.9003546	0.6	0.5882043	14	0.9887	0.9905	0.9937
0.9	0.9000659	0.7	0.7050580	4	0.9286	0.9466	0.9701
0.9	0.9000659	0.7	0.7050580	6	0.9571	0.9670	0.9798
0.9	0.9000659	0.7	0.7050580	8	0.9701	0.9778	0.9846
0.9	0.9000659	0.7	0.7050580	10	0.9775	0.9823	0.9882
0.9	0.9000659	0.7	0.7050580	12	0.9806	0.9850	0.9899
0.9	0.9000659	0.7	0.7050580	14	0.9858	0.9883	0.9914
0.9	0.8927355	0.8	0.8029351	4	0.9280	0.9490	0.9716
0.9	0.8927355	0.8	0.8029351	6	0.9538	0.9636	0.9776
0.9	0.8927355	0.8	0.8029351	8	0.9708	0.9754	0.9847
0.9	0.8927355	0.8	0.8029351	10	0.9784	0.9829	0.9884
0.9	0.8927355	0.8	0.8029351	12	0.9826	0.9858	0.9909
0.9	0.8927355	0.8	0.8029351	14	0.9894	0.9919	0.9942
0.9	0.9005865	0.9	0.8948975	4	0.9329	0.9502	0.9737
0.9	0.9005865	0.9	0.8948975	6	0.9598	0.9672	0.9800
0.9	0.9005865	0.9	0.8948975	8	0.9782	0.9827	0.9882
0.9	0.9005865	0.9	0.8948975	10	0.9787	0.9838	0.9891
0.9	0.9005865	0.9	0.8948975	12	0.9851	0.9880	0.9918
0.9	0.9005865	0.9	0.8948975	14	0.9889	0.9913	0.9938

When there is no correlation between the study variable and the auxiliary variable used at the selection stage, the coverage attained by the 90%, 95%, and 99% intervals are found to be 89.58%, 94.35%, and 98.08% for samples of sizes 6, 8, and 14, respectively, when the auxiliary variable used at the estimation stage has correlation of 0.3. When the value of this correlation coefficient ρ_{xy} increased to 0.9, while keeping ρ_{yz} at zero, the attained coverage becomes 94.94%, 96.11%, and 98.83%, respectively, for samples of sizes 6, 6, and 10. It seems that if there exists zero correlation between the study and the selection variable and there exists another auxiliary variable having high correlation with the study variable, the resultant estimator could perform well for moderate sample sizes.

One can read other results from Table 6.1 in the same way, but let us discuss one more interesting case. Note that if the value of the correlation coefficient between the study variable and the variable used at selection stage is 0.9, and there is another auxiliary variable used at estimation stage that has a correlation coefficient of 0.9 with the study variable, then the attained coverages are 93.29%, 95.02%, and 98.91% for the 90%, 95%, and 99% intervals, respectively, for samples of sizes 4, 4, and 10. We see that when both auxiliary variables are highly correlated with the study variable, the proposed method is very close to the regression type estimator under a PPSWR scheme. Thus, the simulation shows that the Hansen and Hurwitz (1943) PPSWR method works well for the construction of a linear regression type estimator. We observed that as the value of the correlation coefficient increases, the value of the effective sample size required to achieve the nominal coverage of interest decreases.

Thus, we recommend the use of this newly tuned estimator for multicharacter surveys where the variables have very low correlation with the selection variable, and where a moderate sample size is enough. Thus, the newly tuned multicharacter estimator of population total resolves to a certain extent the major problem of estimation of its variance in multicharacter surveys, which is the most cost effective and time-saving scheme in real situations. Again note that in this study we used two variables, one at the selection stage and another at the estimation stage.

6.3.2 R code

The R code, **PUMPKIN61.R**, that was used to study the coverage by the newly tuned multicharacter estimator based on a chi-square type distance function is listed here.

```
#PROGRAM PUMPKIN61.R
library(sampling)
library(SDaA)
set.seed(2013)
for (rhoxy in seq(.3,.9,.2))
 {
  for (rhoyz in seq(.0,.9,.1))
  {
NP<-2000
xe<-rnorm(NP,0,1);ye<-rnorm(NP,0,1);ze<-rnorm(NP,0,1)
y<-2454 + 245*ye
x<-2325 + 232*sqrt((1-rhoxy^2))*xe + rhoxy*232*ye
z<-322 + 52*sqrt((1-rhoyz^2))*ze + rhoyz*52*ye
XPMEAN<-mean(x);ZPMEAN<-mean(z)
RHOXY<-cor(x,y);RHOYZ<-cor(y,z)
TY<-sum(y)
pi<-z/(NP*ZPMEAN)
nreps<-10000
ESTP=rep(0,nreps)
ci1.max=ci1.min=ci2.max=ci2.min=ci3.max=ci3.min=ESTP
vESTP=ESTP
for  (n in seq(4,15,2)  )
 {
 pik<-n*pi
 for  (r in 1:nreps)
```

```
        {
        us<- lahiri.design(pik, n, as.factor(1:NP))
        xs<-x[us]; ys<-y[us]; zs<-z[us]
        ps<-((1+1/NP)^(1-RHOYZ))*((1+pi[us])^(RHOYZ))-1
        ESTX1<-mean(xs/ps);ESTY1<-mean(ys/ps)
        xmult<-(n*ESTX1 - xs/ps)/(n-1)
        ymult<-(n*ESTY1 - ys/ps)/(n-1)
        delt<-n*sum(xmult^2) - sum(xmult)^2
        d<-(n*xmult - sum(xmult))/delt
        wbni<-1/n + d*(NP*XPMEAN-ESTX1)/((n-1)^2)
        ESTi<-n*((n-1)^2*wbni-n+2)*ymult
        ESTP[r]<- sum(ESTi)
        EST_I<-(ESTP[r] - ESTi)/(n-1)
        ESTP[r]<-ESTP[r]/n
        vj<-(wbni^2)*ps*((EST_I - ESTP[r])^2)
        vESTP[r]<-n*((n-1)^3)*sum(vj)
        ci1.max[r]<- ESTP[r]-qt(.05,n-1)*sqrt(vESTP[r])
        ci1.min[r]<- ESTP[r]+qt(.05,n-1)*sqrt(vESTP[r])
        ci2.max[r]<- ESTP[r]-qt(.025,n-1)*sqrt(vESTP[r])
        ci2.min[r]<- ESTP[r]+qt(.025,n-1)*sqrt(vESTP[r])
        ci3.max[r]<- ESTP[r]-qt(.005,n-1)*sqrt(vESTP[r])
        ci3.min[r]<- ESTP[r]+qt(.005,n-1)*sqrt(vESTP[r])
        } # nreps
      round(sum(ci1.min<TY & ci1.max>TY)/nreps,4)->cov1
      round(sum(ci2.min<TY & ci2.max>TY)/nreps,4)->cov2
      round(sum(ci3.min<TY & ci3.max>TY)/nreps,4)->cov3
      cat(rhoxy,RHOXY,rhoyz, RHOYZ, n, cov1,cov2,cov3,'\n')
      }# n
    }# rhoyz
  } # rhoxy
```

6.3.3 Numerical illustration

We explain the method of tuning the estimator of the population total in multicharacter surveys with the following example.

Example 6.1 In a specially designed field, it is easy to take a picture of each one of the pumpkins and record their top size. After taking photographs of the tops of all 200 pumpkins in the field, we selected a PPSWR sample of $n=7$ pumpkins. The circumference (in.), say X, weight (lbs), say Y, and top size (in.), say Z, of the seven selected pumpkins were recorded as follows:

x_i	130.9	67.0	106.5	98.0	115.2	137.1	101.1
y_i	800	800	3084	1042	4500	2500	2397
z_i	105	30	57	32	99	117	85

Apply the multicharacter survey approach with the Bansal and Singh (1985) transformation to estimate the total weight of the pumpkins in the field and construct 90% confidence interval estimates using circumference at the estimation stage and top size at the selection stage. (*Given*: The population total of X is 21,080, total of Z is 14,600, and $\rho_{yz} = 0.4$).

Solution. We can compute the following table:

p_j	p_j^*	$\hat{X}_M(j)$	$\hat{Y}_M(j)$	$\overline{w}_n(j)$	$\hat{Y}_{\text{MTuned cs}(j)}$
0.0071917	0.0058761	21234.39	465818.4	0.143919	390588.1
0.0020547	0.0038208	22024.61	453613.1	0.138279	500631.3
0.0039041	0.0045615	21055.89	375826.8	0.145194	389490.2
0.0021917	0.0038757	20732.93	443700.7	0.147499	328556.3
0.0067808	0.0057119	21585.78	357205.4	0.141411	451167.5
0.008013	0.0062043	21264.28	421352.4	0.143706	403758.6
0.0058219	0.0053286	21785.02	413537.5	0.139989	469900.7

Then a tuned estimate of the total weight of the pumpkins is given by

$$\hat{Y}_{\text{MTuned(cs)}} = \frac{1}{n}\sum_{j \in s} \hat{Y}_{\text{MTuned(cs)}}(j) = 419156.1$$

The standard error of the tuned estimate of the population total is given by

$$SE\left(\hat{Y}_{\text{MTuned(cs)}}\right) = \sqrt{\hat{v}\left(\hat{Y}_{\text{MTuned(cs)}}\right)} = 51840.569$$

Hence, the 90% confidence interval estimate of the total weight of the pumpkins in the field is 318420.6–519891.7 lbs.

6.3.4 R code used for illustration

We used the following R code, **PUMPKIN61EX.R**, to solve the preceding example.

#PROGRAM PUMPKIN61EX.R

```
n<-7
xs<-c(130.9,67,106.5,98,115.2,137.1,101.1)
ys<-c(800,800,3084,1042,4500,2500,2397)
zs<-c(105,30,57,32,99,117,85)
XPMEAN<-105.4;ZPMEAN<-73; RHOYZP<-0.4
NP<-200
pi<-zs/(NP*ZPMEAN)
ps<-((1+1/NP)^(1-RHOYZP))*((1+pi)^(RHOYZP))-1
ESTX1<-mean(xs/ps);ESTY1<-mean(ys/ps)
xmult<-(n*ESTX1 - xs/ps)/(n-1)
```

```
ymult<-(n*ESTY1 - ys/ps)/(n-1)
delt<-n*sum(xmult^2) - sum(xmult)^2
d<-(n*xmult - sum(xmult))/delt
wbni<-1/n + d*(NP*XPMEAN-ESTX1)/((n-1)^2)
ESTi<- n*((n-1)^2*wbni-n+2)*ymult
ESTP<- sum(ESTi)
EST_I<-(ESTP - ESTi)/(n-1)
ESTP<-ESTP/n
vj<-(wbni^2)*ps*((EST_I - ESTP)^2)
vESTP<-n*((n-1)^3)*sum(vj)
L<-ESTP+qt(.05,n-1)*sqrt(vESTP)
U<-ESTP+qt(.95,n-1)*sqrt(vESTP)
cbind(pi,ps,xmult,ymult,wbni,EST_I)
cat("Tuned estimate:", ESTP, "SE: ",vESTP^.5,'\n')
cat("Confidence Interval:"," ", L,"; ", U,'\n')
```

6.4 Tuning of the multicharacter estimator of population total with dual-to-empirical log-likelihood function

The newly tuned multicharacter jackknife dual-to-empirical log-likelihood (dell) estimator of the population total Y is defined as

$$\hat{Y}_{\text{MTuned(dell)}} = \sum_{j \in s} \left[(n-1)^2 \bar{w}_n^*(j) - (n-2) \right] \hat{Y}_M(j) \tag{6.25}$$

where $\bar{w}_n^*(j)$ are the positive jackknife tuned weights, chosen such that the following two constraints are satisfied:

$$\sum_{j \in s} \bar{w}_n^*(j) = 1 \tag{6.26}$$

and

$$\sum_{j \in s} \bar{w}_n^*(j) \Psi_j = 0 \tag{6.27}$$

where

$$\Psi_j = \hat{X}_M(j) - \frac{\left(X - n(2-n)\hat{X}_M \right)}{(n-1)^2} \tag{6.28}$$

Here we suggest the tuning of the jackknife weights $\bar{w}_n^*(j)$ such that the dell distance function defined as

$$\sum_{j \in s} \frac{\ln\left(1 - w_j^*\right)}{n} \tag{6.29}$$

where w_j^* being weights on unit length, or equivalently, the log-likelihood function as

$$\sum_{j \in s} \frac{\ln\left(\bar{w}_n^*(j)\right)}{n} \tag{6.30}$$

is optimum subject to the tuning constraints (6.26) and (6.27).

The Lagrange function is given by

$$L_2 = \sum_{j \in s} \frac{\ln\left(\bar{w}_n^*(j)\right)}{n} - \lambda_0^* \left(\sum_{j \in s} \bar{w}_n^*(j) - 1\right) - \lambda_1^* \left\{\sum_{j \in s} \bar{w}_n^*(j) \Psi_j\right\} \tag{6.31}$$

where λ_0^* and λ_1^* are Lagrange multiplier constants.

On setting:

$$\frac{\partial L_2}{\partial \bar{w}_n^*(j)} = 0$$

we have

$$\bar{w}_n^*(j) = \frac{1}{n\{1 + \lambda_1^* \Psi_j\}} \tag{6.32}$$

Constraints (6.26) and (6.27) yield $\lambda_0^* = 1$, and λ_1^* is a solution to the nonlinear equation

$$\sum_{j \in s} \frac{\Psi_j}{1 + \lambda_1^* \Psi_j} = 0 \tag{6.33}$$

In the following simulation study, by assuming $|\lambda_1^* \Psi_j| < 1$, we approximated the value of λ_1^* as

$$\lambda_1^* \approx \frac{\sum_{j \in s} \Psi_j}{\sum_{j \in s} \Psi_j^2} \tag{6.34}$$

Note that a better solution to the nonlinear equation (6.33) could be used if available.

Thus, under the dell distance function, the newly tuned multicharacter dell-estimator (6.25) of the population total becomes

$$\hat{Y}_{MTuned(dell)} = \sum_{j \in s} \left\{ (n-1)^2 \bar{w}_n^*(j) - (n-2) \right\} \hat{Y}_M(j)$$

$$= \frac{(n-1)^2}{n} \sum_{j \in s} \frac{\hat{Y}_M(j)}{1 + \lambda_1^* \Psi_j} - (n-2) \sum_{j \in s} \hat{Y}_M(j) \quad (6.35)$$

6.4.1 Estimation of variance of multicharacter estimator

An adjusted estimator of the variance of the multicharacter survey estimator $\hat{Y}_{MTuned(dell)}$ is

$$\hat{v}\left(\hat{Y}_{MTuned(dell)}\right) = n(n-1)^3 \sum_{j \in s} \left(\bar{w}_n^*(j) p_j^*\right)^2 \left\{ \hat{Y}_{(j)}^{MTuned(dell)} - \hat{Y}_{MTuned(dell)} \right\}^2 \quad (6.36)$$

Note that the newly tuned doubly jackknifed dell estimator of the population total is given by

$$\hat{Y}_{(j)}^{MTuned(dell)} = \frac{n \hat{Y}_{MTuned(dell)} - n\left((n-1)^2 \bar{w}_n^*(j) - (n-2)\right) \hat{Y}_M(j)}{n-1} \quad (6.37)$$

for $j = 1, 2, \ldots, n$, with

$$\hat{Y}_{MTuned(dell)} = \sum_{j \subset s} \left[(n-1)^2 \bar{w}_n^*(j) - (n-2) \right] \hat{Y}_M(j) \quad (6.38)$$

Again note that in Equation (6.36), instead of using p_j^* the possibility of using

$$\bar{p}^*(j) = \frac{n \bar{p}^* - p_j^*}{n-1} \quad (6.39)$$

where

$$\bar{p}^* = \frac{1}{n} \sum_{j=1}^{n} p_j^*$$

could also be investigated. This has been left an exercise!

The coverage by the $(1-\alpha)100\%$ confidence interval estimates obtained from this newly tuned doubly jackknifed multicharacter dell estimator of the population total is obtained by counting the number of times the true population total Y falls within the interval estimate given by

$$\hat{Y}_{\text{MTuned(dell)}} \mp t_{\alpha/2}(\text{df} = n-1)\sqrt{\hat{v}\left(\hat{Y}_{\text{MTuned(dell)}}\right)} \qquad (6.40)$$

We studied coverage by the 90%, 95%, and 99% confidence interval estimates based on our estimator where the value of the correlation coefficient between the study variable and the auxiliary variable used at the selection stage ranged from 0 to 1. We generated three random variables $y_i^* \sim N(0,1)$, $x_i^* \sim N(0,1)$ and $z_i^* \sim N(0,1)$ for $i = 1, 2, \ldots, N$ from three independent standard normal distribution. For four different values of the correlation coefficient ρ_{xy}, 0.3, 0.5, 0.6, and 0.9, and ten different values of the correlation coefficient ρ_{yz}, 0.0, 0.1, 0.2, 0.3, 0.4, 0.5, 0.6, 0.7, 0.8, and 0.9 we generated the possible populations with three variables Y, X and Z taking values y_i, x_i, and z_i for the ith unit as

$$y_i = \overline{Y} + \sigma_y y_i^* \qquad (6.41)$$

$$x_i = \overline{X} + \sigma_x \sqrt{\left(1 - \rho_{xy}^2\right)} x_i^* + \rho_{xy} \sigma_x x_i^* \qquad (6.42)$$

and

$$z_i = \overline{Z} + \sigma_z \sqrt{\left(1 - \rho_{yz}^2\right)} z_i^* + \rho_{yz} \sigma_z y_i^* \qquad (6.43)$$

Here we fixed $\overline{X} = 2325$, $\overline{Y} = 2454$, $\overline{Z} = 322$, $\sigma_x = 232$, $\sigma_y = 245$, and $\sigma_z = 52$. Note that the variable X is to be used at the estimation stage, and it is constructed to have correlations 0.3, 0.5, 0.7, and 0.9 with the study variable Y because we want to see the effect of low as well as high values of the correlation coefficient. In the same way the variable Z is to be used at the selection stage of sampling, and although this variable is expected to have low correlation with the study variable (e.g., 0.0, 0.1), we considered ten values of ρ_{yz} to investigate how it affects the results. For each of the 40 combinations of ρ_{xy} and ρ_{yz}, we constructed a population of $N = 2000$ units. We used Lahiri (1951) method for selecting a sample from the given population. For each of the sample of sizes $n = 4, 6, 8, 10, 12, 14$, we took 10,000 samples and constructed three confidence intervals for each. Attained coverage was estimated by the proportion of times the true total fell within the interval. The results obtained for various sample sizes are as shown in Table 6.2.

Table 6.2 demonstrates the effect of various values of the correlation coefficient between the study and the auxiliary variable on the coverage of confidence intervals constructed from moderate-sized samples. Among the interesting findings is that if the correlation coefficient between the study variable and the auxiliary variable used at the estimation stage is 0.9 while the study variable and the auxiliary variable used at the selection stage are uncorrelated, then the estimated coverages attained by the

Table 6.2 Performance of the dell tuned multicharacter estimator

ρ_{xy}	RHOXY	ρ_{yz}	RHOYZ	n	90%	95%	99%
0.3	0.2978796	0.0	0.0051677	4	0.6786	0.7530	0.8616
0.3	0.2978796	0.0	0.0051677	6	0.8863	0.9108	0.9477
0.3	0.2978796	0.0	0.0051677	8	0.9288	0.9431	0.9613
0.3	0.2978796	0.0	0.0051677	10	0.9529	0.9624	0.9746
0.3	0.2978796	0.0	0.0051677	12	0.9609	0.9689	0.9774
0.3	0.2978796	0.0	0.0051677	14	0.9658	0.9719	0.9808
0.3	0.2898565	0.1	0.0912027	4	0.7038	0.7728	0.8707
0.3	0.2898565	0.1	0.0912027	6	0.8918	0.9148	0.9465
0.3	0.2898565	0.1	0.0912027	8	0.9293	0.9426	0.9608
0.3	0.2898565	0.1	0.0912027	10	0.9513	0.9602	0.9733
0.3	0.2898565	0.1	0.0912027	12	0.9661	0.9725	0.9809
0.3	0.2898565	0.1	0.0912027	14	0.9720	0.9772	0.9838
0.3	0.2893902	0.2	0.1542838	4	0.6909	0.7623	0.8646
0.3	0.2893902	0.2	0.1542838	6	0.8951	0.9178	0.9508
0.3	0.2893902	0.2	0.1542838	8	0.9335	0.9482	0.9655
0.3	0.2893902	0.2	0.1542838	10	0.9505	0.9607	0.9747
0.3	0.2893902	0.2	0.1542838	12	0.9641	0.9709	0.9792
0.3	0.2893902	0.2	0.1542838	14	0.9698	0.9758	0.9836
0.3	0.2814688	0.3	0.3024993	4	0.6852	0.7614	0.8635
0.3	0.2814688	0.3	0.3024993	6	0.8915	0.9159	0.9474
0.3	0.2814688	0.3	0.3024993	8	0.9332	0.9460	0.9628
0.3	0.2814688	0.3	0.3024993	10	0.9502	0.9581	0.9707
0.3	0.2814688	0.3	0.3024993	12	0.9601	0.9656	0.9756
0.3	0.2814688	0.3	0.3024993	14	0.9701	0.9742	0.9812
0.3	0.2800336	0.4	0.4206022	4	0.6989	0.7688	0.8638
0.3	0.2800336	0.4	0.4206022	6	0.8953	0.9146	0.9487
0.3	0.2800336	0.4	0.4206022	8	0.9326	0.9454	0.9627
0.3	0.2800336	0.4	0.4206022	10	0.9514	0.9620	0.9723
0.3	0.2800336	0.4	0.4206022	12	0.9653	0.9717	0.9786
0.3	0.2800336	0.4	0.4206022	14	0.9740	0.9794	0.9852
0.3	0.2511401	0.5	0.5078852	4	0.7014	0.7713	0.8674
0.3	0.2511401	0.5	0.5078852	6	0.9000	0.9215	0.9514
0.3	0.2511401	0.5	0.5078852	8	0.9374	0.9509	0.9662
0.3	0.2511401	0.5	0.5078852	10	0.9565	0.9658	0.9761
0.3	0.2511401	0.5	0.5078852	12	0.9691	0.9743	0.9822
0.3	0.2511401	0.5	0.5078852	14	0.9774	0.9823	0.9877
0.3	0.3262503	0.6	0.5984581	4	0.7215	0.7862	0.8803
0.3	0.3262503	0.6	0.5984581	6	0.9085	0.9292	0.9564
0.3	0.3262503	0.6	0.5984581	8	0.9413	0.9529	0.9687
0.3	0.3262503	0.6	0.5984581	10	0.9576	0.9662	0.9756
0.3	0.3262503	0.6	0.5984581	12	0.9684	0.9745	0.9819
0.3	0.3262503	0.6	0.5984581	14	0.9736	0.9787	0.9850
0.3	0.3237574	0.7	0.6893118	4	0.7367	0.7979	0.8825
0.3	0.3237574	0.7	0.6893118	6	0.9158	0.9344	0.9582
0.3	0.3237574	0.7	0.6893118	8	0.9503	0.9611	0.9741

Continued

Table 6.2 Continued

ρ_{xy}	RHOXY	ρ_{yz}	RHOYZ	n	90%	95%	99%
0.3	0.3237574	0.7	0.6893118	10	0.9657	0.9720	0.9797
0.3	0.3237574	0.7	0.6893118	12	0.9718	0.9767	0.9832
0.3	0.3237574	0.7	0.6893118	14	0.9772	0.9816	0.9879
0.3	0.3234880	0.8	0.8049675	4	0.7395	0.7965	0.8821
0.3	0.3234880	0.8	0.8049675	6	0.9173	0.9361	0.9611
0.3	0.3234880	0.8	0.8049675	8	0.9531	0.9638	0.9769
0.3	0.3234880	0.8	0.8049675	10	0.9623	0.9693	0.9796
0.3	0.3234880	0.8	0.8049675	12	0.9736	0.9801	0.9865
0.3	0.3234880	0.8	0.8049675	14	0.9772	0.9816	0.9867
0.3	0.3039484	0.9	0.8964122	4	0.7634	0.8140	0.8910
0.3	0.3039484	0.9	0.8964122	6	0.9365	0.9508	0.9686
0.3	0.3039484	0.9	0.8964122	8	0.9602	0.9666	0.9774
0.3	0.3039484	0.9	0.8964122	10	0.9714	0.9759	0.9831
0.3	0.3039484	0.9	0.8964122	12	0.9792	0.9830	0.9879
0.3	0.3039484	0.9	0.8964122	14	0.9843	0.9874	0.9918
0.5	0.4833906	0.0	0.0026272	4	0.7085	0.7769	0.8692
0.5	0.4833906	0.0	0.0026272	6	0.8977	0.9210	0.9528
0.5	0.4833906	0.0	0.0026272	8	0.9355	0.9477	0.9656
0.5	0.4833906	0.0	0.0026272	10	0.9516	0.9592	0.9708
0.5	0.4833906	0.0	0.0026272	12	0.9638	0.9713	0.9779
0.5	0.4833906	0.0	0.0026272	14	0.9734	0.9788	0.9857
0.5	0.5064535	0.1	0.0992117	4	0.7046	0.7775	0.8738
0.5	0.5064535	0.1	0.0992117	6	0.9004	0.9207	0.9484
0.5	0.5064535	0.1	0.0992117	8	0.9357	0.9471	0.9630
0.5	0.5064535	0.1	0.0992117	10	0.9537	0.9621	0.9735
0.5	0.5064535	0.1	0.0992117	12	0.9662	0.9740	0.9813
0.5	0.5064535	0.1	0.0992117	14	0.9724	0.9767	0.9834
0.5	0.5167572	0.2	0.1951615	4	0.7023	0.7730	0.8731
0.5	0.5167572	0.2	0.1951615	6	0.8974	0.9199	0.9502
0.5	0.5167572	0.2	0.1951615	8	0.9381	0.9509	0.9682
0.5	0.5167572	0.2	0.1951615	10	0.9598	0.9681	0.9794
0.5	0.5167572	0.2	0.1951615	12	0.9650	0.9721	0.9802
0.5	0.5167572	0.2	0.1951615	14	0.9730	0.9769	0.9833
0.5	0.5171975	0.3	0.3007205	4	0.7183	0.7843	0.8759
0.5	0.5171975	0.3	0.3007205	6	0.9016	0.9230	0.9505
0.5	0.5171975	0.3	0.3007205	8	0.9380	0.9512	0.9685
0.5	0.5171975	0.3	0.3007205	10	0.9575	0.9670	0.9772
0.5	0.5171975	0.3	0.3007205	12	0.9705	0.9753	0.9821
0.5	0.5171975	0.3	0.3007205	14	0.9733	0.9779	0.9847
0.5	0.5010488	0.4	0.4293076	4	0.7099	0.7770	0.8719
0.5	0.5010488	0.4	0.4293076	6	0.9034	0.9267	0.9558
0.5	0.5010488	0.4	0.4293076	8	0.9435	0.9543	0.9694
0.5	0.5010488	0.4	0.4293076	10	0.9564	0.9655	0.9760
0.5	0.5010488	0.4	0.4293076	12	0.9677	0.9737	0.9820
0.5	0.5010488	0.4	0.4293076	14	0.9747	0.9803	0.9858

Continued

Table 6.2 Continued

ρ_{xy}	RHOXY	ρ_{yz}	RHOYZ	n	90%	95%	99%
0.5	0.5044690	0.5	0.5025602	4	0.7282	0.7970	0.8809
0.5	0.5044690	0.5	0.5025602	6	0.9142	0.9346	0.9585
0.5	0.5044690	0.5	0.5025602	8	0.9429	0.9537	0.9691
0.5	0.5044690	0.5	0.5025602	10	0.9599	0.9682	0.9778
0.5	0.5044690	0.5	0.5025602	12	0.9702	0.9759	0.9828
0.5	0.5044690	0.5	0.5025602	14	0.9769	0.9814	0.9860
0.5	0.4796009	0.6	0.6105799	4	0.7371	0.7968	0.8832
0.5	0.4796009	0.6	0.6105799	6	0.9132	0.9324	0.9582
0.5	0.4796009	0.6	0.6105799	8	0.9495	0.9595	0.9729
0.5	0.4796009	0.6	0.6105799	10	0.9654	0.9714	0.9800
0.5	0.4796009	0.6	0.6105799	12	0.9718	0.9768	0.9823
0.5	0.4796009	0.6	0.6105799	14	0.9782	0.9820	0.9870
0.5	0.4774821	0.7	0.6929410	4	0.7285	0.7923	0.8822
0.5	0.4774821	0.7	0.6929410	6	0.9199	0.9391	0.9634
0.5	0.4774821	0.7	0.6929410	8	0.9532	0.9629	0.9742
0.5	0.4774821	0.7	0.6929410	10	0.9649	0.9720	0.9806
0.5	0.4774821	0.7	0.6929410	12	0.9714	0.9773	0.9840
0.5	0.4774821	0.7	0.6929410	14	0.9778	0.9823	0.9865
0.5	0.4933499	0.8	0.8119242	4	0.7424	0.8052	0.8870
0.5	0.4933499	0.8	0.8119242	6	0.9174	0.9370	0.9611
0.5	0.4933499	0.8	0.8119242	8	0.9591	0.9674	0.9778
0.5	0.4933499	0.8	0.8119242	10	0.9706	0.9751	0.9817
0.5	0.4933499	0.8	0.8119242	12	0.9767	0.9811	0.9866
0.5	0.4933499	0.8	0.8119242	14	0.9821	0.9856	0.9893
0.5	0.5033757	0.9	0.9006591	4	0.7755	0.8261	0.8969
0.5	0.5033757	0.9	0.9006591	6	0.9364	0.9525	0.9696
0.5	0.5033757	0.9	0.9006591	8	0.9613	0.9697	0.9792
0.5	0.5033757	0.9	0.9006591	10	0.9694	0.9749	0.9824
0.5	0.5033757	0.9	0.9006591	12	0.9802	0.9841	0.9891
0.5	0.5033757	0.9	0.9006591	14	0.9834	0.9868	0.9899
0.7	0.7068314	0.0	−0.0153304	4	0.7309	0.7968	0.8847
0.7	0.7068314	0.0	−0.0153304	6	0.9143	0.9347	0.9584
0.7	0.7068314	0.0	−0.0153304	8	0.9468	0.9590	0.9733
0.7	0.7068314	0.0	−0.0153304	10	0.9591	0.9673	0.9775
0.7	0.7068314	0.0	−0.0153304	12	0.9744	0.9787	0.9847
0.7	0.7068314	0.0	−0.0153304	14	0.9748	0.9801	0.9863
0.7	0.7037323	0.1	0.0930306	4	0.7322	0.7949	0.8786
0.7	0.7037323	0.1	0.0930306	6	0.9163	0.9358	0.9605
0.7	0.7037323	0.1	0.0930306	8	0.9537	0.9644	0.9746
0.7	0.7037323	0.1	0.0930306	10	0.9618	0.9699	0.9786
0.7	0.7037323	0.1	0.0930306	12	0.9710	0.9772	0.9839
0.7	0.7037323	0.1	0.0930306	14	0.9765	0.9806	0.9869
0.7	0.7091130	0.2	0.2079550	4	0.7358	0.7975	0.8846
0.7	0.7091130	0.2	0.2079550	6	0.9163	0.9339	0.9611
0.7	0.7091130	0.2	0.2079550	8	0.9481	0.9590	0.9732

Continued

Table 6.2 Continued

ρ_{xy}	RHOXY	ρ_{yz}	RHOYZ	n	90%	95%	99%
0.7	0.7091130	0.2	0.2079550	10	0.9619	0.9685	0.9783
0.7	0.7091130	0.2	0.2079550	12	0.9712	0.9761	0.9861
0.7	0.7091130	0.2	0.2079550	14	0.9773	0.9817	0.9868
0.7	0.7228749	0.3	0.2823218	4	0.7385	0.8007	0.8844
0.7	0.7228749	0.3	0.2823218	6	0.9217	0.9405	0.9648
0.7	0.7228749	0.3	0.2823218	8	0.9538	0.9637	0.9753
0.7	0.7228749	0.3	0.2823218	10	0.9643	0.9714	0.9807
0.7	0.7228749	0.3	0.2823218	12	0.9729	0.9786	0.9840
0.7	0.7228749	0.3	0.2823218	14	0.9790	0.9829	0.9873
0.7	0.7043535	0.4	0.4168755	4	0.7377	0.7976	0.8799
0.7	0.7043535	0.4	0.4168755	6	0.9149	0.9335	0.9586
0.7	0.7043535	0.4	0.4168755	8	0.9477	0.9586	0.9717
0.7	0.7043535	0.4	0.4168755	10	0.9649	0.9716	0.9801
0.7	0.7043535	0.4	0.4168755	12	0.9712	0.9769	0.9831
0.7	0.7043535	0.4	0.4168755	14	0.9751	0.9792	0.9843
0.7	0.6969780	0.5	0.5394821	4	0.7482	0.8093	0.8884
0.7	0.6969780	0.5	0.5394821	6	0.9256	0.9419	0.9630
0.7	0.6969780	0.5	0.5394821	8	0.9487	0.9598	0.9732
0.7	0.6969780	0.5	0.5394821	10	0.9623	0.9701	0.9785
0.7	0.6969780	0.5	0.5394821	12	0.9709	0.9766	0.9836
0.7	0.6969780	0.5	0.5394821	14	0.9789	0.9827	0.9875
0.7	0.7091886	0.6	0.5960213	4	0.7486	0.8056	0.8891
0.7	0.7091886	0.6	0.5960213	6	0.9284	0.9442	0.9652
0.7	0.7091886	0.6	0.5960213	8	0.9545	0.9652	0.9774
0.7	0.7091886	0.6	0.5960213	10	0.9701	0.9748	0.9835
0.7	0.7091886	0.6	0.5960213	12	0.9744	0.9797	0.9850
0.7	0.7091886	0.6	0.5960213	14	0.9768	0.9809	0.9867
0.7	0.7066548	0.7	0.7085645	4	0.7619	0.8170	0.8939
0.7	0.7066548	0.7	0.7085645	6	0.9282	0.9466	0.9675
0.7	0.7066548	0.7	0.7085645	8	0.9579	0.9657	0.9770
0.7	0.7066548	0.7	0.7085645	10	0.9712	0.9767	0.9839
0.7	0.7066548	0.7	0.7085645	12	0.9770	0.9822	0.9875
0.7	0.7066548	0.7	0.7085645	14	0.9819	0.9847	0.9893
0.7	0.6880201	0.8	0.8009049	4	0.7574	0.8127	0.8881
0.7	0.6880201	0.8	0.8009049	6	0.9312	0.9466	0.9668
0.7	0.6880201	0.8	0.8009049	8	0.9568	0.9645	0.9766
0.7	0.6880201	0.8	0.8009049	10	0.9686	0.9742	0.9827
0.7	0.6880201	0.8	0.8009049	12	0.9751	0.9804	0.9860
0.7	0.6880201	0.8	0.8009049	14	0.9806	0.9832	0.9871
0.7	0.6878207	0.9	0.8979229	4	0.7663	0.8151	0.8908
0.7	0.6878207	0.9	0.8979229	6	0.9412	0.9541	0.9737
0.7	0.6878207	0.9	0.8979229	8	0.9672	0.9743	0.9822
0.7	0.6878207	0.9	0.8979229	10	0.9743	0.9787	0.9843
0.7	0.6878207	0.9	0.8979229	12	0.9804	0.9829	0.9881
0.7	0.6878207	0.9	0.8979229	14	0.9826	0.9861	0.9900

Continued

Table 6.2 Continued

ρ_{xy}	RHOXY	ρ_{yz}	RHOYZ	n	90%	95%	99%
0.9	0.8976998	0.0	0.0042819	4	0.7890	0.8333	0.9049
0.9	0.8976998	0.0	0.0042819	6	0.9429	0.9560	0.9727
0.9	0.8976998	0.0	0.0042819	8	0.9667	0.9736	0.9819
0.9	0.8976998	0.0	0.0042819	10	0.9783	0.9825	0.9883
0.9	0.8976998	0.0	0.0042819	12	0.9843	0.9868	0.9908
0.9	0.8976998	0.0	0.0042819	14	0.9866	0.9887	0.9921
0.9	0.8999227	0.1	0.1200173	4	0.7854	0.8336	0.9056
0.9	0.8999227	0.1	0.1200173	6	0.9463	0.9580	0.9737
0.9	0.8999227	0.1	0.1200173	8	0.9663	0.9731	0.9818
0.9	0.8999227	0.1	0.1200173	10	0.9790	0.9830	0.9885
0.9	0.8999227	0.1	0.1200173	12	0.9824	0.9865	0.9905
0.9	0.8999227	0.1	0.1200173	14	0.9854	0.9889	0.9915
0.9	0.8966213	0.2	0.1928654	4	0.7990	0.8463	0.9106
0.9	0.8966213	0.2	0.1928654	6	0.9476	0.9587	0.9748
0.9	0.8966213	0.2	0.1928654	8	0.9694	0.9744	0.9835
0.9	0.8966213	0.2	0.1928654	10	0.9757	0.9815	0.9878
0.9	0.8966213	0.2	0.1928654	12	0.9801	0.9834	0.9891
0.9	0.8966213	0.2	0.1928654	14	0.9867	0.9886	0.9917
0.9	0.8977677	0.3	0.2522655	4	0.7944	0.8413	0.9070
0.9	0.8977677	0.3	0.2522655	6	0.9460	0.9579	0.9751
0.9	0.8977677	0.3	0.2522655	8	0.9694	0.9742	0.9820
0.9	0.8977677	0.3	0.2522655	10	0.9773	0.9820	0.9881
0.9	0.8977677	0.3	0.2522655	12	0.9818	0.9856	0.9902
0.9	0.8977677	0.3	0.2522655	14	0.9889	0.9908	0.9934
0.9	0.9045316	0.4	0.3869980	4	0.7914	0.8413	0.9048
0.9	0.9045316	0.4	0.3869980	6	0.9490	0.9608	0.9763
0.9	0.9045316	0.4	0.3869980	8	0.9686	0.9750	0.9836
0.9	0.9045316	0.4	0.3869980	10	0.9795	0.9832	0.9884
0.9	0.9045316	0.4	0.3869980	12	0.9825	0.9856	0.9907
0.9	0.9045316	0.4	0.3869980	14	0.9852	0.9886	0.9917
0.9	0.8984540	0.5	0.4761073	4	0.7837	0.8289	0.9008
0.9	0.8984540	0.5	0.4761073	6	0.9490	0.9607	0.9767
0.9	0.8984540	0.5	0.4761073	8	0.9677	0.9738	0.9822
0.9	0.8984540	0.5	0.4761073	10	0.9779	0.9813	0.9869
0.9	0.8984540	0.5	0.4761073	12	0.9829	0.9861	0.9905
0.9	0.8984540	0.5	0.4761073	14	0.9859	0.9883	0.9914
0.9	0.9003546	0.6	0.5882043	4	0.7966	0.8404	0.9070
0.9	0.9003546	0.6	0.5882043	6	0.9509	0.9623	0.9751
0.9	0.9003546	0.6	0.5882043	8	0.9706	0.9772	0.9856
0.9	0.9003546	0.6	0.5882043	10	0.9780	0.9826	0.9876
0.9	0.9003546	0.6	0.5882043	12	0.9852	0.9880	0.9912
0.9	0.9003546	0.6	0.5882043	14	0.9887	0.9905	0.9937
0.9	0.9000659	0.7	0.7050580	4	0.7871	0.8336	0.9003
0.9	0.9000659	0.7	0.7050580	6	0.9473	0.9606	0.9763
0.9	0.9000659	0.7	0.7050580	8	0.9696	0.9778	0.9846

Continued

Table 6.2 Continued

ρ_{xy}	RHOXY	ρ_{yz}	RHOYZ	n	90%	95%	99%
0.9	0.9000659	0.7	0.7050580	10	0.9775	0.9823	0.9882
0.9	0.9000659	0.7	0.7050580	12	0.9806	0.9850	0.9899
0.9	0.9000659	0.7	0.7050580	14	0.9858	0.9883	0.9914
0.9	0.8927355	0.8	0.8029351	4	0.7966	0.8425	0.9068
0.9	0.8927355	0.8	0.8029351	6	0.9444	0.9577	0.9753
0.9	0.8927355	0.8	0.8029351	8	0.9708	0.9754	0.9847
0.9	0.8927355	0.8	0.8029351	10	0.9785	0.9829	0.9884
0.9	0.8927355	0.8	0.8029351	12	0.9826	0.9858	0.9909
0.9	0.8927355	0.8	0.8029351	14	0.9894	0.9919	0.9942
0.9	0.9005865	0.9	0.8948975	4	0.8014	0.8456	0.9077
0.9	0.9005865	0.9	0.8948975	6	0.9500	0.9603	0.9758
0.9	0.9005865	0.9	0.8948975	8	0.9780	0.9824	0.9880
0.9	0.9005865	0.9	0.8948975	10	0.9787	0.9838	0.9891
0.9	0.9005865	0.9	0.8948975	12	0.9851	0.9881	0.9918
0.9	0.9005865	0.9	0.8948975	14	0.9889	0.9913	0.9938

90%, 95%, and 99% intervals were 94.29%, 95.60%, and 99.08% for samples of sizes 6, 6, and 12, respectively.

If the value of the correlation coefficient between the study variable and each of the auxiliary variables is 0.9, then the attained coverage of the 90%, 95%, and 99% intervals are estimated to be 80.14%, 96.03%, and 98.91% for samples of sizes 4, 6, and 10, respectively. If the study variable and the auxiliary variable used at the selection stage are uncorrelated while the study variable and auxiliary variable used at the estimation stage have correlation coefficient 0.3, then the coverages for the 90%, 95%, and 99% intervals are estimated as 88.63%, 94.31%, and 98.06% for samples of sizes 6, 8, and 14 units, respectively.

Therefore, in the case of multicharacter surveys, the newly tuned estimators based on the dell function seem to perform well when sample size is small and correlation between the study and auxiliary variables is low. Thus, the newly tuned multicharacter dell estimator resolves, to a certain extent, the problem of estimation of variance in case of multicharacter surveys, which is the most cost effective and time-saving scheme in real situations.

6.4.2 R code

We used the following R code, **PUMPKIN62.R**, to study the coverage by the newly tuned dell estimator.

```
#PROGRAM PUMPKIN62.R
library(sampling)
library(SDaA)
set.seed(2013)
```

```
for (rhoxy in seq(.3,.9,.2))
 {
  for (rhoyz in seq(.0,.9,.1))
  {
  NP<-2000
  xe<-rnorm(NP,0,1);ye<-rnorm(NP,0,1);ze<-rnorm(NP,0,1)
  y<-2454 + 245*ye
  x<-2325 + 232*sqrt((1-rhoxy^2))*xe + rhoxy*232*ye
  z<-322 + 52*sqrt((1-rhoyz^2))*ze + rhoyz*52*ye
  XPMEAN<-mean(x);ZPMEAN<-mean(z)
  RHOXY<-cor(x,y);RHOYZ<-cor(y,z)
  TY<-sum(y)
  pi<-z/(NP*ZPMEAN)
  nreps<-10000
  ESTP=rep(0,nreps)
  ci1.max=ci1.min=ci2.max=ci2.min=ci3.max=ci3.min=ESTP
  vESTP=ESTP
  for (n in seq(4,15,2) )
   {
   pik<-n*pi
   for (r in 1:nreps)
    {
    us<- lahiri.design(pik, n, as.factor(1:NP))
    xs<-x[us]; ys<-y[us]; zs<-z[us]
    ps<-((1+ 1/NP)^(1-RHOYZ))*((1+pi[us])^(RHOYZ))-1
    ESTX1<-mean(xs/ps);ESTY1<-mean(ys/ps)
    xmult<-(n*ESTX1 - xs/ps)/(n-1)
    ymult<-(n*ESTY1 - ys/ps)/(n-1)
    shi<-xmult - (NP*XPMEAN -n*(2-n)*ESTX1)/((n-1)^2)
    wbni<-(1/(1+ (sum(shi)/sum(shi^2))*shi))/n
    ESTi<- n*((n-1)^2*wbni-n+2)*ymult
    ESTP[r]<- sum(ESTi)
    EST_I<-(ESTP[r] - ESTi)/(n-1)
    ESTP[r]<-ESTP[r]/n
    vj<-(wbni^2)*ps*((EST_I - ESTP[r])^2)
    vESTP[r]<-n*((n-1)^3)*sum(vj)
    ci1.max[r]<- ESTP[r]-qt(.05,n-1)*sqrt(vESTP[r])
    ci1.min[r]<- ESTP[r]+qt(.05,n-1)*sqrt(vESTP[r])
    ci2.max[r]<- ESTP[r]-qt(.025,n-1)*sqrt(vESTP[r])
    ci2.min[r]<- ESTP[r]+qt(.025,n-1)*sqrt(vESTP[r])
    ci3.max[r]<- ESTP[r]-qt(.005,n-1)*sqrt(vESTP[r])
    ci3.min[r]<- ESTP[r]+qt(.005,n-1)*sqrt(vESTP[r])
    }# nreps
   round(sum(ci1.min<TY & ci1.max>TY)/nreps,4)->cov1
   round(sum(ci2.min<TY & ci2.max>TY)/nreps,4)->cov2
```

```
        round(sum(ci3.min<TY & ci3.max>TY)/nreps,4)->cov3
        cat(rhoxy,RHOXY,rhoyz, RHOYZ, n, cov1,cov2,cov3,'\n')
        }#  n
    }#  rhoyz
}#  rhoxy
```

6.4.3 Numerical illustration

The following example gives a numerical illustration of the use of steps in computing the estimator of total, standard error, and confidence interval estimate by using the proposed tuned estimator of population total for multicharacter surveys.

Example 6.2 Use data from Section 6.3.3 and apply the empirically tuned multicharacter survey method to estimate the total weight and construct a 90% confidence interval estimate.

Solution. We compute the following results:

p_j	p_j^*	$\hat{X}_M(j)$	$\hat{Y}_M(j)$	$\overline{w}_n^*(j)$	$\hat{Y}_{MTuned(dell)(j)}$
0.0071917	0.0058761	21234.39	465818.4	0.1438664	391891.5
0.0020547	0.0038208	22024.61	453613.1	0.1383667	499220.6
0.0039041	0.0045615	21055.89	375826.8	0.1451698	390130.2
0.0021917	0.0038757	20732.93	443700.7	0.1475890	327142.9
0.0067808	0.0057119	21585.78	357205.4	0.1413678	452081.9
0.0080136	0.0062043	21264.28	421352.4	0.1436504	405008.3
0.005821918	0.0053286	21785.02	413537.5	0.1399893	470160.5

Then a dell tuned estimate of the total weight of pumpkins is given by

$$\hat{Y}_{MTuned(dell)} = \frac{1}{n}\sum_{j \in s} \hat{Y}_{MTuned(dell)(j)} = 419376.6$$

The standard error of the tuned estimate of the population total is given by

$$SE\left(\hat{Y}_{MTuned(dell)}\right) = \sqrt{\hat{v}\left(\hat{Y}_{MTuned(dell)}\right)} = 51827.29$$

Hence, the 90% confidence interval estimate of the total weight of the pumpkins in the field is 318666.8–520086.3 lbs.

6.4.4 R code used for illustration

We used the following R code, **PUMPKIN62EX.R**, in the example.

#PROGRAM PUMPKIN62EX.R
```
n<-7
xs<-c(130.9,67,106.5,98,115.2,137.1,101.1)
ys<-c(800,800,3084,1042,4500,2500,2397)
zs<-c(105,30,57,32,99,117,85)
```

```
XPMEAN<-105.4;ZPMEAN<-73; RHOXY<-0.4
NP<-200
pi<-zs/(NP*ZPMEAN)
ps<-((1+1/NP)^(1-RHOXY))*((1+pi)^(RHOXY))-1
ESTX1<-mean(xs/ps);ESTY1<-mean(ys/ps)
xmult<-(n*ESTX1 - xs/ps)/(n-1)
ymult<-(n*ESTY1 - ys/ps)/(n-1)
shi<-xmult - (NP*XPMEAN -n*(2-n)*ESTX1)/((n-1)^2)
delt<-n*sum(xmult^2) - sum(xmult)^2
d<-(n*xmult - sum(xmult))/delt
wbni<-(1/(1+(sum(shi)/sum(shi^2))*shi))/n
ESTi<- n*((n-1)^2*wbni-n+2)*ymult
ESTP<- sum(ESTi)
EST_I<-(ESTP - ESTi)/(n-1)
ESTP<-ESTP/n
vj<-(wbni^2)*ps*((EST_I - ESTP)^2)
vESTP<-n*((n-1)^3)*sum(vj)
L<-ESTP+qt(.05,n-1)*sqrt(vESTP)
U<-ESTP+qt(.95,n-1)*sqrt(vESTP)
cbind(pi,ps,xmult,ymult,wbni,EST_I)
cat("Tuned estimate:",ESTP,"SE:",vESTP^.5,'\n')
cat("Confidence Interval:"," ", L,"; ", U,'\n')
```

6.5 Exercises

Exercise 6.1 Consider a newly tuned estimator of the population total Y in multicharacter surveys as

$$\hat{Y}_{\mathrm{MTuned}} = \sum_{j \in s} \left[\left\{ (n-1)^2 \bar{w}_n(j) - (n-2) \right\} \hat{Y}_M(j) \right] \tag{6.44}$$

where

$$\hat{Y}_M(j) = \frac{\sum_{i=1}^{n} \frac{y_i}{p_{ik}^*} - \frac{y_j}{p_{jk}^*}}{n-1} \tag{6.45}$$

for $k = 0, 1, 2, 3, 4$, is the estimator of the population total Y obtained by removing the jth unit from the sample s. The transformations p_{ik}^* are given by

$$p_{i0}^* = \frac{1}{N} \quad \text{(Rao, 1966)}$$

$$p_{i1}^* = \left(1 + \frac{1}{N}\right)^{(1-\rho_{yz})} (1+p_i)^{\rho_{yz}} - 1 \quad \text{(Bansal \& Singh, 1985)}$$

$$p_{i2}^* = \frac{(1-\rho_{yz})}{N} + \rho_{yz} p_i \quad \text{(Amahia, Chaubey, \& Rao, 1989)}$$

$$p_{i3}^* = \left[N(1-\rho_{yz}) + \frac{\rho_{yz}}{p_i} \right]^{-1} \quad \text{(Amahia, Chaubey, \& Rao, 1989)}$$

and

$$p_{i4}^* = \left(\frac{1}{N}\right)^{(1-\rho_{yz})} (p_i)^{\rho_{yz}} \quad \text{(Mangat \& Singh, 1992-93)}$$

where

$$p_i = \frac{z_i}{Z}, \quad i = 1, 2, \ldots, N \tag{6.46}$$

where $Z = \sum_{i=1}^{N} z_i$ is the known population total for the auxiliary variable used at the selection stage.

Consider the problem of tuning the jackknife weights $\bar{w}_n(j)$ such that the following distance functions:

$$D_1 = (2^{-1}n) \sum_{j \in s} q_j^{-1} \left(1 - (n-1)\bar{w}_n(j) - n^{-1}\right)^2 \tag{6.47}$$

$$D_2 = \sum_{j \in s} \frac{\ln(\bar{w}_n(j))}{n}, \quad \bar{w}_n(j) > 0 \tag{6.48}$$

and

$$D_3 = \frac{1}{n} \sum_{j \in s} \tanh^{-1} \left(\frac{\{\bar{w}_n(j)\}^2 - 1}{\{\bar{w}_n(j)\}^2 + 1} \right) \tag{6.49}$$

are optimized, where q_j are suitably chosen weights that form different types of estimators, $\tanh^{-1}()$ is the hyperbolic tangent function, and the jackknife tuned weights $\bar{w}_n(j)$ are chosen such that the following two constraints are satisfied:

$$\sum_{j \in s} \bar{w}_n(j) = 1 \tag{6.50}$$

and

$$\sum_{j \in s} \bar{w}_n(j) \hat{X}_M(j) = \frac{(X - n(2-n)\hat{X}_M)}{(n-1)^2} \tag{6.51}$$

where

$$\hat{X}_M(j) = \frac{n\hat{X}_M - x_j/p_{jk}^*}{n-1} \qquad (6.52)$$

and

$$\hat{X}_M = \frac{1}{n}\sum_{i \in s}(x_i/p_{ik}^*) \qquad (6.53)$$

is the estimator of the known population total X of the auxiliary variable used at the estimation stage. Write code in any scientific programming language to study these distance functions for different choices of p_{jk}^*. Discuss the nature of the tuned weights in each situation. Construct the 90%, 95%, and 99% confidence interval estimates in each situation by estimating the variance using the method discussed in this chapter. Simulate the distribution of $-2D_2$. Discuss the result in each situation, and conclude which transformation p_{jk}^* is more sensible for the situation you are dealing with. Extend the results to the case of random nonresponse by following Singh, Rueda, and Sanchez-Borrego (2010).

Exercise 6.2 Consider a newly tuned estimator of the population total Y in multicharacter surveys as

$$\hat{Y}_{MTuned} = \sum_{j \in s}\left[\left\{(n-1)^2 \bar{w}_n(j) - (n-2)\right\}\hat{Y}_M(j)\right] \qquad (6.54)$$

with

$$\hat{Y}_M(j) = \frac{n\hat{Y}_P - y_j/p_j^*}{n-1} \qquad (6.55)$$

where an estimator of the population total Y is

$$\hat{Y}_P = \frac{1}{n}\sum_{i=1}^{n}\frac{y_i}{p_i^*} \qquad (6.56)$$

for

$$p_i^* = (p_i^+)^{\rho_{yz}(1+\rho_{yz})/2}(p_i^-)^{-\rho_{yz}(1-\rho_{yz})/2}\left(\frac{1}{N}\right)^{(1-\rho_{yz})(1+\rho_{yz})} \qquad (6.57)$$

and ρ_{yz} is the population correlation coefficient between the study variable y and the auxiliary variable z used at the sample selection stage with $p_i^+ = z_i/Z$ and $p_i^- = (Z - nz_i)/\{(N-n)Z\}$ with known population total $Z = \sum_{i=1}^{N} z_i$. Discuss the cases for $\rho_{yz}=1$, $\rho_{yz}=0$, and $\rho_{yz}=-1$. Consider the problem of tuning the jackknife weights $\bar{w}_n(j)$ such that the following distance functions

$$D_1 = (2^{-1}n)\sum_{j\in s} q_j^{-1}\left(1-(n-1)\bar{w}_n(j)-n^{-1}\right)^2 \tag{6.58}$$

$$D_2 = \sum_{j\in s} \frac{\ln(\bar{w}_n(j))}{n}, \quad \bar{w}_n(j) > 0 \tag{6.59}$$

and

$$D_3 = \frac{1}{n}\sum_{j\in s} \tanh^{-1}\left(\frac{\{\bar{w}_n(j)\}^2 - 1}{\{\bar{w}_n(j)\}^2 + 1}\right) \tag{6.60}$$

are optimized, where q_j are suitably chosen weights that form different types of estimators, $\tanh^{-1}()$ is the hyperbolic tangent function, and the jackknife tuned weights $\bar{w}_n(j)$ are chosen such that the following two tuning constraints are satisfied:

$$\sum_{j\in s} \bar{w}_n(j) = 1 \tag{6.61}$$

and

$$\sum_{j\in s} \bar{w}_n(j)\hat{X}_M(j) = \frac{\left(X - n(2-n)\hat{X}_M\right)}{(n-1)^2} \tag{6.62}$$

where

$$\hat{X}_M(j) = \frac{n\hat{X}_M - x_j/p_j^*}{n-1} \tag{6.63}$$

and

$$\hat{X}_M = \frac{1}{n}\sum_{i\in s} \frac{x_i}{p_i^*} \tag{6.64}$$

is the estimator of the known population total X of the auxiliary variable used at the estimation stage.

Write codes in any scientific programming language, such as R, to study these distance functions. Discuss the nature of the tuned weights in each situation. Construct the 90%, 95%, and 99% confidence interval estimates in each situation by estimating the variance using the method discussed in this chapter. Simulate the distribution of $-2D_2$ and D_3.

Hint: Singh and Horn (1998).

Exercise 6.3 Tuning of sensitive variables in multicharacter surveys

Let the value y_i of a sensitive variable Y, defined on a finite survey population of N identifiable and labeled persons, be assumed to be unavailable through a direct response survey. Suppose that one wants to estimate the population total Y by choosing a sample s from the population with a probability $p(s)$ according to design p. Instead of direct response, suppose a randomized response r_i is available in an independent manner from the respective persons i, done in such a way that their expectations, variance, and covariance (E_R, V_R, C_R), respectively, satisfy $E_R(r_i) = y_i$, $V_R(r_i) = \alpha_i y_i^2 + \beta_i y_i + \theta_i = V_i^2$ (say), $C_R(r_i, r_j) = 0$, for $i \neq j$ where $\alpha_i > 0$, β_i, and θ_i are known for every unit in the population. Consider a newly tuned estimator of the population total Y in multicharacter surveys as

$$\hat{Y}_{\text{MTuned}} = \sum_{j \in s} \left[\left\{ (n-1)^2 \bar{w}_n(j) - (n-2) \right\} \hat{Y}_M(j) \right] \tag{6.65}$$

where

$$\hat{Y}_M(j) = \frac{n \hat{y}_{\text{mc}} - r_j / p_{jk}^*}{n-1} \tag{6.66}$$

with

$$\hat{y}_{\text{mc}} = \frac{1}{n} \sum_{i=1}^{n} \frac{r_i}{p_{ik}^*} \tag{6.67}$$

for $k = 0, 1, 2, 3, 4$, is the estimator of the population total Y obtained by removing the jth scrambled response unit from the sample s, and

$$p_{i0}^* = \frac{1}{N} \quad \text{(Rao, 1966)}$$

$$p_{i1}^* = \left(1 + \frac{1}{N}\right)^{(1-\rho_{yz})} (1+p_i)^{\rho_{yz}} - 1 \quad \text{(Bansal \& Singh, 1985)}$$

$$p_{i2}^* = \frac{(1-\rho_{yz})}{N} + \rho_{yz} p_i \quad \text{(Amahia, Chaubey, \& Rao, 1989)}$$

$$p_{i3}^* = \left[N(1-\rho_{yz}) + \frac{\rho_{yz}}{p_i} \right]^{-1} \quad \text{(Amahia, Chaubey, \& Rao, 1989)}$$

$$p_{i4}^* = \left(\frac{1}{N}\right)^{(1-\rho_{yz})} (p_i)^{\rho_{yz}} \quad \text{(Mangat \& Singh, 1992–93)}$$

where

$$p_i = \frac{z_i}{Z}, \quad i = 1, 2, \ldots, N \tag{6.68}$$

and $Z = \sum_{i=1}^{N} z_i$ is the known total of the auxiliary variable used at the selection stage.
Consider the problem of tuning the jackknife weights $\bar{w}_n(j)$ such that the following distance functions:

$$D_1 = (2^{-1}n) \sum_{j \in s} q_j^{-1} \left(1 - (n-1)\bar{w}_n(j) - n^{-1}\right)^2 \tag{6.69}$$

$$D_2 = \sum_{j \in s} \frac{\ln(\bar{w}_n(j))}{n}, \quad \bar{w}_n(j) > 0 \tag{6.70}$$

and

$$D_3 = \frac{1}{n} \sum_{j \in s} \tanh^{-1}\left(\frac{\{\bar{w}_n(j)\}^2 - 1}{\{\bar{w}_n(j)\}^2 + 1}\right) \tag{6.71}$$

are optimized, where q_j are suitably chosen weights that form different types of estimators, $\tanh^{-1}()$ is the hyperbolic tangent function, and the tuned jackknife weights $\bar{w}_n(j)$ are chosen such that the following two tuning constraints are satisfied:

$$\sum_{j \in s} \bar{w}_n(j) = 1 \tag{6.72}$$

and

$$\sum_{j \in s} \bar{w}_n(j) \hat{X}_M(j) = \frac{(X - n(2-n)\hat{X}_M)}{(n-1)^2} \tag{6.73}$$

where

$$\hat{X}_M(j) = \frac{n\hat{X}_M - x_j/p_{jk}^*}{n-1} \tag{6.74}$$

and

$$\hat{X}_M = \frac{1}{n} \sum_{i \in s} \frac{x_i}{p_{ik}^*} \tag{6.75}$$

is the estimator of the known population total X of the auxiliary variable used at the estimation stage. Write code in any scientific programming language to study these distance functions for different choices of p_{jk}^*. Discuss the nature of the tuned weights in each situation. Construct the 90%, 95%, and 99% confidence interval estimates in each situation by estimating the variance using the method discussed in this chapter. Simulate the distribution of $-2D_2$ and D_3. Discuss the results in each situation, and conclude which transformation p_{jk}^* is more suitable for the situation you are dealing with. *Hint*: Bansal, Singh, and Singh (1994).

Exercise 6.4 Consider the optimization of a new compromised chi-squared type distance function defined as

$$D = \rho_{yz} \left[\sum_{i \in s} \frac{(p_i^* - p_i)^2}{h_i p_i} \right] + (1 - \rho_{yz}) \left[\sum_{i \in s} \frac{(p_i^* - (1/N))^2}{l_i (1/N)} \right] \quad (6.76)$$

subject to the condition:

$$\sum_{i \in s} p_i^* = \rho_{yz} \sum_{i \in s} p_i + (1 - \rho_{yz})(n/N) \quad (6.77)$$

where h_i and l_i are some choice of weights with respect to the required transformed variable p_i^*. Note that if $\rho_{yz} \to 0$, then the transformed probability p_i^* in Equation (6.77) becomes $1/N$, and if $\rho_{yz} \to 1$ then p_i^* becomes $p_i = z_i/Z$. Then consider an estimator of the population total Y as

$$\hat{Y}_{new} = \frac{1}{n} \sum_{i \in s} \frac{y_i}{p_i^*} \quad (6.78)$$

Investigate the estimator $\hat{Y}_{new}$, from the points of view of bias and variance through a simulation study, and compare your findings with the kth estimator, for $k = 0, 1, 2, 3, 4$, given by

$$\hat{Y}_k = \frac{1}{n} \sum_{i \in s} \frac{y_i}{p_{ik}^*} \quad (6.79)$$

where

$$p_{i0}^* = \frac{1}{N} \quad (6.80)$$

$$p_{i1}^* = \left(1 + \frac{1}{N}\right)^{(1 - \rho_{yz})} (1 + p_i)^{\rho_{yz}} - 1 \quad (6.81)$$

$$p_{i2}^* = \frac{(1 - \rho_{yz})}{N} + \rho_{yz} p_i \quad (6.82)$$

$$p_{i3}^* = \left[N(1-\rho_{yz}) + \frac{\rho_{yz}}{p_i}\right]^{-1} \quad (6.83)$$

$$p_{i4}^* = \left(\frac{1}{N}\right)^{(1-\rho_{yz})} (p_i)^{\rho_{yz}} \quad (6.84)$$

In addition, suggest a calibrated estimator of the population total in the presence of another auxiliary variable, X, to be used at the estimation stage to improve the suggested new estimator $\hat{Y}_{new}$.

Exercise 6.5 In a specially designed field, it is easy to take a picture of each one of the pumpkins and record their top size. After taking photographs of the tops of all the 200 pumpkins in the field, we selected a PPSWR sample of $n=7$ pumpkins. The weight (lbs), say Y, circumference (in.), say X, and top size (in.), say Z, of the seven selected pumpkins were recorded as follows:

x_i	130	67	108	98	116	137	101
y_i	800	800	3084	1042	4500	2500	2397
z_i	105	30	57	32	99	117	85

Apply the multicharacter survey approach with the following five transformations:

$$p_{i0}^* = \frac{1}{N} \quad (6.85)$$

$$p_{i1}^* = \left(1 + \frac{1}{N}\right)^{(1-\rho_{yz})} (1+p_i)^{\rho_{yz}} - 1 \quad (6.86)$$

$$p_{i2}^* = \frac{(1-\rho_{yz})}{N} + \rho_{yz} p_i \quad (6.87)$$

$$p_{i3}^* = \left[N(1-\rho_{yz}) + \frac{\rho_{yz}}{p_i}\right]^{-1} \quad (6.88)$$

$$p_{i4}^* = \left(\frac{1}{N}\right)^{(1-\rho_{yz})} (p_i)^{\rho_{yz}} \quad (6.89)$$

to estimate the total weight of the pumpkins in the field and construct 90% confidence interval estimates using circumference at the estimation stage and top size at the selection stage using two tuning methods discussed in this chapter. Discuss your findings. (*Given*: The population total of X is 21,080, total of Z is 14,600, and $\rho_{yz} = 0.3$).

Tuning of the Horvitz–Thompson estimator 7

7.1 Introduction

In this chapter, we consider the problem of estimation of population total by tuning the design weights in the Horvitz and Thompson (1952) estimator. The tuning of the design weights using both the chi-square type distance function and a new displacement function are investigated using simulation study. At the end of the chapter, a few unsolved exercises are also included for future investigation.

7.2 Jackknifed weights in the Horvitz–Thompson estimator

Let $d_j = \pi_j^{-1}$, with π_j being the first-order inclusion probability, denote the jth design weight. Then, the well-known Horvitz and Thompson (1952) estimator of the population total Y is defined as

$$\hat{Y}_{\text{HT}} = \sum_{j \in s} d_j y_j \tag{7.1}$$

Let $w_j^{\diamond}$ be any calibrated (or arbitrary) weights, following Singh (2003), such that

$$\sum_{j \in s} w_j^{\diamond} = \sum_{j \in s} d_j \quad \text{or} \quad \bar{w}_n^{\diamond} = \bar{d} \tag{7.2}$$

where

$$\bar{w}_n^{\diamond} = \frac{1}{n} \sum_{j \in s} w_j^{\diamond} \quad \text{and} \quad \bar{d} = \frac{1}{n} \sum_{j \in s} d_j$$

Obviously, the weight $w_j^{\diamond}$ can be written as

$$w_j^{\diamond} = n\bar{w}_n^{\diamond} - (n-1)\bar{w}_n^{\diamond}(j) \tag{7.3}$$

or

$$w_j^{\diamond} = n\bar{d} - (n-1)\bar{w}_n^{\diamond}(j) \tag{7.4}$$

A New Concept for Tuning Design Weights in Survey Sampling. http://dx.doi.org/10.1016/B978-0-08-100594-1.00007-3
Copyright © 2016 Elsevier Ltd. All rights reserved.

where

$$\bar{w}_n^\diamond(j) = \frac{1}{n-1} \sum_{i \neq j \in s} w_i^\diamond \qquad (7.5)$$

is the jackknifed average weight after removing the jth weight.

Therefore, using Equations (7.2) and (7.4), we have

$$\sum_{j \in s} w_j^\diamond = n\bar{d} \quad \text{or} \quad \sum_{j \in s} \left[n\bar{d} - (n-1)\bar{w}_n^\diamond(j) \right] = n\bar{d}$$

or

$$\sum_{j \in s} \bar{w}_n^\diamond(j) = \sum_{j \in s} d_j \qquad (7.6)$$

In the same way one can see

$$\sum_{j \in s} \bar{w}_n^\diamond(j) = \sum_{j \in s} \bar{d}(j) = \sum_{j \in s} w_j^\diamond = \sum_{j \in s} d_j \qquad (7.7)$$

where

$$\bar{d}(j) = (n\bar{d} - d_j)/(n-1) \qquad (7.8)$$

7.3 Tuning with a chi-square distance function while using jackknifed sample means

The newly tuned estimator of the population total Y is defined as

$$\hat{Y}_{\text{HTuned(cs)}} = \sum_{j \in s} \left[(n-1)^2 \bar{w}_n^\diamond(j) - n(n-2)\bar{d} \right] \bar{y}_n(j) \qquad (7.9)$$

where

$$\bar{y}_n(j) = \frac{n\bar{y}_n - y_j}{n-1} \qquad (7.10)$$

with

$$\bar{y}_n = \frac{1}{n} \sum_{i \in s} y_i \qquad (7.11)$$

and $\bar{w}_n^\Diamond(j)$ are the jackknife weights such that

$$\sum_{j\in s}\bar{w}_n^\Diamond(j) = \sum_{j\in s}\bar{d}(j) \tag{7.12}$$

and

$$\sum_{j\in s}\bar{w}_n^\Diamond(j)\bar{x}_n(j) = \frac{X-(2-n)n^2\bar{d}\bar{x}_n}{(n-1)^2} \tag{7.13}$$

where $X = \sum_{i\in\Omega} x_i$ denotes the known population total of the auxiliary variable, and

$$\bar{x}_n(j) = \frac{n\bar{x}_n - x_j}{n-1} \tag{7.14}$$

with

$$\bar{x}_n = \frac{1}{n}\sum_{i\in s} x_i \tag{7.15}$$

Note that constraint (7.13) ensures that the calibrated weight $w_j^\Diamond$ satisfies the calibration constraint:

$$\sum_{j\in s} w_j^\Diamond x_j = X \tag{7.16}$$

with

$$x_j = n\bar{x}_n - (n-1)\bar{x}_n(j) \tag{7.17}$$

The chi-square distance between the calibrated weights $w_j^\Diamond$ and the design weights d_j is given by

$$D = \frac{1}{2}\sum_{j\in s}\frac{\left(w_j^\Diamond - d_j\right)^2}{d_j q_j} = \frac{(n-1)^2}{2}\sum_{j\in s}\frac{\left(\bar{d}(j) - \bar{w}_n^\Diamond(j)\right)^2}{d_j q_j} \tag{7.18}$$

The Lagrange function is then given by

$$L = \frac{(n-1)^2}{2}\sum_{j\in s}\frac{\left(\bar{d}(j) - \bar{w}_n^\Diamond(j)\right)^2}{d_j q_j} - \lambda_0\left[\sum_{j\in s}\bar{w}_n^\Diamond(j) - \sum_{j\in s}\bar{d}(j)\right]$$
$$- \lambda_1\left[\sum_{j\in s}\bar{w}_n^\Diamond(j)\bar{x}_n(j) - \frac{X-(2-n)n^2\bar{x}_n\bar{d}}{(n-1)^2}\right] \tag{7.19}$$

On setting

$$\frac{\partial L}{\partial \bar{w}_n^{\diamond}(j)} = 0 \tag{7.20}$$

we have

$$\bar{w}_n^{\diamond}(j) = \bar{d}(j) + \frac{1}{(n-1)^2} \left(\lambda_0 d_j q_j + \lambda_1 d_j q_j \bar{x}_n(j) \right) \tag{7.21}$$

Note that

$$\sum_{j \in s} \bar{d}(j) \bar{x}_n(j) = \frac{1}{(n-1)^2} \sum_{j \in s} \left(n\bar{d} - d_j \right) \left(n\bar{x}_n - x_j \right)$$

$$= \frac{1}{(n-1)^2} \left[\bar{d}\bar{x}_n n^2 (n-2) + \hat{X}_{HT} \right] \tag{7.22}$$

and

$$(n-1)^2 \left[\frac{X - (2-n)n^2 \bar{d}\bar{x}_n}{(n-1)^2} - \frac{\bar{d}\bar{x}_n n^2 (n-2) + \hat{X}_{HT}}{(n-1)^2} \right] = \left(X - \hat{X}_{HT} \right) \tag{7.23}$$

where

$$\hat{X}_{HT} = \sum_{i \in s} d_i x_i \tag{7.24}$$

The values of λ_0 and λ_1 are obtained by solving the normal equations given by

$$\begin{bmatrix} \sum_{j \in s} d_j q_j, & \sum_{j \in s} d_j q_j \bar{x}_n(j) \\ \sum_{j \in s} d_j q_j \bar{x}_n(j), & \sum_{j \in s} d_j q_j \{\bar{x}_n(j)\}^2 \end{bmatrix} \begin{bmatrix} \lambda_0 \\ \lambda_1 \end{bmatrix} = \begin{bmatrix} 0 \\ \left(X - \hat{X}_{HT} \right) \end{bmatrix} \tag{7.25}$$

Thus, the optimal jackknifed tuning weights are given by

$$\bar{w}_n^{\diamond}(j) = \bar{d}(j) + \frac{\Delta_j}{(n-1)^2} \left(X - \hat{X}_{HT} \right) \tag{7.26}$$

where

$$\Delta_j = \frac{(d_j q_j \bar{x}_n(j)) \left(\sum_{j \in s} d_j q_j \right) - (d_j q_j) \left(\sum_{j \in s} d_j q_j \bar{x}_n(j) \right)}{\left(\sum_{j \in s} d_j q_j \right) \left(\sum_{j \in s} d_j q_j \{\bar{x}_n(j)\}^2 \right) - \left\{ \sum_{j \in s} d_j q_j \bar{x}_n(j) \right\}^2} \tag{7.27}$$

Using these weights, the newly tuned estimator of the population total Y becomes

$$\hat{Y}_{\text{HTuned(cs)}} = \sum_{j \in s} \left[(n-1)^2 \bar{w}_n^\diamond(j) - n(n-2) \bar{d} \right] \bar{y}_n(j) \qquad (7.28)$$
$$= \hat{Y}_{\text{HT}} + \hat{\beta}_1 \left(X - \hat{X}_{\text{HT}} \right)$$

where

$$\hat{\beta}_1 = \frac{\left(\sum_{j \in s} d_j q_j \right) \left(\sum_{j \in s} d_j q_j \bar{x}_n(j) \bar{y}_n(j) \right) - \left(\sum_{j \in s} d_j q_j \bar{y}_n(j) \right) \left(\sum_{j \in s} d_j q_j \bar{x}_n(j) \right)}{\left(\sum_{j \in s} d_j q_j \right) \left(\sum_{j \in s} d_j q_j \{ \bar{x}_n(j) \}^2 \right) - \left\{ \sum_{j \in s} d_j q_j \bar{x}_n(j) \right\}^2} \qquad (7.29)$$

which is a new estimator of the regression coefficient.

7.3.1 Estimation of variance and coverage

An adjusted estimator of variance of the tuned Horvitz and Thompson (1952) estimator $\hat{Y}_{\text{HTuned(cs)}}$ is

$$\hat{v}\left(\hat{Y}_{\text{HTuned(cs)}} \right) = \sum_{j \in s} \left\{ f_j \bar{w}_n^\diamond(j) / \bar{d}(j) \right\}^2 \left\{ \hat{Y}_{\text{HTuned(cs)}(j)} - \hat{Y}_{\text{HTuned(cs)}} \right\}^2 \qquad (7.30)$$

Note that each newly tuned Horvitz and Thompson (1952) doubly jackknifed estimator of the population total Y is given by

$$\hat{Y}_{\text{HTuned(cs}(j))} = \frac{n \hat{Y}_{\text{HTuned(cs)}} - \left\{ (n-1)^2 \bar{w}_n^\diamond(j) - n(n-2) \bar{d} \right\} \bar{y}_n(j)}{n-1} \qquad (7.31)$$

for $j = 1, 2, \ldots, n$, and

$$f_j = 1 - ((N-1)/N) \left(\pi_j - (n-1)/(N-1) \right)/p_j \qquad (7.32)$$

The coverage by the $(1-\alpha)100\%$ confidence interval estimates using this newly tuned regression type Horvitz and Thompson (1952) estimator of the population total is obtained by observing the proportion of times the true population total Y falls within the interval estimate given by

$$\hat{Y}_{\text{HTuned(cs)}} \mp t_{\alpha/2}(\text{df} = n-1) \sqrt{\hat{v}\left(\hat{Y}_{\text{HTuned(cs)}} \right)} \qquad (7.33)$$

We studied the nominal 90%, 95%, and 99% intervals by selecting 10,000 random samples from a population consisting of three related variables Y, X, and Z, which take values y_i, x_i, and z_i for the ith unit in the population as follows:

$$y_i = \bar{Y} + \sigma_y y_i^* \tag{7.34}$$

$$x_i = \bar{X} + x_i^* \sigma_x \sqrt{1 - \rho_{xy}^2} + \rho_{xy} \sigma_x y_i^* \tag{7.35}$$

and

$$z_i = \bar{Z} + z_i^* \sigma_z \sqrt{1 - \rho_{yz}^2} + \rho_{yz} \sigma_z y_i^* \tag{7.36}$$

where $x_i^* \sim N(0, 1)$, $y_i^* \sim N(0, 1)$, and $z_i^* \sim N(0, 1)$ are three standard normal variables with mean zero and standard deviation one. As an example, the variable Y represents the unknown weight of pumpkins, the variable Z represents the known top size of pumpkins, and the variable X represents the known circumference of pumpkins. For this illustration, we made the following choice of parameters: $\rho_{xy} = 0.80$, $\rho_{yz} = 0.70$, $\bar{X} = 2320$, $\bar{Y} = 2450$, $\bar{Z} = 322$, $\sigma_x = 50$, $\sigma_y = 50$, $\sigma_z = 52$, and $N = 2000$. Thus, our problem is to estimate the population total $Y = N\bar{Y}$ of the study variable Y, using known population totals $X = N\bar{X}$ and $Z = N\bar{Z}$ of the two auxiliary variables X and Z, respectively. We used the Z variable at the selection stage by taking $d_i = \pi_i^{-1}$, where

$$\pi_i = \left(\frac{N-n}{N-1}\right) p_i + \left(\frac{n-1}{N-1}\right) \tag{7.37}$$

with $p_i = z_i/Z$, $Z = N\bar{Z}$ and $n = 3\text{--}31$.

The variable X is used at the tuning stage of the estimator of the population total of the Y variable. For each sample of size n (from 3 to 31), we generated 10,000 samples to study the coverages. The results obtained for various sample sizes are shown in Table 7.1.

Table 7.1 shows that the coverages by the intervals formed from the newly tuned estimator of the population total under the probability proportional to size and without replacement (PPSWOR) sampling scheme perform much better than in earlier cases. We used the PPSWOR scheme proposed independently by Midzuno (1951) and Sen (1952). The Midzuno–Sen sampling scheme is easy to implement in practice; it requires the selection of the first units in a sample s of size n using the PPSWR scheme and the selection of the remaining $(n-1)$ units using the SRSWOR scheme. Any other PPSWOR sampling scheme available in the literature could be investigated in this context. It is very interesting to note that the attained coverage of the nominal 90% confidence interval is estimated to be 89.62% for a sample size of 23 units, for the 95% interval attained coverage is 95.03% when the sample size is 23, and for the 99% interval attained coverage is 98.96% when the sample size is 21. Our overall observation regarding the proposed tuned estimator is that it shows nominal coverages for moderate sample sizes.

Table 7.1 **Performance of the newly tuned Horvitz and Thompson type estimator of population total**

Sample size (n)	90% coverage	95% coverage	99% coverage
3	0.3041	0.4108	0.7022
5	0.3282	0.4175	0.6304
7	0.4452	0.5367	0.733
9	0.551	0.6483	0.8204
11	0.619	0.7206	0.8769
13	0.6923	0.7846	0.9132
15	0.7435	0.8358	0.9443
17	0.8008	0.8805	0.9684
19	0.8349	0.9083	0.9797
21	0.8796	0.9393	0.9896
23	0.8962	0.9503	0.9935
25	0.9239	0.9672	0.9965
27	0.9465	0.9808	0.9989
29	0.9618	0.9875	0.9988
31	0.9707	0.9909	0.9998

7.3.2 R code

The following R code, **PUMPKIN71.R**, was used to study the coverage by the newly tuned Horvitz and Thompson (1952) estimator.

```
#PROGRAM PUMPKIN71.R
library("sampling")
set.seed(2013)
rhoxy<-0.8;rhoyz<-0.7;NP<-2000;
  xe<-rnorm(NP,0,1);ye<-rnorm(NP,0,1);ze<-rnorm(NP,0,1);
  y<-2450+50*ye
  x<-2320+50*sqrt(1-rhoxy^2)*xe + rhoxy*50*ye
  z<-322+52*sqrt(1-rhoyz^2)*ze + rhoyz*52*ye
XPMEAN<-mean(x);ZPMEAN<-mean(z); TY<-sum(y)
cor(x,y);cor(z,y);sqrt(var(x));sqrt(var(y));sqrt(var(z))
nreps<-10000
ESTP=rep(0,nreps)
ci1.max=ci1.min=ci2.max=ci2.min=ci3.max=ci3.min=ESTP
vESTP=ESTP
for (n in seq(3,50,2) )
 {
 inclusionprobabilities(z,n)->pik
 for (r in 1:nreps)
  {
  ese<-UPmidzuno(pik)
  (1:length(pik))[ese==1]->us
```

```
#us<-sample(NP,n)
xs<-x[us]; ys<-y[us]; zs<-z[us]
xmj<-(sum(xs)-xs)/(n-1)
ymj<-(sum(ys)-ys)/(n-1)
p<-zs/(NP*ZPMEAN)
pi<-(NP-n)*p/(NP-1) + (n-1)/(NP-1); di<-1/pi
Fi<-1-((NP-1)/NP)*(pi-(n-1)/(NP-1))/p
YHT<-sum(di*ys); XHT<-sum(di*xs);
DB<-mean(di)
dj<-(sum(di)-di)/(n-1)
delta<-sum(di)*sum(di*xmj^2) - sum(di*xmj)^2
del<-(sum(di)*di*xmj - di*sum(di*xmj))/delta
wbni<-dj + del*(NP*XPMEAN-XHT)/((n-1)^2)
ESTi<- ((n-1)^2*wbni-n*DB*(n-2))*ymj
ESTP[r]<-sum(ESTi)
EST_I<-(n*ESTP[r] - ESTi)/(n-1)
vj<-(di*Fi/wbni)^2*(EST_I-ESTP[r])^2
vESTP[r]<-sum(vj)
ci1.min[r]<- ESTP[r]+qt(.05,n-1)*sqrt(vESTP[r])
ci1.max[r]<- ESTP[r]+qt(.95,n-1)*sqrt(vESTP[r])
ci2.max[r]<- ESTP[r]+qt(.975,n-1)*sqrt(vESTP[r])
ci2.min[r]<- ESTP[r]+qt(.025,n-1)*sqrt(vESTP[r])
ci3.min[r]<- ESTP[r]+qt(.005,n-1)*sqrt(vESTP[r])
ci3.max[r]<- ESTP[r]+qt(.995,n-1)*sqrt(vESTP[r])
}# nreps
round(sum(ci1.min<TY & ci1.max>TY)/nreps,4)->cov1
round(sum(ci2.min<TY & ci2.max>TY)/nreps,4)->cov2
round(sum(ci3.min<TY & ci3.max>TY)/nreps,4)->cov3
cat(n, cov1,cov2,cov3,'\n')
}# n
```

7.3.3 Numerical illustration

The following numerical illustration is used to explain all steps involved in the use of the newly tuned Horvitz and Thompson (1952) estimator.

Example 7.1 In a specially designed field, it is easy to take a picture of each one of the $N = 200$ pumpkins and record their top sizes. After taking the photographs, Midzuno–Sen sampling was applied to select a sample of $n = 7$ pumpkins. The weight (lbs), say Y, circumference (in.), say X, and top size (in.), say Z, of the seven selected pumpkins are recorded as follows:

x_i	130.9	67.0	106.5	98.0	115.2	137.1	101.1
y_i	800	800	3084	1042	4500	2500	2397
z_i	105	30	56	30	97	117	87

Assume it is known that the population average circumference is $\bar{X} = 105.40$ in. and the population average top size is $\bar{Z} = 61$ in.. Apply the tuned Horvitz and Thompson (1952) estimator to estimate the total weight Y of the pumpkins and construct the 95% confidence interval estimate.

Solution. One can easily compute the following:

$\bar{x}_n(j)$	$\bar{y}_n(j)$	$\bar{w}_n^\diamond(j)$	$\hat{Y}_{\text{HTuned(cs }(j))}$
104.1500	2387.167	27.85886	478049.4
114.8000	2387.167	28.01937	475750.4
108.2167	2006.500	27.69873	481722.0
109.6333	2346.833	27.52766	482897.9
106.7667	1770.500	27.99467	479433.2
103.1167	2103.833	27.89032	478951.0
109.1167	2121.000	28.09405	476276.3

where

$$\hat{Y}_{\text{HTuned(cs}(j))} = \frac{n\hat{Y}_{\text{HTuned(cs)}} - \left\{(n-1)^2 \bar{w}_n^\diamond(j) - n(n-2)\bar{d}\right\} \bar{y}_n(j)}{n-1}$$

We can then compute

$$\hat{Y}_{\text{HTuned(cs)}(j)} = \frac{1}{n} \sum_{j \in s} \hat{Y}_{\text{HTuned(cs)}(j)} = 419135.0$$

and

$$SE\left(\hat{Y}_{\text{HTuned(cs)}}\right) = \sqrt{\hat{v}\left(\hat{Y}_{\text{HTuned(cs)}}\right)} = 5548.925$$

Thus, the 95% confidence interval estimate of the total weight of all pumpkins in the field is 405557.3–432712.8 lbs.

7.3.4 R code used for illustration

We used the following R code, **PUMPKIN71EX.R**, to solve the preceding numerical illustration.

```
#PROGRAM PUMPKIN71EX.R
n<-7
xs<-c(130.9,67,106.5,98,115.2,137.1,101.1)
ys<-c(800,800,3084,1042,4500,2500,2397)
zs<-c(105,30,56,30,97,117,87)
XPMEAN<-105.4;ZPMEAN<-61; RHOYZP<-0.4
NP<-200
```

```
xmj<-(sum(xs)-xs)/(n-1)
ymj<-(sum(ys)-ys)/(n-1)
p<-zs/(NP*ZPMEAN)
pi<-(NP-n)*p/(NP-1) + (n-1)/(NP-1); di<-1/pi
Fi<-1-((NP-1)/NP)*(pi-(n-1)/(NP-1))/p
YHT<-sum(di*ys); XHT<-sum(di*xs)
DB<-mean(di)
dj<-(sum(di)-di)/(n-1)
delta<-sum(di)*sum(di*xmj^2) - sum(di*xmj)^2
del<-(sum(di)*di*xmj - di*sum(di*xmj))/delta
wbni<-dj + del*(NP*XPMEAN-XHT)/((n-1)^2)
ESTi<- ((n-1)^2*wbni-n*DB*(n-2))*ymj
ESTP<- sum(ESTi)
EST_I<-(n*ESTP - ESTi)/(n-1)
vj<-((Fi*wbni/dj)^2)*((EST_I - ESTP)^2)
vESTP<-sum(vj)
L<-ESTP+qt(.025,n-1)*sqrt(vESTP)
U<-ESTP+qt(.975,n-1)*sqrt(vESTP)
cbind(xmj,ymj,wbni,EST_I)
cat("Tuned estimate:", ESTP, "SE: ",vESTP^.5 ,'\n')
cat("Confidence Interval:"," ", L,"; ", U,'\n')
```

7.4 Tuning of the Horvitz–Thompson estimator with a displacement function

The newly tuned Horvitz and Thompson (1952) jackknife estimator of the population total Y is defined as

$$\hat{Y}_{\text{HTuned(new)}} = \sum_{j \in s} \left[(n-1)^2 \bar{w}_n^\circ(j) - n(n-2)\bar{d} \right] \bar{y}_n(j) \tag{7.38}$$

where $\bar{w}_n^\circ(j)$ are the tuned weights constructed so that the following two constraints are satisfied:

$$\sum_{j \in s} \bar{w}_n^\circ(j) = \sum_{j \in s} \bar{d}(j) \tag{7.39}$$

and

$$\sum_{j \in s} \bar{w}_n^\circ(j) \Psi_j = 0 \tag{7.40}$$

where

$$\Psi_j = \bar{x}_n(j) - \frac{X - (2-n)n^2 \bar{d} \, \bar{x}_n}{n\bar{d}(n-1)^2} \tag{7.41}$$

We consider tuning the weights $\bar{w}_n^\circ(j)$ such that an alternative to weighted displacement function due to Singh (2012), defined as

$$\sum_{j\in s} \bar{d}(j) \tanh^{-1}\left[\frac{\{\bar{w}_n^\circ(j)\}^2 - 1}{\{\bar{w}_n^\circ(j)\}^2 + 1}\right] \tag{7.42}$$

is optimum, subject to tuning constraints (7.39) and (7.40).

The Lagrange function becomes

$$L_2 = \sum_{j\in s} \bar{d}(j) \tanh^{-1}\left[\frac{\{\bar{w}_n^\circ(j)\}^2 - 1}{\{\bar{w}_n^\circ(j)\}^2 + 1}\right] - \lambda_0^* \left\{\sum_{j\in s} \bar{w}_n^\circ(j) - \sum_{j\in s} \bar{d}(j)\right\}$$
$$- \lambda_1^* \left\{\sum_{j\in s} \bar{w}_n^\circ(j) \Psi_j\right\} \tag{7.43}$$

where λ_0^* and λ_1^* are the Lagrange multiplier constants.

On setting

$$\frac{\partial L_2}{\partial \bar{w}_n^\circ(j)} = 0 \tag{7.44}$$

we have

$$\bar{w}_n^\circ(j) = \frac{\bar{d}(j)}{1 + \lambda_1^* \Psi_j} \tag{7.45}$$

Constraints (7.39) and (7.40) yield $\lambda_0^* = 1$, and λ_1^* is a solution to the single parametric equation

$$\sum_{j\in s} \frac{\bar{d}(j) \Psi_j}{1 + \lambda_1^* \Psi_j} = 0 \tag{7.46}$$

Thus, under the alternative to displacement function, the newly tuned Horvitz and Thompson (1952) jackknife estimator (7.38) of the population total becomes

$$\hat{Y}_{\text{HTuned(new)}} = \sum_{j\in s} \left[\frac{(n-1)^2 \bar{d}(j)}{1 + \lambda_1^* \Psi_j} - n(n-2)\bar{d}\right] \bar{y}_n(j) \tag{7.47}$$

7.4.1 Estimation of variance of the Horvitz–Thompson type estimator

An adjusted estimator to estimate the variance of the newly tuned Horvitz and Thompson (1952) estimator $\hat{Y}_{\text{HTuned(new)}}$ is

$$\hat{v}\left(\hat{Y}_{\text{HTuned(new)}}\right) = \sum_{j\in s} \left(\frac{f_j \bar{w}_n^\circ(j)}{\bar{d}(j)}\right)^2 \left(\hat{Y}_{\text{HTuned(new)}(j)} - \hat{Y}_{\text{HTuned(new)}}\right)^2 \tag{7.48}$$

Note that each newly tuned Horvitz and Thompson (1952) jackknifed estimator of the population total is given by

$$\hat{Y}_{\text{HTuned(new)}(j)} = \frac{n\hat{Y}_{\text{HTuned(new)}} - \left\{(n-1)^2 \bar{w}_n^\circ(j) - n(n-2)\bar{d}\right\} \bar{y}_n(j)}{n-1} \qquad (7.49)$$

for $j = 1, 2, \ldots, n$. In the simulation, we approximated the value of λ_1^* by

$$\lambda_1^* \approx \frac{\sum_{j \in s} \bar{d}(j) \Psi_j}{\sum_{j \in s} \bar{d}(j) \Psi_j^2} \qquad (7.50)$$

A better solution to the nonlinear equation (7.46) could be used, if available. The coverage by the $(1-\alpha)100\%$ confidence interval estimates constructed using the newly tuned Horvitz and Thompson (1952) estimator of the population total is estimated by observing the proportion of times the true population total Y falls within the interval estimate given by

$$\hat{Y}_{\text{HTuned(new)}} \mp t_{\alpha/2}(\text{df} = n-1)\sqrt{\hat{v}(\hat{Y}_{\text{HTuned(new)}})} \qquad (7.51)$$

We studied coverage by the nominal 90%, 95%, and 99% intervals by selecting 10,000 random samples from a population consisting of three related variables Y, X, and Z taking values y_i, x_i, and z_i for the ith unit in the population as

$$y_i = \bar{Y} + \sigma_y y_i^* \qquad (7.52)$$

$$x_i = \bar{X} + x_i^* \sigma_x \sqrt{1 - \rho_{xy}^2} + \rho_{xy} \sigma_x y_i^* \qquad (7.53)$$

and

$$z_i = \bar{Z} + z_i^* \sigma_z \sqrt{1 - \rho_{yz}^2} + \rho_{yz} \sigma_z y_i^* \qquad (7.54)$$

where $x_i^* \sim N(0, 1)$, $y_i^* \sim N(0, 1)$, and $z_i^* \sim N(0, 1)$ are three standard normal variables with mean 0 and standard deviation 1.

The value y_i of the variable Y can be imagined as the weight of the ith pumpkin, the value z_i of the variable Z can be imagined as the top size of the ith pumpkin measured using remote sensing images, and value x_i of the variable X can be imagined as the circumference of the ith pumpkin in the population. Again, for this illustration, we made the following choice of parameters: $\rho_{xy}=0.80$, $\rho_{yz}=0.70$, $\bar{X}=2320$, $\bar{Y}=2450$, $\bar{Z}=322$, $\sigma_x=50$, $\sigma_y=50$, $\sigma_z=52$, and $N=2000$. Thus, our problem is to estimate the population total $Y=N\bar{Y}$ of the study variable Y, using known population totals $X=N\bar{X}$ and $Z=N\bar{Z}$ of the two auxiliary variables X and Z, respectively. We used the variable Z at the selection stage such that $d_i = \pi_i^{-1}$, where

$$\pi_i = \left(\frac{N-n}{N-1}\right)p_i + \left(\frac{n-1}{N-1}\right) \tag{7.55}$$

with $p_i = z_i/(N\bar{Z})$ and $n=3$–31. This is an example of the Midzuno–Sen sampling scheme. The variable X is used at the tuning stage of the estimator of the population total of the study variable, Y. For each sample of size n, with n varying from 3 to 31, we generated 10,000 samples and determined the attained coverage by the interval estimators. The results obtained for the various sample sizes are shown in Table 7.2.

Table 7.2 shows the coverages of intervals constructed using the newly tuned estimator of the population total under the PPSWOR scheme. We used the Midzuno–Sen sampling scheme, but any sampling scheme available in the literature could be investigated in this way. Our overall observation about the proposed tuned estimator is that it shows good results for small sample sizes. In particular, the nominal 90% coverage is estimated as 89.84% for a sample of size 23, the nominal 95% coverage is estimated as 95.18% for a sample of size 23, and the nominal 99% coverage is estimated as 99.02% for a sample of size 21.

Table 7.2 Performance of the newly tuned Horvitz–Thompson interval estimator of the population total

Sample size (n)	90% coverage	95% coverage	99% coverage
3	0.4966	0.6170	0.8579
5	0.4442	0.5451	0.7606
7	0.4823	0.5822	0.7826
9	0.5730	0.6697	0.8427
11	0.6330	0.7350	0.8891
13	0.7005	0.7923	0.9194
15	0.7517	0.8421	0.9477
17	0.8045	0.8852	0.9711
19	0.8393	0.9118	0.9804
21	0.8834	0.9415	0.9902
23	0.8984	0.9518	0.9939
25	0.9257	0.9679	0.9967
27	0.9477	0.9814	0.9989
29	0.9626	0.9879	0.9989
31	0.9708	0.9916	0.9998

7.4.2 R code

The following R code, **PUMPKIN72.R**, was used to study the coverage by newly the tuned Horvitz and Thompson (1952) interval estimator based on optimizing the displacement function.

```
#PROGRAM PUMPKIN72.R
library("sampling")
set.seed(2013)
rhoxy<-0.8;rhoyz<-0.7;NP<-2000;
  xe<-rnorm(NP,0,1);ye<-rnorm(NP,0,1);ze<-rnorm(NP,0,1);
  y<-2450 + 50*ye
  x<-2320 + 50*sqrt(1-rhoxy^2)*xe + rhoxy*50*ye
  z<-322 + 52*sqrt(1-rhoyz^2)*ze + rhoyz*52*ye
XPMEAN<-mean(x);ZPMEAN<-mean(z); TY<-sum(y)
cor(x,y);cor(z,y);sqrt(var(x));sqrt(var(y));sqrt(var(z))
nreps<-10000
ESTP=rep(0,nreps)
ci1.max=ci1.min=ci2.max=ci2.min=ci3.max=ci3.min=ESTP
vESTP=ESTP
for (n in seq(3,31,2) )
 {
 inclusionprobabilities(z,n)->pik
 for (r in 1:nreps)
  {
  ese<-UPmidzuno(pik)
  (1:length(pik))[ese==1]->us
  #us<-sample(NP,n)
  xs<-x[us]; ys<-y[us]; zs<-z[us]
  xmj<-(sum(xs)-xs)/(n-1)
  ymj<-(sum(ys)-ys)/(n-1)
  p<-zs/(NP*ZPMEAN)
  pi<-(NP-n)*p/(NP-1) + (n-1)/(NP-1); di<-1/pi
  Fi<-1-((NP-1)/NP)*(pi-(n-1)/(NP-1))/p
  YHT<-sum(di*ys); XHT<-sum(di*xs)
  DB<-mean(di); dj<-(sum(di)-di)/(n-1)
  shi<-(NP*XPMEAN-(2-n)*(n^2)*DB*mean(xs))/(n*DB*((n-1)^2))
  shi<-xmj-shi
  wbni<-dj/(1+(sum(dj*shi)/sum(dj*shi^2))*shi)
  ESTi<- ((n-1)^2*wbni-n*DB*(n-2))*ymj
  ESTP[r]<- sum(ESTi)
  EST_I<-(n*ESTP[r] - ESTi)/(n-1)
  vj<-(di*Fi/wbni)^2*(EST_I-ESTP[r])^2
  vESTP[r]<-sum(vj)
  ci1.max[r]<- ESTP[r]-qt(.05,n-1)*sqrt(vESTP[r])
  ci1.min[r]<- ESTP[r]+qt(.05,n-1)*sqrt(vESTP[r])
```

```
ci2.max[r]<- ESTP[r]-qt(.025,n-1)*sqrt(vESTP[r])
ci2.min[r]<- ESTP[r]+qt(.025,n-1)*sqrt(vESTP[r])
ci3.max[r]<- ESTP[r]-qt(.005,n-1)*sqrt(vESTP[r])
ci3.min[r]<- ESTP[r]+qt(.005,n-1)*sqrt(vESTP[r])
}# nreps
round(sum(ci1.min<TY & ci1.max>TY)/nreps,4)->cov1
round(sum(ci2.min<TY & ci2.max>TY)/nreps,4)->cov2
round(sum(ci3.min<TY & ci3.max>TY)/nreps,4)->cov3
cat(n, cov1,cov2,cov3,'\n')
}# n
```

7.4.3 Numerical illustration

In the following example, we explain the various steps involved in tuning the Horvitz and Thompson (1952) type estimator using the displacement function.

Example 7.2 Using the data in Example 7.1, apply the newly tuned Horvitz and Thompson (1952) estimator $\hat{Y}_{\text{HTuned(new)}}$ to estimate the total weight Y of the pumpkins in the field and construct the 95% confidence interval estimate.

Solution. One can easily compute the following:

$\bar{x}_n(j)$	$\bar{y}_n(j)$	$\bar{w}_n^\circ(j)$	$\hat{Y}_{\text{HTuned(new)}(j)}$
104.1500	2387.167	27.84847	477663.6
114.8000	2387.167	27.99474	475568.6
108.2167	2006.500	27.71860	480948.2
109.6333	2346.833	27.53564	482250.9
106.7667	1770.500	28.00524	478786.3
103.1167	2103.833	27.86208	478772.8
109.1167	2121.000	28.11904	475423.7

where

$$\hat{Y}_{\text{HTuned(new)}(j)} = \frac{n\hat{Y}_{\text{HTuned(new)}} - \left\{(n-1)^2 \bar{w}_n^\circ(j) - n(n-2)\bar{d}\right\}\bar{y}_n(j)}{n-1}$$

From these we compute

$$\hat{Y}_{\text{HTuned(new)}} = \frac{1}{n}\sum_{j \in s} \hat{Y}_{\text{HTuned(new)}(j)} = 418676.8$$

and

$$SE(\hat{Y}_{\text{HTuned(new)}}) = \sqrt{\hat{v}(\hat{Y}_{\text{HTuned(new)}})} = 5572.19$$

Thus, the 95% confidence interval estimate of the total weight of all pumpkins in the field is 405042.1–432311.4 lbs.

7.4.4 R code used for illustration

We used the following R code, **PUMPKIN72EX.R**, to solve the preceding illustration.

#PROGRAM PUMPKIN72EX.R
```
n<-7
xs<-c(130.9,67,106.5,98,115.2,137.1,101.1)
ys<-c(800,800,3084,1042,4500,2500,2397)
zs<-c(105,30,56,30,97,117,87)
XPMEAN<-105.4;ZPMEAN<-61; RHOYZP<-0.4
NP<-200
xmj<-(sum(xs)-xs)/(n-1)
ymj<-(sum(ys)-ys)/(n-1)
p<-zs/(NP*ZPMEAN)
pi<-(NP-n)*p/(NP-1) + (n-1)/(NP-1); di<-1/pi
Fi<-1-((NP-1)/NP)*(pi-(n-1)/(NP-1))/p
YHT<-sum(di*ys); XHT<-sum(di*xs)
DB<-mean(di)
dj<-(sum(di)-di)/(n-1)
shi<-xmj-(NP*XPMEAN-(2-n)*(n^2)*DB* mean(xs)) /(n*DB*((n-1)^2))
wbni<-dj/(1+(sum(dj*shi)/sum(dj*shi^2))*shi)
ESTi<- ((n-1)^2*wbni-n*DB*(n-2))*ymj
ESTP<- sum(ESTi)
EST_I<-(n*ESTP - ESTi)/(n-1)
vj<-(di*Fi/wbni)^2*(EST_I-ESTP)^2
vESTP<-sum(vj)
L<-ESTP+qt(.025,n-1)*sqrt(vESTP)
U<-ESTP+qt(.975,n-1)*sqrt(vESTP)
cbind(xmj,ymj,wbni,EST_I)
cat("Tuned estimate:", ESTP, "SE: ",vESTP^.5 ,'\n')
cat("Confidence Interval:"," ", L,"; ", U,'\n')
```

7.5 Exercises

Exercise 7.1 Consider the problem of estimating the population total Y with an estimator defined by

$$\hat{Y}_{\text{HTuned}(1)} = \frac{1}{(n-1)} \sum_{j \in s} \bar{w}_n^{\Diamond}(j)[\bar{d}(j)]^{-1} \hat{Y}_{\text{HT}}(j) \qquad (7.56)$$

where

$$\hat{Y}_{\text{HT}}(j) = \hat{Y}_{\text{HT}} - d_j y_j \qquad (7.57)$$

denotes the jth jackknifed Horvitz and Thompson (1952) estimator of the population total and $\bar{w}_n^\diamond(j)$ are the tuned weights constructed so that the following two constraints are satisfied:

$$\sum_{j \in s} \bar{w}_n^\diamond(j) = \sum_{j \in s} \bar{d}(j) \tag{7.58}$$

and

$$(n-1)^{-1} \sum_{j \in s} \bar{w}_n^\diamond(j) [\bar{d}(j)]^{-1} \hat{X}_{HT}(j) = X \tag{7.59}$$

where $\hat{X}_{HT}(j) = \hat{X}_{HT} - d_j x_j$ has its usual meaning.

By optimizing the distance function

$$\frac{1}{2} \sum_{j \in s} \frac{\{\bar{w}_n^\diamond(j) - \bar{d}(j)\}^2}{q_j \bar{d}(j)} \tag{7.60}$$

Show that the tuned estimator of the population total is given by

$$\hat{Y}_{HTuned(1)} = \hat{Y}_{HT} + \hat{b}(X + \hat{X}_{HT}) \tag{7.61}$$

where $\hat{b}$ is given by

$$b = \frac{\left(\sum_{j \in s} q_j \bar{d}(j)\right) \sum_{j \in s} q_j [\bar{d}(j)]^{-1} \hat{X}_{HT}(j) \hat{Y}_{HT}(j) - \left(\sum_{j \in s} q_j \hat{Y}_{HT}(j)\right) \left(\sum_{j \in s} q_j \hat{X}_{HT}(j)\right)}{\left(\sum_{j \in s} q_j \bar{d}(j)\right) \left(\sum_{j \in s} q_j (\bar{d}(j))^{-1} (\hat{X}_{HT}(j))^2\right) - \left(\sum_{j \in s} q_j \hat{X}_{HT}(j)\right)^2} \tag{7.62}$$

Develop a doubly jackknifed estimator of variance to estimate the variance of the tuned estimator of population total given in Equation (7.56). Write R code to study the coverage by the 90%, 95%, and 99% confidence intervals of this estimator suggested to you using the Midzuno–Sen sampling scheme.

Exercise 7.2 Consider the problem of estimating the population total Y with an estimator defined by

$$\hat{Y}_{HTuned(2)} = \frac{1}{(n-1)} \sum_{j \in s} \bar{w}_n^\diamond(j) [\bar{d}(j)]^{-1} \hat{Y}_{HT}(j) \tag{7.63}$$

where

$$\hat{Y}_{HT}(j) = \hat{Y}_{HT} - d_j y_j \tag{7.64}$$

denotes the jth jackknifed Horvitz and Thompson (1952) estimator of the population total and $\bar{w}_n^{\diamond}(j)$ are the tuned weights constructed so that the following two constraints are satisfied:

$$\sum_{j \in s} \bar{w}_n^{\diamond}(j) = \sum_{j \in s} \bar{d}(j) \tag{7.65}$$

and

$$(n-1)^{-1} \sum_{j \in s} \bar{w}_n^{\diamond}(j)[\bar{d}(j)]^{-1} \hat{X}_{HT}(j) = X \tag{7.66}$$

where $\hat{X}_{HT}(j) = \hat{X}_{HT} - d_j x_j$ has its usual meaning. Optimize each one of the following distance/displacement functions:

$$D_1 = \frac{1}{2} \sum_{j \in s} \frac{\{\bar{w}_n^{\diamond}(j) - \bar{d}(j)\}^2}{q_j \bar{d}(j)} \tag{7.67}$$

$$D_2 = \sum_{j \in s} \frac{(\bar{w}_n^{\diamond}(j) - \bar{d}(j))^2}{2 \bar{w}_n^{\diamond}(j)}, \quad \bar{w}_n^{\diamond}(j) > 0 \tag{7.68}$$

$$D_3 = \sum_{j \in s} \bar{d}(j) \tanh^{-1}\left(\frac{\{\bar{w}_n^{\diamond}(j)\}^2 - 1}{\{\bar{w}_n^{\diamond}(j)\}^2 + 1}\right) \tag{7.69}$$

and

$$D_4 = \frac{1}{2} \sum_{j \in s} \frac{(\bar{w}_n^{\diamond}(j) - \bar{d}(j))^2}{q_j \bar{d}(j)} + \frac{1}{2} \sum_{j \in s} \frac{\Phi_j \{\bar{w}_n^{\diamond}(j)\}^2}{q_j \bar{d}(j)} \tag{7.70}$$

where q_j are suitably chosen weights that form different types of estimators, Φ_j is a penalty, and $\tanh^{-1}()$ is the hyperbolic tangent function, subject to the two tuning constraints (7.65) and (7.66). Write code in any scientific programming language to study these distance functions. Discuss the nature of the tuned weights. Construct the 90%, 95%, and 99% confidence interval estimates in each situation by estimating the variance using the method discussed here.

Exercise 7.3 Consider the problem of estimating the population total Y with an estimator defined by

$$\hat{Y}_{HTuned(cs)} = \sum_{j \in s} \left[(n-1)^2 \bar{w}_n^{\diamond}(j) - n(n-2)\bar{d}\right] \bar{y}_n(j) \tag{7.71}$$

where

$$\bar{y}_n(j) = \frac{n \bar{y}_n - y_j}{n-1} \tag{7.72}$$

with $\bar{y}_n = \frac{1}{n}\sum_{j\in s} y_j$, and $\bar{w}_n^\diamond(j)$ are the jackknife weights constructed so that

$$\sum_{j\in s} \bar{w}_n^\diamond(j) = \sum_{j\in s} \bar{d}(j) \qquad (7.73)$$

and

$$\sum_{j\in s} \bar{w}_n^\diamond(j)\bar{x}_n(j) = \frac{X-(2-n)n^2\bar{d}\bar{x}_n}{(n-1)^2} \qquad (7.74)$$

Optimize the penalized chi-square distance between the jackknifed calibrated weights $\bar{w}_n^\diamond(j)$ and the jackknifed design weights $\bar{d}(j)$ as given by

$$D = \frac{(n-1)^2}{2}\sum_{j\in s}\frac{(\bar{d}(j)-\bar{w}_n^\diamond(j))^2}{d_j q_j} + \sum_{j\in s}\frac{\{n\bar{d}-(n-1)\bar{w}_n^\diamond(j)\}^2 \Phi_j}{d_j q_j} \qquad (7.75)$$

where q_j are suitably chosen weights that form different types of estimators and Φ_j is a penalty, subject to the two tuning constraints (7.73) and (7.74). Write code in any scientific programming language to study these distance functions. Examine 90%, 95%, and 99% confidence interval estimates in each situation by estimating the variance using the method discussed here.

Exercise 7.4 Consider the problem of estimating the population total Y with an estimator defined by

$$\hat{Y}_{\text{HTuned(cs)}} = \sum_{j\in s}\left[(n-1)^2\bar{w}_n^\diamond(j) - n(n-2)\bar{d}\right]\bar{y}_n(j) \qquad (7.76)$$

where

$$\bar{y}_n(j) = \frac{n\bar{y}_n - y_j}{n-1} \qquad (7.77)$$

with $\bar{y}_n = \frac{1}{n}\sum_{j\in s} y_j$, and $\bar{w}_n^\diamond(j)$ are the jackknife weights so that

$$\sum_{j\in s} \bar{w}_n^\diamond(j) = \sum_{j\in s} \bar{d}(j) \qquad (7.78)$$

and

$$\sum_{j\in s} \bar{w}_n^\diamond(j)\{\hat{G}_x(j)\}^{(1-n)} = \frac{1}{(n-1)}\left[n\bar{d}\sum_{j\in s}\{\hat{G}_x(j)\}^{(1-n)} - X(\hat{G}_x)^{-n}\right] \qquad (7.79)$$

where $\hat{G}_x(j) = \left(\prod_{\substack{i=1 \\ i \neq j}}^{n} x_i\right)^{1/(n-1)}$ is the jth jackknifed sample geometric mean

$\hat{G}_x = \left(\prod_{i=1}^{n} x_i\right)^{1/n}$, $X = \sum_{i=1}^{N} x_i$ is the known population total of the auxiliary variable, and $\bar{d}(j) = (n\bar{d} - d_j)/(n-1)$ are the jackknifed design weights. Subject to the preceding two constraints, optimize the weights $\bar{w}_n^{\diamond}(j)$ in the following distance function

$$D = \frac{(n-1)^2}{2} \sum_{j \in s} \frac{\left(\bar{d}(j) - \bar{w}_n^{\diamond}(j)\right)^2}{d_j q_j} \tag{7.80}$$

Develop a doubly jackknifed estimator of variance for the tuned estimator of population total given in Equation (7.76). Write R code to study the coverage by 90%, 95%, and 99% confidence intervals of the estimator suggested to you using the Midzuno–Sen sampling scheme.

Exercise 7.5 In Exercise 7.4, replace constraint (7.79) with

$$\sum_{j \in s} \frac{\bar{w}_n^{\diamond}(j) \hat{H}_x(j)}{n\hat{H}_x(j) - (n-1)\hat{H}_x} = \frac{1}{(n-1)} \left[\sum_{j \in s} \frac{n\bar{d}\,\hat{H}_x(j)}{n\hat{H}_x(j) - (n-1)\hat{H}_x} - \frac{X}{\hat{H}_x}\right] \tag{7.81}$$

where $\hat{H}_x(j) = (n-1)\left(\sum_{\substack{i=1 \\ i \neq j}}^{n} x_i^{-1}\right)^{-1}$ is the jth jackknifed sample harmonic mean, $\hat{H}_x = n\left(\sum_{i=1}^{n} x_i^{-1}\right)^{-1}$ and $X = \sum_{i=1}^{N} x_i$ is the known population total of the auxiliary variable.

Tuning in stratified sampling

8.1 Introduction

In this chapter, we extend the new methodology of tuning the jackknife technique in survey sampling to the case of stratified random sampling. This will provide new estimators of the population mean and will provide estimates of variance of these estimators. The new methodology is illustrated with a model used to estimate the weight of pumpkins with the help of known circumferences. Finally, an extension of the work to the case of multistage sampling is suggested in the exercises.

8.2 Stratification

Stratification means the population of N units is first divided into homogeneous and mutually exclusive groups called strata. Then an independent random sample of the required size is selected from each stratum. In short, the hth stratum in stratified sampling design consists of N_h units, where $h = 1, 2, \ldots, L$ so that

$$\sum_{h=1}^{L} N_h = N \tag{8.1}$$

Let y_{hi} be the value of the study variable for the ith unit in the hth stratum, $i = 1, 2, \ldots, N_h$, so the hth stratum population mean is given by

$$\bar{Y}_h = N_h^{-1} \sum_{i=1}^{N_h} y_{hi}, \quad \text{for } h = 1, 2, \ldots, L \tag{8.2}$$

Obviously, using the concept of weighted average, the true mean of the whole population can be written as

$$\begin{aligned}
\bar{Y} &= \frac{N_1 \bar{Y}_1 + N_2 \bar{Y}_2 + \cdots + N_L \bar{Y}_L}{N_1 + N_2 + \cdots + N_L} \\
&= \frac{N_1 \bar{Y}_1 + N_2 \bar{Y}_2 + \cdots + N_L \bar{Y}_L}{N} \\
&= \left(\frac{N_1}{N}\right) \bar{Y}_1 + \left(\frac{N_2}{N}\right) \bar{Y}_2 + \cdots + \left(\frac{N_L}{N}\right) \bar{Y}_L \\
&= \Omega_1 \bar{Y}_1 + \Omega_2 \bar{Y}_2 + \cdots + \Omega_L \bar{Y}_L \\
&= \sum_{h=1}^{L} \Omega_h \bar{Y}_h
\end{aligned} \tag{8.3}$$

A New Concept for Tuning Design Weights in Survey Sampling. http://dx.doi.org/10.1016/B978-0-08-100594-1.00008-5
Copyright © 2016 Elsevier Ltd. All rights reserved.

where

$$\Omega_h = \frac{N_h}{N} \qquad (8.4)$$

is the known proportion of population units falling within the hth stratum. Consider drawing a sample s_h of size n_h using a simple random sampling (SRS) scheme from the hth stratum consisting of N_h units, such that

$$\sum_{h=1}^{L} n_h = n \qquad (8.5)$$

is the required sample size.

Assume the value of the ith unit of the study variable selected from the hth stratum is denoted by y_{hi} and $i = 1, 2, ..., n_h$. An unbiased estimator of population mean $\bar{Y}$ is given by

$$\bar{y}_{st} = \sum_{h=1}^{L} \Omega_h \bar{y}_h \qquad (8.6)$$

where

$$\bar{y}_h = n_h^{-1} \sum_{i=1}^{n_h} y_{hi} \qquad (8.7)$$

denotes the hth stratum sample mean.

Also assume the value of the auxiliary variable for the ith unit selected from the hth stratum is denoted by x_{hi}, where $i = 1, 2, ..., n_h$. An unbiased estimator of the population mean

$$\bar{X} = \sum_{h=1}^{L} \Omega_h \bar{X}_h \qquad (8.8)$$

is given by

$$\bar{x}_{st} = \sum_{h=1}^{L} \Omega_h \bar{x}_h \qquad (8.9)$$

where

$$\bar{x}_h = n_h^{-1} \sum_{i=1}^{n_h} x_{hi} \qquad (8.10)$$

denotes the hth stratum sample mean, and

$$\bar{X}_h = N_h^{-1} \sum_{i=1}^{N_h} x_{hi} \qquad (8.11)$$

denotes the known hth stratum population mean of the auxiliary variable. For more information about stratified random sampling, one could refer to Neyman (1934).

8.3 Tuning with a chi-square distance function using stratum-level known population means of an auxiliary variable

The newly tuned estimator of the population mean $\bar{Y}$ for stratified random sampling is defined as

$$\bar{y}_{\text{StTuned(cs)}} = \sum_{h=1}^{L} \Omega_h \sum_{j \in s_h} \left[(n_h - 1)^2 \bar{w}_h(j) - (n_h - 2) \right] \bar{y}_h(j) \qquad (8.12)$$

where $\bar{w}_h(j)$ is the tuned stratum weight such that in each stratum the following two constraints are satisfied:

$$\sum_{j \in s_h} \bar{w}_h(j) = 1 \qquad (8.13)$$

and

$$\sum_{j \in s_h} \bar{w}_h(j) \bar{x}_h(j) = \frac{\bar{X}_h - n_h(2 - n_h) \bar{x}_h}{(n_h - 1)^2} \qquad (8.14)$$

where

$$\bar{y}_h(j) = \frac{n_h \bar{y}_h - y_{hj}}{n_h - 1} \qquad (8.15)$$

is the hth stratum sample mean of the study variable obtained by removing the jth unit from the sample s_h,

$$\bar{x}_h(j) = \frac{n_h \bar{x}_h - x_{hj}}{n_h - 1} \qquad (8.16)$$

is the hth stratum sample mean of the auxiliary variable obtained by removing the jth unit from the sample s_h, and

$$\bar{w}_h(j) = \frac{1 - w_{hj}}{n_h - 1} \tag{8.17}$$

is the tuned jackknife weight of the calibrated weights w_{hj} constructed so that in each stratum

$$\sum_{j \in s_h} w_{hj} = 1 \tag{8.18}$$

and

$$\sum_{j \in s_h} w_{hj} x_{hj} = \bar{X}_h \tag{8.19}$$

We suggest the tuning of the weights w_{hj} such that in the hth stratum, the chi-square type distance function, defined as

$$(2^{-1} n_h) \sum_{j \in s_h} q_{hj}^{-1} \left(w_{hj} - n_h^{-1} \right)^2 = (2^{-1} n_h) \sum_{j \in s_h} q_{hj}^{-1} \left\{ 1 - (n_n - 1) \bar{w}_h(j) - n_h^{-1} \right\}^2 \tag{8.20}$$

is optimal, subject to the tuning constraints (8.13) and (8.14), and q_{hj} is a choice of weights in the hth stratum.

The hth stratum Lagrange function is then given by

$$L_{h1} = (2^{-1} n_h) \sum_{j \in s_h} q_{hj}^{-1} \left\{ 1 - (n_h - 1) \bar{w}_h(j) - n_h^{-1} \right\}^2 - \lambda_{h0} \left\{ \sum_{j \in s_h} \bar{w}_h(j) - 1 \right\}$$

$$- \lambda_{h1} \left\{ \sum_{j \in s_h} \bar{w}_h(j) \bar{x}_h(j) - \frac{\bar{X}_h - n_h(2 - n_h) \bar{x}_h}{(n_h - 1)^2} \right\} \tag{8.21}$$

where λ_{h0} and λ_{h1} are Lagrange multiplier constants.

On setting:

$$\frac{\partial L_{h1}}{\partial \bar{w}_h(j)} = 0 \tag{8.22}$$

we have

$$\bar{w}_h(j) = \frac{1}{n_h} \left\{ 1 + \frac{1}{(n_h - 1)^2} \left(q_{hj} \lambda_{h0} + \lambda_{h1} q_{hj} \bar{x}_h(j) \right) \right\} \tag{8.23}$$

Using Equation (8.23) in Equations (8.13) and (8.14), a set of normal equations to find the optimal values of λ_{h0} and λ_{h1} is found to be

$$\begin{bmatrix} \sum_{j \in s_h} q_{hj}, & \sum_{j \in s_h} q_{hj}\bar{x}_h(j) \\ \sum_{j \in s_h} q_{hj}\bar{x}_h(j), & \sum_{j \in s_h} q_{hj}\{\bar{x}_h(j)\}^2 \end{bmatrix} \begin{bmatrix} \lambda_{h0} \\ \lambda_{h1} \end{bmatrix}$$

$$= \begin{bmatrix} 0 \\ (n_h-1)^2 \left\{ \dfrac{n_h(\bar{X}_h - n_h(2-n_h)\bar{x}_h)}{(n_h-1)^2} - \sum_{j \in s_h} \bar{x}_h(j) \right\} \end{bmatrix} \quad (8.24)$$

Note that

$$(n_h-1)^2 \left\{ \dfrac{n_h(\bar{X}_h - n_h(2-n_h)\bar{x}_h)}{(n_h-1)^2} - \sum_{j \in s_h} \bar{x}_h(j) \right\}$$

$$= (n_h-1)^2 \left\{ \dfrac{n_h(\bar{X}_h - n_h(2-n_h)\bar{x}_h) - (n_h-1)^2 n_h \bar{x}_h}{(n_h-1)^2} \right\}$$

$$= n_h \left[(\bar{X}_h - n_h(2-n_h)\bar{x}_h) - (n_h-1)^2 \bar{x}_h \right] \quad (8.25)$$

$$= n_h \left[\bar{X}_h - 2n_h\bar{x}_h + n_h^2\bar{x}_h - (n_h^2 + 1 - 2n_h)\bar{x}_h \right]$$

$$= n_h \left[\bar{X}_h - 2n_h\bar{x}_h + n_h^2\bar{x}_h - n_h^2\bar{x}_h - \bar{x}_h + 2n_h\bar{x}_h \right]$$

$$= n_h [\bar{X}_h - \bar{x}_h]$$

Thus, we have

$$\begin{bmatrix} \sum_{j \in s_h} q_{hj}, & \sum_{j \in s_h} q_{hj}\bar{x}_h(j) \\ \sum_{j \in s_h} q_{hj}\bar{x}_h(j), & \sum_{j \in s_h} q_{hj}\{\bar{x}_h(j)\}^2 \end{bmatrix} \begin{bmatrix} \lambda_{h0} \\ \lambda_{h1} \end{bmatrix} = \begin{bmatrix} 0 \\ n_h(\bar{X}_h - \bar{x}_h) \end{bmatrix} \quad (8.26)$$

The jackknifed tuned weight $\bar{w}_h(j)$ then becomes:

$$\bar{w}_h(j) = \dfrac{1}{n_h} \left[1 + \dfrac{\Delta_{hj}}{(n_h-1)^2} \{\bar{X}_h - \bar{x}_h\} \right] \quad (8.27)$$

with

$$\Delta_{hj} = q_{hj} \left\{ \frac{\bar{x}_h(j)\left(\sum_{j\in s_h} q_{hj}\right) - \sum_{j\in s_h} q_{hj}\bar{x}_h(j)}{\left(\sum_{j\in s_h} q_{hj}\right)\left\{\sum_{j\in s_h} q_{hj}(\bar{x}_h(j))^2\right\} - \left\{\sum_{j\in s_h} q_{hj}\bar{x}_h(j)\right\}^2} \right\} \quad (8.28)$$

Thus, under the chi-square (cs) type distance function, the newly tuned jackknife estimator (8.12) of the population mean $\bar{Y}$, for the stratified random sampling becomes

$$\begin{aligned}\bar{y}_{\text{StTuned(cs)}} &= \sum_{h=1}^{L} \frac{\Omega_h}{n_h} \left[\sum_{j\in s_h} \bar{y}_h(j) + \hat{\beta}_h \left\{ n_h \bar{X}_h - \sum_{j\in s_h} \bar{x}_h(j) \right\} \right] \\ &= \sum_{h=1}^{L} \Omega_h \left[\bar{y}_h + \hat{\beta}_h \{\bar{X}_h - \bar{x}_h\} \right] \end{aligned} \quad (8.29)$$

where

$$\hat{\beta}_h = \left[\frac{\left(\sum_{j\in s_h} q_{hj}\right)\left(\sum_{j\in s_h} q_{hj}\bar{x}_h(j)\bar{y}_h(j)\right) - \left(\sum_{j\in s_h} q_{hj}\bar{y}_h(j)\right)\left(\sum_{j\in s_h} q_{hj}\bar{x}_h(j)\right)}{\left(\sum_{j\in s_h} q_{hj}\right)\left(\sum_{j\in s_h} q_{hj}(\bar{x}_h(j))^2\right) - \left(\sum_{j\in s_h} q_{hj}\bar{x}_h(j)\right)^2} \right] \quad (8.30)$$

8.3.1 Estimation of variance and coverage

We suggest the following adjusted estimator of the variance of the estimator $\bar{y}_{\text{StTuned(cs)}}$:

$$\hat{v}_{\text{StTuned(cs)}} = \sum_{h=1}^{L} \Omega_h^2 n_h (n_h - 1)^3 \sum_{j\in s_h} \{\bar{w}_h(j)\}^2 \left\{ \bar{y}_{\text{StTuned(cs)}(j)}^{(h)} - \bar{y}_{\text{StTuned(cs)}}^{(h)} \right\}^2 \quad (8.31)$$

Note that each newly tuned estimator of the hth stratum population mean is

$$\bar{y}_{\text{StTuned(cs)}(j)}^{(h)} = \left[\frac{n_h \bar{y}_{1h} - n_h \{(n_h-1)^2 \bar{w}_h(j) - (n_h-2)\} \bar{y}_h(j)}{n_h - 1} \right] \quad (8.32)$$

where

$$\bar{y}_{lh} = \sum_{j \in s_h} \left[(n_h - 1)^2 \bar{w}_h(j) - (n_h - 2)\right] \bar{y}_h(j) \tag{8.33}$$

for $h = 1, 2, \ldots, L$; $j = 1, 2, \ldots, n_h$.

Remember that the strata are mutually exclusive, and samples are drawn independently from each stratum, thus, an estimator of variance may be developed by taking the sum, over all strata, of the estimators of variance in each stratum. Attained coverage by the $(1 - \alpha)100\%$ confidence interval estimates obtained from this newly tuned estimator of the population mean is estimated by observing the proportion of times the true population mean $\bar{Y}$ falls within the interval estimates given by

$$\bar{y}_{\text{StTuned(cs)}} \pm t_{\alpha/2}(\text{df} = n - L) \sqrt{\hat{v}_{\text{StTuned(cs)}}} \tag{8.34}$$

We use degree of freedom df $= n - L$ because one population mean is being estimated in each stratum. Here we considered the problem of estimating the average weight of three types of pumpkins, namely *Sumbo Pumpkins* of small size, *Mumbo Pumpkins* of medium size, and *Jumbo Pumpkins* of large size, the size of the pumpkins being based on their circumferences. We generated six variables, two variables in each stratum, with different means, different standard deviations, and different correlations between weight and circumference as follows:

$$x_{hi} = \bar{X}_h + \sigma_{hx} x_{hi}^* \tag{8.35}$$

and

$$y_{hi} = \bar{Y}_h + \sigma_{hy} y_{hi}^* \sqrt{1 - \rho_{hxy}^2} + \sigma_{hy} \rho_{hxy} x_{hi}^* \tag{8.36}$$

where $x_{hi}^* \sim N(0,1)$ and $y_{hi}^* \sim N(0,1)$ for $h = 1, 2, 3$.

We studied the attained coverages by the nominally 90%, 95%, and 99% confidence intervals, based on the newly tuned estimator of the population mean, by selecting 10,000 random samples from each of the populations consisting of the three types of pumpkins. We set the values of the correlation coefficients as $\rho_{1xy} = 0.6$, $\rho_{2xy} = 0.7$, and $\rho_{3xy} = 0.4$. The known average values of the circumference (inches) in the three strata were taken to be $\bar{X}_1 = 20$, $\bar{X}_2 = 300$, and $\bar{X}_3 = 1400$, respectively. We assumed the average weights (lbs) of the pumpkins in the three strata were $\bar{Y}_1 = 25$, $\bar{Y}_2 = 400$, and $\bar{Y}_3 = 8000$. We also set $\sigma_{1x} = 12$, $\sigma_{2x} = 280$, $\sigma_{3x} = 329$, $\sigma_{1y} = 15$, $\sigma_{2y} = 200$, and $\sigma_{3y} = 6000$. For various small sample sizes n_1, n_2, and n_3 are taken from the three independent strata consisting of $N_1 = 800$ *Sumbo*, $N_2 = 2000$ *Mumbo*, and $N_3 = 6000$ *Jumbo* pumpkins, we found the attained coverages shown in Table 8.1.

Table 8.1 Attained coverage of the newly tuned jackknife estimator

n_1	n_2	n_3	Expected coverage			n_1	n_2	n_3	Expected coverage		
			90%	95%	99%				90%	95%	99%
2	2	2	0.9190	0.9416	0.9707	4	2	2	0.9035	0.9244	0.9553
		3	0.9700	0.9810	0.9931			3	0.9655	0.9759	0.9880
		4	0.9850	0.9910	0.9964			4	0.9837	0.9902	0.9972
		5	0.9922	0.9965	0.9992			5	0.9925	0.9959	0.9988
	3	2	0.9004	0.9236	0.9575		3	2	0.8839	0.9102	0.9408
		3	0.9609	0.9739	0.9890			3	0.9582	0.9718	0.9870
		4	0.9812	0.9898	0.9966			4	0.9831	0.9889	0.9952
		5	0.9907	0.9952	0.9979			5	0.9905	0.9939	0.9979
	4	2	0.8885	0.9105	0.9444		4	2	0.8845	0.9056	0.9376
		3	0.9574	0.9712	0.9875			3	0.9569	0.9698	0.9850
		4	0.9829	0.9897	0.9962			4	0.9816	0.9891	0.9948
		5	0.9921	0.9951	0.9982			5	0.9881	0.9924	0.9966
	5	2	0.8888	0.9128	0.9487		5	2	0.8888	0.9123	0.9431
		3	0.9590	0.9724	0.9863			3	0.9579	0.9714	0.9844
		4	0.9808	0.9875	0.9952			4	0.9803	0.9874	0.9952
		5	0.9914	0.9943	0.9978			5	0.9889	0.9937	0.9977
3	2	2	0.9062	0.9295	0.9580	5	2	2	0.8986	0.9203	0.9507
		3	0.9660	0.9773	0.9904			3	0.9619	0.9747	0.9887
		4	0.9848	0.9913	0.9963			4	0.9822	0.9884	0.9958
		5	0.9921	0.9958	0.9988			5	0.9899	0.9939	0.9983
	3	2	0.8844	0.9133	0.9482		3	2	0.8800	0.9069	0.9404
		3	0.9582	0.9711	0.9858			3	0.9569	0.9709	0.9852
		4	0.9844	0.9907	0.9965			4	0.9817	0.9886	0.9944
		5	0.9910	0.9950	0.9986			5	0.9903	0.9938	0.9981

Continued

4	2	0.8903	0.9140	0.9445	4	2	0.8773	0.9004	0.9347
	3	0.9596	0.9723	0.9858		3	0.9565	0.9693	0.9841
	4	0.9808	0.9875	0.9946		4	0.9821	0.9893	0.9959
	5	0.9902	0.9942	0.9977		5	0.9902	0.9937	0.9977
5	2	0.8809	0.9055	0.9441	5	2	0.8791	0.9033	0.9364
	3	0.9546	0.9671	0.9827		3	0.9517	0.9650	0.9826
	4	0.9818	0.9891	0.9951		4	0.9774	0.9844	0.9936
	5	0.9907	0.9947	0.9977		5	0.9889	0.9942	0.9980

Table 8.1 shows that the attained coverage by the newly tuned stratified interval estimator of the population mean remains quite appreciable for small sample sizes. In particular, note that the attained coverage by the 90% interval is estimated to be 90.04% for a sample of two *Sumbo*, three *Mumbo*, and two *Jumbo* pumpkins, the attained coverage by the 95% interval is estimated to be 94.16% for a sample of two *Sumbo*, two *Mumbo*, and two *Jumbo* pumpkins, and attained coverage for the 99% interval is estimated to be 98.98% for a sample of two *Sumbo*, three *Mumbo*, and four *Jumbo* pumpkins. Thus, the newly tuned estimator of the population mean weight of the pumpkins shows quite good coverage in the case of small sample sizes.

8.3.2 R code

The following R code, **PUMPKIN81.R**, was used to study the coverage by the newly tuned stratified estimator based on the chi-square type distance function.

```
#PROGRAM PUMPKIN81.R
set.seed(2013)
rnorm(800,0,1)->x1s
rnorm(2000,0,1)->x2s
rnorm(6000,0,1)->x3s
x1<-20 +12*x1s; x2<-300+280*x2s; x3<-1400+329*x3s
y1<-25 + 15* rnorm(800,0,1)*sqrt(1-0.6^2) + 15*0.6*x1s
y2<-400 + 200* rnorm(2000,0,1)*sqrt(1-0.7^2) + 200*0.7*x2s
y3<-8000+6000* rnorm(6000,0,1)*sqrt(1-0.4^2) + 6000*0.4*x3s
Nh<-c(800,2000,6000);N<-sum(Nh)
Wh<-Nh/N
xmean<- c(mean(x1),mean(x2),mean(x3))
xmh<-rep(0,3);deltah=sest=esh=varh=ymh=xmh
ymean<-mean(c(y1,y2,y3))
nreps<-10000
ESTP=rep(0,nreps)
ci1.max=ci1.min=ci2.max=ci2.min=ci3.max=ci3.min=vESTP=ESTP
for (i1 in 2:5)
{for (i2 in 2:5)
  {for (i3 in 2:5)
    {nh<-c(i1,i2,i3);n<-sum(nh)
        x<-matrix(ncol=2,nrow=n)
        y<-matrix(ncol=2,nrow=n)
        rep((1:3),nh)->x[,2]
        y[,2]=x[,2]
        for (r in 1:nreps) {
           s1<-sample(800,i1)
           s2<-sample(2000,i2)
           s3<-sample(6000,i3)
           x[,1]<-c(x1[s1],x2[s2],x3[s3])
```

```
                y[,1]<-c(y1[s1],y2[s2],y3[s3])
                for(h in 1:3)
                  {
                  xmj<-rep(0,nh[h])
                  ymj=di=dif=wi=xmj
                  x[(x[,2]==h),1]->xsh; y[(y[,2]==h),1]->ysh
                  xmj<-(sum(xsh)-xsh)/(nh[h]-1)
                  ymj<-(sum(ysh)-ysh)/(nh[h]-1)
                  delta<-nh[h]*sum(xmj^2) - sum(xmj)^2
                  di<-(nh[h]*xmj-sum(xmj))/delta
                  dif <- nh[h]*(xmean[h]-mean(xsh))/((nh[h]-1)^2)
                  wi <- (1+di*dif )/nh[h]
                  est_i = nh[h]*(((nh[h]-1)^2)*wi-(nh[h]-2))*ymj
                  sest = sum(est_i)
                  sest_j = (sest-est_i)/(nh[h]-1)
                  esh[h] = sest/nh[h]
                  vj<-(wi)^2*(sest_j-esh[h])^2
                  varh[h] = sum((Wh[h]^2)*(nh[h]*(nh[h]-1)^3)*vj)
                  }
                ESTP[r] <- sum(Wh*esh)
                vESTP[r] <- sum(varh)
                ci1.max[r]<- ESTP[r]+qt(.95,n-3)*sqrt(vESTP[r])
                ci1.min[r]<- ESTP[r]+qt(.05,n-3)*sqrt(vESTP[r])
                ci2.max[r]<- ESTP[r]+qt(.975,n-3)*sqrt(vESTP[r])
                ci2.min[r]<- ESTP[r]+qt(.025,n-3)*sqrt(vESTP[r])
                ci3.max[r]<- ESTP[r]+qt(.995,n-3)*sqrt(vESTP[r])
                ci3.min[r]<- ESTP[r]+qt(.005,n-3)*sqrt(vESTP[r])
              }# nreps
              round(sum(ci1.min<ymean & ci1.max>ymean)/nreps,4)->cov1
              round(sum(ci2.min<ymean & ci2.max>ymean)/nreps,4)->cov2
              round(sum(ci3.min<ymean & ci3.max>ymean)/nreps,4)->cov3
              cat(i1,i2,i3,cov1,cov2,cov3,'\n')
            }#i1
          }#i2
      }#i
```

8.3.3 Numerical illustration

The following example explains the computational steps required to implement the proposed estimator in a stratified random sampling scheme.

Example 8.1 Suppose that there are 800 *Sumbo*, 2000 *Mumbo*, and 6000 *Jumbo* pumpkins in a field. The farmer applied the proportional allocation method to select a sample of 3 *Sumbo*, 7 *Mumbo*, and 20 *Jumbo* pumpkins to estimate the weight of the pumpkins in the entire field using the newly tuned stratified random sampling estimator. The farmer knows that $\bar{X}_1 = 20$, $\bar{X}_2 = 282$, and $\bar{X}_3 = 1403$ are the true average circumferences of the three types of pumpkins. The farmer collected the following sample information on circumference (x) in inches and weight (y) in pounds:

Sumbo pumpkins	
x_{1i}	y_{1i}
19.88	33.96
15.00	17.62
25.95	25.74

Mumbo pumpkins	
x_{2i}	y_{2i}
433.58	471.16
366.66	458.00
240.93	295.52
142.63	302.31
101.75	291.96
188.78	367.71
502.58	493.79

Jumbo pumpkins	
x_{3i}	y_{3i}
1245.83	5961.43
1389.43	5024.06
1549.95	5257.64
1218.13	6261.65
1396.49	5177.95
1605.36	2863.82
1197.08	5034.54
1603.43	3364.57
1506.00	1247.42
1428.66	2683.35
1329.38	6264.21
1236.34	5061.31
1604.52	4123.96
1348.69	4007.46
1340.51	8835.86
1480.90	7225.51
1500.06	8798.72
1109.26	9087.72
1599.05	1056.17
1374.50	2404.53

The farmer now wishes to construct 90%, 95%, and 99% confidence interval estimates of the average weight of the pumpkins.

Solution. Using the tuned estimator for stratified random sampling the 90%, 95%, and 99% confidence interval estimates of the average weight of the pumpkins in the field are (2468.29, 4530.96), (2257.24, 4742.01), and (1821.97, 5177.27), respectively.

8.3.4 R code used for illustration

We used the following R code, **PUMPKIN81EX.R**, in the preceding illustration.

PROGRAM PUMPKIN81EX.R
```
nh<-c(3,7,20);n<-sum(nh)
x<-matrix(ncol=2,nrow=n)
y<-matrix(ncol=2,nrow=n)
rep((1:3),c(3,7,20))->x[,2]
y[,2]=x[,2]
x[,1]<-c(
19.88, 15.00, 25.95,
433.58,366.66,240.93,142.63,101.75,
188.78,502.58,
1245.83,1389.43,1549.95,1218.13,1396.49,
1605.36,1197.08,1603.43,1506.00,1428.66,1329.38,
1236.34,1604.52,1348.69,1340.51,1480.90,1500.06,
1109.26,1599.05,1374.5)
y[,1]<-c(133.96, 117.62, 125.74,
471.16,458.00,295.52,302.31,291.96,
367.71,493.79,
5961.43,5024.06,5257.64,6261.65,
5177.95,2863.82,5034.54,3364.57,1247.42,
2683.35,6264.21,5061.31,4123.96,4007.46,
8835.86,7225.51,8798.72,9087.72,1056.17,
2404.53)
Nh<-c(800,2000,6000);N<-sum(Nh)
Wh<-Nh/N
#xmean<- as.vector(tapply(x[,1],x[,2],mean))
xmean<-c(20,282,1403)
xmh<-rep(0,3);deltah=sest=esh=varh=ymh=xmh
for(h in 1:3)
   {
   xmj<-rep(0,nh[h])
   ymj=di=dif=wi=xmj
   x[(x[,2]==h),1]->xsh; y[(y[,2]==h),1]->ysh
   xmj<-(sum(xsh)-xsh)/(nh[h]-1)
   ymj<-(sum(ysh)-ysh)/(nh[h]-1)
```

```
    delta<-nh[h]*sum(xmj^2) - sum(xmj)^2
    di<-(nh[h]*xmj-sum(xmj))/delta
    dif <- nh[h]*(xmean[h]-mean(xsh))/((nh[h]-1)^2)
    wi <- (1+di*dif)/nh[h]
    est_i = nh[h]*((nh[h]-1)^2*wi-(nh[h]-2))*ymj
    sest = sum(est_i)
    sest_j = (sest-est_i)/(nh[h]-1)
    esh[h] = sest/nh[h]
    vj<-(wi)^2*(sest_j-esh[h])^2
    varh[h] = sum(Wh[h]^2*(nh[h]*(nh[h]-1)^3)*vj)
    }
es <- sum(Wh*esh)
var <- sum(varh)
cat("Tuned estimate:", es, "SE: ",var^.5 ,'\n')
#cat("Confidence Interval:"," ", L,"; ", U,'\n')
cat("90%:",es+qt(.05,n-3)*sqrt(var),
            es+qt(.95,n-3)*sqrt(var),'\n')
cat("95%:",es+qt(.025,n-3)*sqrt(var),
            es+qt(.975,n-3)*sqrt(var),'\n')
cat("99%:",es+qt(.005,n-3)*sqrt(var),
            es+qt(.995,n-3)*sqrt(var),'\n')
```

8.4 Tuning with dual-to-empirical log-likelihood function using stratum-level known population means of an auxiliary variable

The newly tuned jackknife estimator of the population mean $\bar{Y}$ is defined by

$$\bar{y}_{\text{StTuned(dell)}} = \sum_{h=1}^{L} \Omega_h \sum_{j \in s_h} \left[(n_h - 1)^2 \bar{w}_h^*(j) - (n_h - 2) \right] \bar{y}_h(j) \tag{8.37}$$

where $\bar{w}_h^*(j)$ are the stratum level tuned positive weights such that in each stratum the following two constraints are satisfied:

$$\sum_{j \in s_h} \bar{w}_h^*(j) = 1 \tag{8.38}$$

and

$$\sum_{j \in s_h} \bar{w}_h^*(j) \left\{ \bar{x}_h(j) - \frac{\bar{X}_h - n_h(2 - n_h)\bar{x}_h}{(n_h - 1)^2} \right\} = 0 \tag{8.39}$$

where:

$$w_{hj}^* = 1 - (n_h - 1)\bar{w}_h^*(j) \tag{8.40}$$

are some arbitrary weights of unit length in each stratum.

Here, we suggest tuning the weights $\bar{w}_h^*(j)$ so that the weighted dual-to-empirical log-likelihood (dell) distance function in each stratum, defined as

$$\sum_{j \in s_h} \frac{\log\left(1 - w_{hj}^*\right)}{n_h} = \sum_{j \in s_h} \frac{\log\left(\bar{w}_h^*(j)\right)}{n_h} \tag{8.41}$$

is optimal, subject to the tuning constraints (8.38) and (8.39).

The Lagrange function becomes

$$L_{h2} = \sum_{j \in s_h} \frac{\log\left(\bar{w}_h^*(j)\right)}{n_h} - \lambda_{h0}^* \left\{ \sum_{j \in s_h} \bar{w}_h^*(j) - 1 \right\} - \lambda_{h1}^* \left\{ \sum_{j \in s_h} \bar{w}_h^*(j) \Psi_{hj} \right\} \tag{8.42}$$

where λ_{h0}^* and λ_{h1}^* are Lagrange multiplier constants, and

$$\Psi_{hj} = \left\{ \bar{x}_h(j) - \left(\frac{\bar{X}_h - n_h(2 - n_h)\bar{x}_h}{(n_h - 1)^2} \right) \right\} \tag{8.43}$$

On setting

$$\frac{\partial L_{h2}}{\partial \bar{w}_h^*(j)} = 0 \tag{8.44}$$

we have

$$\bar{w}_h^*(j) = \frac{1}{n_h\left(1 + \lambda_{h1}^* \Psi_{hj}\right)} \tag{8.45}$$

Constraints (8.38) and (8.39) yield $\lambda_{h0}^* = 1$, and λ_{h1}^* is a solution to the parametric equation

$$\sum_{j \in s_h} \frac{\Psi_{hj}}{\left(1 + \lambda_{h1}^* \Psi_{hj}\right)} = 0 \tag{8.46}$$

Thus, under the dell distance function, the newly tuned jackknife estimator (8.37) of the population mean in stratified random sampling becomes

$$\bar{y}_{StTuned(dell)} = \sum_{h=1}^{L} \Omega_h \sum_{j \in s_h} \left[\frac{(n_h - 1)^2}{n_h\left(1 + \lambda_{h1}^* \Psi_{hj}\right)} - (n_h - 2) \right] \bar{y}_h(j) \tag{8.47}$$

8.4.1 Estimation of variance and coverage

An adjusted estimator of the variance of the estimator $\bar{y}_{\text{StTuned(dell)}}$ is given by

$$\hat{v}_{\text{StTuned(dell)}} = \sum_{h=1}^{L} \Omega_h^2 n_h (n_h - 1)^3 \sum_{j \in s_h} \{\bar{w}_h^*(j)\}^2 \left\{ \bar{y}_{\text{StTuned(dell)}(j)}^{(h)} - \bar{y}_{\text{StTuned(dell)}}^{(h)} \right\}^2 \tag{8.48}$$

Note that each newly tuned estimator of the hth stratum population mean is

$$\bar{y}_{\text{StTuned(dell)}(j)}^{(h)} = \left[\frac{n_h \bar{y}_{\text{lh}} - n_h \left\{ (n_h - 1)^2 \bar{w}_h^*(j) - (n_h - 2) \right\} \bar{y}_h(j)}{n_h - 1} \right] \tag{8.49}$$

and

$$\bar{y}_{\text{lh}} = \sum_{j \in s_h} \left[(n_h - 1)^2 \bar{w}_h^*(j) - (n_h - 2) \right] \bar{y}_h(j) \tag{8.50}$$

for $h = 1, 2, \ldots, L;\ j = 1, 2, \ldots, n_h$.

Attained coverage by the $(1-\alpha)100\%$ confidence interval estimates obtained from this newly tuned dell estimator of the population mean is estimated by observing the proportion of times the true population mean $\bar{Y}$ falls within the interval estimates given by

$$\bar{y}_{\text{StTuned(dell)}} \mp t_{\alpha/2}(\text{df} = n - L)\sqrt{\hat{v}_{\text{StTuned(dell)}}} \tag{8.51}$$

In the simulation study, under certain mild assumptions, we approximate

$$\lambda_{h1}^* \approx \frac{\sum_{j \in s_h} \Psi_{hj}}{\sum_{j \in s_h} \Psi_{hj}^2} \tag{8.52}$$

Here we consider the problem of estimating the average weight (lbs) of three types of pumpkins, namely the *Sumbo Pumpkins*, *Mumbo Pumpkins*, and *Jumbo Pumpkins*, with sizes based on their circumferences (in.). In the simulation study, we considered three different strata based on the circumferences of the pumpkins. We generated six variables, two variables in each stratum, with different means, different standard deviations, and different correlations between weight and circumference as follows:

$$x_{hi} = \bar{X}_h + \sigma_{hx} x_{hi}^* \tag{8.53}$$

and

$$y_{hi} = \bar{Y}_h + \sigma_{hy} y_{hi}^* \sqrt{1 - \rho_{hxy}^2} + \sigma_{hy} \rho_{hxy} x_{hi}^* \tag{8.54}$$

here $x_{hi}^* \sim N(0, 1)$ and $y_{hi}^* \sim N(0, 1)$ for $h = 1, 2, 3$.

Again, recall that the strata are mutually exclusive and samples are drawn independently, so again we estimated the variance in each stratum and then took the sum over all strata. We studied the attained coverage of the nominally 90%, 95%, and 99% confidence intervals, constructed using the newly tuned estimator of the population mean by selecting 10,000 random samples from the each of the three types of pumpkins. We set the values of correlation coefficients as $\rho_{1xy} = 0.6$, $\rho_{2xy} = 0.7$, and $\rho_{3xy} = 0.4$. The known average values of the circumferences (in.) in the three strata were taken to be $\bar{X}_1 = 20$, $\bar{X}_2 = 300$, and $\bar{X}_3 = 1400$, respectively. We assumed the average weights (lbs) of the pumpkins in the three strata were $\bar{Y}_1 = 25$, $\bar{Y}_2 = 400$, and $\bar{Y}_3 = 8000$, respectively. We also set $\sigma_{1x} = 12$, $\sigma_{2x} = 280$, $\sigma_{3x} = 329$, $\sigma_{1y} = 15$, $\sigma_{2y} = 200$, and $\sigma_{3y} = 6000$. We selected n_1 Sumbo, n_2 Mumbo, and n_3 Jumbo pumpkins from the three independent strata consisting of $N_1 = 800$ Sumbo, $N_2 = 2000$ Mumbo, and $N_3 = 6000$ Jumbo pumpkins. The results obtained for various sample sizes are shown in Table 8.2.

Table 8.2 shows that the coverage of intervals based on the newly tuned dell estimator of the population mean in stratified random sampling remains as good as the estimator based on the chi-square distance function. In particular, we note that attained coverage of the nominally 90% confidence interval is estimated to be 89.14% for a sample of two Sumbo, two Mumbo, and three Jumbo pumpkins, attained coverage of the 95% intervals is estimated to be 92.01% for a sample two Sumbo, two Mumbo, and three Jumbo pumpkins, and attained coverage of the 99% interval is estimated to be 99.09% for a sample five Sumbo, five Mumbo, and four Jumbo pumpkins. Thus, intervals based on the newly tuned dell estimator of the population mean weight in stratified random sampling show lower coverage than those based on estimators developed from the chi-square type distance function, in case of small sample sizes, but shows almost equal coverage for reasonably large sample sizes.

8.4.2 R code

To study the true coverage of the newly tuned estimator based on the dell distance function, we used the following R code, **PUMPKIN82.R**.

PROGRAM PUMPKIN82.R

```
set.seed(2013)
rnorm(800,0,1)->x1s
rnorm(2000,0,1)->x2s
rnorm(6000,0,1)->x3s
x1<-20 +12*x1s; x2<-300+280*x2s; x3<-1400+329*x3s
y1<-25 + 15* rnorm(800,0,1)*sqrt(1-0.6^2) + 15*0.6*x1s
```

Table 8.2 Performance of confidence intervals based on the newly tuned dual-to-empirical log-likelihood (dell) estimator with stratified random sampling

n_1	n_2	n_3	Expected coverage			n_1	n_2	n_3	Expected coverage		
			90%	95%	99%				90%	95%	99%
2	2	2	0.5632	0.6365	0.7741	4	2	2	0.5394	0.5972	0.7044
		3	0.8914	0.9201	0.9598			3	0.8807	0.9071	0.9433
		4	0.9747	0.9840	0.9937			4	0.9719	0.9825	0.9937
		5	0.9894	0.9949	0.9986			5	0.9892	0.9941	0.9984
	3	2	0.5554	0.6225	0.7415		3	2	0.5307	0.5905	0.6944
		3	0.8776	0.9058	0.9460			3	0.8753	0.9026	0.9386
		4	0.9732	0.9841	0.9947			4	0.9745	0.9821	0.9917
		5	0.9888	0.9940	0.9974			5	0.9893	0.9933	0.9975
	4	2	0.5637	0.6276	0.7489		4	2	0.5519	0.6094	0.7062
		3	0.8738	0.9042	0.9442			3	0.8841	0.9085	0.9415
		4	0.9758	0.9848	0.9945			4	0.9762	0.9852	0.9929
		5	0.9907	0.9945	0.9980			5	0.9874	0.9917	0.9966
	5	2	0.5710	0.6377	0.7484		5	2	0.5712	0.6268	0.7245
		3	0.8842	0.9132	0.9490			3	0.8768	0.9062	0.9392
		4	0.9740	0.9829	0.9934			4	0.9718	0.9822	0.9922
		5	0.9910	0.9940	0.9974			5	0.9885	0.9932	0.9976
3	2	2	0.5559	0.6191	0.7318	5	2	2	0.5310	0.5851	0.6787
		3	0.8839	0.9099	0.9486			3	0.8743	0.9031	0.9397
		4	0.9750	0.9856	0.9936			4	0.9729	0.9818	0.9932
		5	0.9890	0.9933	0.9975			5	0.9872	0.9923	0.9976
	3	2	0.5461	0.6082	0.7135		3	2	0.5243	0.5818	0.6731
		3	0.8799	0.9060	0.9452			3	0.8757	0.9022	0.9377
		4	0.9779	0.9862	0.9949			4	0.9743	0.9827	0.9916
		5	0.9892	0.9938	0.9980			5	0.9888	0.9928	0.9974

Continued

	n					n			
4	2	0.5576	0.6148	0.7288	4	2	0.5382	0.5978	0.6996
	3	0.8819	0.9097	0.9446		3	0.8784	0.9011	0.9359
	4	0.9731	0.9827	0.9920		4	0.9751	0.9840	0.9929
	5	0.9895	0.9935	0.9976		5	0.9890	0.9926	0.9976
5	2	0.5669	0.6268	0.7342	5	2	0.5606	0.6202	0.7163
	3	0.8774	0.9047	0.9403		3	0.8714	0.8995	0.9377
	4	0.9760	0.9837	0.9924		4	0.9703	0.9798	0.9909
	5	0.9902	0.9946	0.9975		5	0.9886	0.9939	0.9978

```
y2<-400 + 200* rnorm(2000,0,1)*sqrt(1-0.7^2) + 200*0.7*x2s
y3<-8000+6000* rnorm(6000,0,1)*sqrt(1-0.4^2) + 6000*0.4*x3s
Nh<-c(800,2000,6000);N<-sum(Nh)
Wh<-Nh/N
xmean<- c(mean(x1),mean(x2),mean(x3))
xmh<-rep(0,3);deltah=sest=esh=varh=ymh=xmh
ymean<-mean(c(y1,y2,y3))
nreps<-10000
ESTP=rep(0,nreps)
ci1.max=ci1.min=ci2.max=ci2.min=ci3.max=ci3.min=vESTP=ESTP
for (i1 in 2:5)
   {for (i2 in 2:5)
      {for (i3 in 2:5)
         {nh<-c(i1,i2,i3);n<-sum(nh)
          x<-matrix(ncol=2,nrow=n)
          y<-matrix(ncol=2,nrow=n)
          rep((1:3),nh)->x[,2]
          y[,2]=x[,2]
          for (r in 1:nreps) {
             s1<-sample(800,i1)
             s2<-sample(2000,i2)
             s3<-sample(6000,i3)
             x[,1]<-c(x1[s1],x2[s2],x3[s3])
             y[,1]<-c(y1[s1],y2[s2],y3[s3])
             for(h in 1:3)
                {
                xmj<-rep(0,nh[h])
                ymj=di=dif=wi=xmj
                x[(x[,2]==h),1]->xsh; y[(y[,2]==h),1]->ysh
                xmj<-(sum(xsh)-xsh)/(nh[h]-1)
                ymj<-(sum(ysh)-ysh)/(nh[h]-1)
                phi<- xmj - (xmean[h] - nh[h]*(2-nh[h])*mean(xsh))/
                   ((nh[h]-1)^2)
                al<-sum(phi)/sum(phi^2)
                wi<-(1/(1+al*phi))/nh[h]
                est_i = nh[h]*(((nh[h]-1)^2)*wi-(nh[h]-2))*ymj
                sest = sum(est_i)
                sest_j = (sest-est_i)/(nh[h]-1)
                esh[h] = sest/nh[h]
                vj<-(wi)^2*(sest_j-esh[h])^2
                varh[h] = sum((Wh[h]^2)*(nh[h]*(nh[h]-1)^3)*vj)
                }
             ESTP[r] <- sum(Wh*esh)
             vESTP[r] <- sum(varh)
             ci1.max[r]<- ESTP[r]+qt(.95,n-3)*sqrt(vESTP[r])
```

```
            ci1.min[r]<- ESTP[r]+qt(.05,n-3)*sqrt(vESTP[r])
            ci2.max[r]<- ESTP[r]+qt(.975,n-3)*sqrt(vESTP[r])
            ci2.min[r]<- ESTP[r]+qt(.025,n-3)*sqrt(vESTP[r])
            ci3.max[r]<- ESTP[r]+qt(.995,n-3)*sqrt(vESTP[r])
            ci3.min[r]<- ESTP[r]+qt(.005,n-3)*sqrt(vESTP[r])
        }# nreps
        round(sum(ci1.min<ymean & ci1.max>ymean)/nreps,4)->cov1
        round(sum(ci2.min<ymean & ci2.max>ymean)/nreps,4)->cov2
        round(sum(ci3.min<ymean & ci3.max>ymean)/nreps,4)->cov3
        cat(i1,i2,i3,cov1,cov2,cov3,'\n')
    }#i1
  }#i2
}#i3
```

8.4.3 Numerical illustration

We explain the use of the dell estimator in stratified random sampling with the following example.

Example 8.2 Suppose there are 800 *Sumbo*, 2000 *Mumbo*, and 6000 *Jumbo* pumpkins in a field. The farmer applied the proportional allocation method to select a sample of 3 *Sumbo*, 7 *Mumbo*, and 20 *Jumbo* pumpkins to estimate average weigh of the pumpkins in the whole field using the dell estimator tuned for stratified random sampling. The farmer collected the following information on circumference (x) in inches and weight (y) in lbs.

Sumbo pumpkins	
x_{1i}	y_{1i}
19.88	33.96
15.00	17.62
25.95	25.74

Mumbo pumpkins	
x_{2i}	y_{2i}
433.58	471.16
366.66	458.00
240.93	295.52
142.63	302.31
101.75	291.96
188.78	367.71
502.58	493.79

Jumbo pumpkins	
x_{3i}	y_{3i}
1245.83	5961.43
1389.43	5024.06
1549.95	5257.64
1218.13	6261.65
1396.49	5177.95
1605.36	2863.82
1197.08	5034.54
1603.43	3364.57
1506.00	1247.42
1428.66	2683.35
1329.38	6264.21
1236.34	5061.31
1604.52	4123.96
1348.69	4007.46
1340.51	8835.86
1480.90	7225.51
1500.06	8798.72
1109.26	9087.72
1599.05	1056.17
1374.50	2404.53

The farmer knows that $\bar{X}_1 = 20$, $\bar{X}_2 = 282$, and $\bar{X}_3 = 1403$ are the average circumferences of the three types of pumpkins, and he wishes to construct 90%, 95%, and 99% confidence interval estimates.

Solution. Using the tuned estimator in stratified random sampling the 90%, 95%, and 99% confidence interval estimates of the average weight of pumpkins in the field are (2468.29, 4530.96), (2257.25, 4742.01), and (1821.98, 5177.27), respectively. It appears that the tuned dell estimator in stratified random sampling works as well as the chi-square distance estimator for this particular situation.

8.4.4 R code used for illustration

We used the following R code, **PUMPKIN82EX.R**, in the preceding illustration.

PROGRAM PUMPKIN82EX.R
```
nh<-c(3,7,20);n<-sum(nh)
x<-matrix(ncol=2,nrow=n)
y<-matrix(ncol=2,nrow=n)
rep((1:3),c(3,7,20))->x[,2]
y[,2]=x[,2]
x[,1]<-c(
19.88, 15.00, 25.95,
```

```
    433.58,366.66,240.93,142.63,101.75,
    188.78,502.58,
    1245.83,1389.43,1549.95,1218.13,1396.49,
    1605.36,1197.08,1603.43,1506.00,1428.66,1329.38,
    1236.34,1604.52,1348.69,1340.51,1480.90,1500.06,
    1109.26,1599.05,1374.5)
    y[,1]<-c(133.96, 117.62, 125.74,
    471.16,458.00,295.52,302.31,291.96,
    367.71,493.79,
    5961.43,5024.06,5257.64,6261.65,
    5177.95,2863.82,5034.54,3364.57,1247.42,
    2683.35,6264.21,5061.31,4123.96,4007.46,
    8835.86,7225.51,8798.72,9087.72,1056.17,
    2404.53)
    Nh<-c(800,2000,6000);N<-sum(Nh)
    Wh<-Nh/N
    #xmean<- as.vector(tapply(x[,1],x[,2],mean))
    xmean<-c(20,282,1403)
    xmh<-rep(0,3);deltah=sest=esh=varh=ymh=xmh
    for(h in 1:3)
      {
      xmj<-rep(0,nh[h])
      ymj=di=dif=wi=xmj
      x[(x[,2]==h),1]->xsh; y[(y[,2]==h),1]->ysh
      xmj<-(sum(xsh)-xsh)/(nh[h]-1)
      ymj<-(sum(ysh)-ysh)/(nh[h]-1)
      phi<- xmj - (xmean[h] - nh[h]*(2-nh[h])*mean(xsh))/((nh[h]-1)^2)
      al<-sum(phi)/sum(phi^2)
      wi<-(1/(1+al*phi))/nh[h]
      est_i = nh[h]*(((nh[h]-1)^2)*wi-(nh[h]-2))*ymj
      sest = sum(est_i)
      sest_j = (sest-est_i)/(nh[h]-1)
      esh[h] = sest/nh[h]
      vj<-(wi)^2*(sest_j-esh[h])^2
      varh[h] = sum(Wh[h]^2*nh[h]*(nh[h]-1)^  3*vj)
      }
    es <- sum(Wh*esh); var <- sum(varh)
    cat("Tuned estimate:", es, "SE: ",var^.5 ,'\n')
    cat("90%:",es+qt(.05,n-3)*sqrt(var),
                es+qt(.95,n-3)*sqrt(var),'\n')
    cat("95%:",es+qt(.025,n-3)*sqrt(var),
                es+qt(.975,n-3)*sqrt(var),'\n')
    cat("99%:",es+qt(.005,n-3)*sqrt(var),
                es+qt(.995,n-3)*sqrt(var),'\n')
```

8.5 Exercises

Exercise 8.1 Consider the newly tuned estimator of the population mean $\bar{Y}$ in stratified random sampling, defined by

$$\bar{y}_{\text{StTuned(cs)}} = \sum_{h=1}^{L} \Omega_h \sum_{j \in s_h} \left[(n_h - 1)^2 \bar{w}^*_{n_h}(j) - (n_h - 2) \right] \bar{y}_h(j) \tag{8.55}$$

where

$$\bar{y}_h(j) = \frac{n_h \bar{y}_h - y_{hj}}{n_h - 1} \tag{8.56}$$

is the hth stratum sample mean of the study variable obtained by removing the jth unit from the stratum sample s_h, and $\Omega_h = N_h/N$ are the hth stratum weights. The tuned weights $\bar{w}^*_{n_h}(j)$ are computed independently in the h strata such that the chi-squared type distance function, defined as

$$D_h = \frac{n_h}{2} \sum_{j \in s_h} q_{hj}^{-1} \left[1 - (n_h - 1)\bar{w}^*_{n_h}(j) - n_h^{-1} \right]^2 \tag{8.57}$$

is optimized, subject to the following two tuning constraints:

$$\sum_{j \in s_h} \bar{w}^*_{n_h}(j) = 1 \tag{8.58}$$

and

$$\sum_{j \in s_h} \bar{w}^*_{n_h}(j) \bar{x}_h(j) = \frac{\bar{X}_h - n_h(2 - n_h)\bar{x}_h}{(n_h - 1)^2} \tag{8.59}$$

where

$$\bar{x}_h(j) = \frac{n_h \bar{x}_h - x_{hj}}{n_h - 1} \tag{8.60}$$

is the jth jackknifed estimator obtained from the hth stratum sample mean $\bar{x}_h = n_h^{-1} \sum_{i \in s_h} x_{hi}$ by removing the jth unit. The stratum population means of the auxiliary variable $\bar{X}_h$ are assumed to be known. Find the optimal tuning weights $\bar{w}^*_{n_h}(j)$ for the estimator $\bar{y}_{\text{StTuned(cs)}}$. Suggest doubly jackknifed estimators of variance and confidence interval. Create a hypothetical pumpkin farm where the pumpkins can be divided into three mutually exclusive strata, say *Sumbo*, *Mumbo*, and *Jumbo* pumpkins, based on their known circumferences. Select a sample of size n using the method of proportional allocation. Study the nominally 90%, 95%, and 99% coverages by the confidence interval estimators you suggested. Discuss the difficulties you encountered, if any.

Exercise 8.2 In Exercises 8.1, consider an additional constraint given by

$$\sum_{j \in s_h} \bar{w}_{n_h}^*(j) \hat{\sigma}_{hx}^2(j) = \frac{\sigma_{hx}^2 - n_h(2-n_h)\hat{\sigma}_{hx}^2}{(n_h-1)^2} \qquad (8.61)$$

where

$$\hat{\sigma}_{hx}^2(j) = \frac{n_h \hat{\sigma}_{hx}^2 - (x_{hj} - \bar{x}_h)^2}{n_h - 1} \qquad (8.62)$$

and

$$\hat{\sigma}_{hx}^2 = n_n^{-1} \sum_{i \in s_h} (x_{hi} - \bar{x}_h)^2 \qquad (8.63)$$

Note that $\hat{\sigma}_{hx}^2$ is the maximum likelihood estimator of the known finite population variance $\sigma_{hx}^2 = N_h^{-1} \sum_{i=1}^{N_h} (x_{h_i} - \bar{X}_h)^2$ of the auxiliary variable in the hth stratum. Also, $\hat{\sigma}_{hx}^2(j)$ is a partial jth jackknifed estimator of variance obtained by dropping the jth squared deviation about the hth stratum sample mean from the total sum of squares from the sample s_h of the auxiliary variable divided by $(n_h - 1)$. Discuss and report the changes observed in the results.

Exercise 8.3 In Exercise 8.1, replace the second constraint with the following constraint:

$$\sum_{j \in s_h} \bar{w}_{n_h}(j) \left[\hat{G}_{hx}(j)\right]^{(1-n_h)} = \frac{1}{(n_h-1)} \left[\sum_{j \in s_h} \left(\hat{G}_{hx}(j)\right)^{(1-n_h)} - \bar{X}_h \left(\hat{G}_{hx}\right)^{-n_h}\right] \qquad (8.64)$$

where $\bar{X}_h = N_h^{-1} \sum_{i=1}^{N_h} x_{hi}$ denotes the known population arithmetic mean of the auxiliary variable in the hth stratum, and

$$\hat{G}_{hx}(j) = \left(\prod_{i \neq j=1}^{n_h} x_{hi}\right)^{1/(n_h-1)}, \quad j = 1, 2, \ldots, n_h \qquad (8.65)$$

is the jth jackknifed estimator of the geometric mean of the auxiliary variable obtained by dropping the jth unit from the usual estimator of the geometric mean of the auxiliary variable in the hth stratum which is given by

$$\hat{G}_{hx} = \left(\prod_{i=1}^{n_h} x_i\right)^{1/n_h}, \quad h = 1, 2, \ldots, L \tag{8.66}$$

Exercise 8.4 In Exercise 8.1, replace the second constraint with the following constraint:

$$\sum_{j \in s_h} \frac{\bar{w}_{n_h}(j) \hat{H}_{hx}(j)}{n_h \hat{H}_{hx}(j) - (n_h - 1)\hat{H}_{hx}} = \frac{1}{(n_h - 1)} \left[\sum_{j \in s_h} \frac{\hat{H}_{hx}(j)}{n_h \hat{H}_{hx}(j) - (n_h - 1)\hat{H}_{hx}} - \frac{\bar{X}_h}{\hat{H}_{hx}} \right] \tag{8.67}$$

where $\bar{X}_h = N_h^{-1} \sum_{i=1}^{N_h} x_{hi}$ denotes the known population arithmetic mean of the auxiliary variable in the hth stratum, and

$$\hat{H}_{hx}(j) = (n_h - 1) \left(\sum_{i \neq j = 1}^{n_h} x_{hi}^{-1} \right)^{-1}, \quad j = 1, 2, \ldots, n_h \tag{8.68}$$

is the jth jackknifed estimator of the harmonic mean of the auxiliary variable obtained by dropping jth unit from the usual estimator of the harmonic mean of the auxiliary variable in the hth stratum, which is given by

$$\hat{H}_{hx} = n_h \left(\sum_{i=1}^{n_h} x_{hi}^{-1} \right)^{-1} \tag{8.69}$$

Exercise 8.5 Consider the newly tuned estimator of the population mean $\bar{Y}$ for stratified random sampling defined by

$$\bar{y}_{\text{StTuned(dell)}} = \sum_{h=1}^{L} \Omega_h \sum_{j \in s_h} \left[(n_h - 1)^2 \bar{w}_{n_h}^*(j) - (n_h - 2) \right] \bar{y}_h(j) \tag{8.70}$$

where

$$\bar{y}_h(j) = \frac{n_h \bar{y}_h - y_{hj}}{n_h - 1} \tag{8.71}$$

is the hth stratum sample mean of the study variable obtained by removing the jth unit from the sample s_h, and $\Omega_h = N_h/N$ is the hth stratum weight. The tuned weights $\bar{w}_{n_h}^*(j)$ are computed independently in the h strata such that the dell distance function defined as

$$D_h = \frac{1}{n_h} \sum_{j \in s_h} \ln\left(\bar{w}_{n_h}^*(j)\right) \qquad (8.72)$$

is optimized, subject to the following two tuning constraints:

$$\sum_{j \in s_h} \bar{w}_{n_h}^*(j) = 1 \qquad (8.73)$$

and

$$\sum_{j \in s_h} \bar{w}_{n_h}^*(j) \Psi_{h_j} = 0 \qquad (8.74)$$

where

$$\Psi_{h_j} = \bar{x}_h(j) - \frac{\bar{X}_h - n_h(2-n_h)\bar{x}_h}{(n_h - 1)^2} \qquad (8.75)$$

with

$$\bar{x}_h(j) = \frac{n_h \bar{x}_h - x_{hj}}{n_h - 1} \qquad (8.76)$$

being the jth jackknifed estimator obtained from the hth stratum sample mean $\bar{x}_h = n_h^{-1} \sum_{i \in s_h} x_{hi}$ by removing the jth unit. Assume that the stratum population means of the auxiliary variable $\bar{X}_h$ are known. For the optimal tuning weights $\bar{w}_{n_h}^*(j)$, find the estimator $\bar{y}_{\text{StTuned(dell)}}$. Suggest doubly jackknifed estimators of variance and a confidence interval estimator. Create a hypothetical pumpkin farm where the pumpkins can be divided into three mutually exclusive strata, say *Sumbo*, *Mumbo*, and *Jumbo*, pumpkins, based on their known circumferences. Select a sample of size n using the method of proportional allocation in each stratum. Study the coverages by the nominally 90%, 95%, and 99% confidence interval estimators you suggested. Discuss the difficulties you encountered, if any.

Exercise 8.6 In Exercise 8.5 consider an additional constraint given by

$$\sum_{j \in s_h} \bar{w}_{n_h}^*(j) \left[\hat{\sigma}_{hx}^2(j) - \frac{\hat{\sigma}_{hx}^2 - n_h(2-n_h)\hat{\sigma}_{hx}^2}{(n_h - 1)^2}\right] = 0 \qquad (8.77)$$

where

$$\hat{\sigma}_{hx}^2(j) = \frac{n_h \hat{\sigma}_{hx}^2 - (x_{hj} - \bar{x}_h)^2}{n_h - 1} \qquad (8.78)$$

and

$$\hat{\sigma}_{hx}^2 = n_h^{-1} \sum_{i \in s_h} (x_{hi} - \bar{x}_h)^2. \tag{8.79}$$

Note that $\hat{\sigma}_{hx}^2$ is the maximum likelihood estimator of the known finite population variance $\sigma_{hx}^2 = N_h^{-1} \sum_{i=1}^{N_h} (x_{hi} - \bar{X}_h)^2$ of the auxiliary variable in the hth stratum. Also, $\hat{\sigma}_{hx}^2(j)$ is a partial jth jackknifed estimator of the variance obtained by dropping the jth squared deviation about the hth stratum sample mean from the total sum of squares from the sample s_h of the auxiliary variable divided by $(n_h - 1)$. Discuss and report the changes observed in the results.

Exercise 8.7 Consider a newly tuned estimator of the population mean $\bar{Y}$ for combined stratified random sampling defined as

$$\bar{y}_{\text{StTuned(cs)}}^{(c)} = \sum_{h=1}^{L} \Omega_h \sum_{j \in s_n} \left\{ (n_h - 1)^2 \bar{w}_{n_h}^*(j) - (n_h - 2) \right\} \bar{y}_h(j) \tag{8.80}$$

where

$$\bar{y}_h(j) = \frac{n_h \bar{y}_h - y_{hj}}{n_h - 1} \tag{8.81}$$

is the hth stratum sample mean of the study variable obtained by removing the jth unit from the sample s_h, and $\Omega_h = N_h/N$ are the hth stratum weights. The tuned weights $\bar{w}_{n_h}^*(j)$ are computed using pooled information across all strata such that a new weighted chi-squared type distance function defined by

$$D_c = \sum_{h=1}^{L} \Omega_h \left[\frac{n_h}{2} \sum_{j \in s_h} q_{hj}^{-1} \left\{ 1 - (n_h - 1) \bar{w}_{n_h}^*(j) - n_h^{-1} \right\}^2 \right] \tag{8.82}$$

is optimized, subject to the following two tuning constraints:

$$\sum_{h=1}^{L} \Omega_h (n_h - 1) \sum_{j \in s_h} \bar{w}_{n_h}(j) = \left(\sum_{h=1}^{L} n_h \Omega_h - 1 \right) \tag{8.83}$$

and

$$\sum_{h=1}^{L} \Omega_h (n_h - 1)^2 \sum_{j \in s_h} \bar{w}_{n_h}(j) \bar{x}_h(j) = \bar{X} - \sum_{h=1}^{L} \Omega_h n_h (2 - n_h) \bar{x}_h \tag{8.84}$$

where

$$\bar{x}_h(j) = \frac{n_h \bar{x}_h - x_{hj}}{n_h - 1} \tag{8.85}$$

is the jth jackknifed estimator obtained from the hth stratum sample mean $\bar{x}_h = n_h^{-1} \sum_{i \in s_h} x_{hi}$ by removing the jth unit. Assume that the combined population mean $\bar{X} = \sum_{h=1}^{L} \Omega_h \bar{X}_h$ of the auxiliary variable is known, while at the stratum level population means, $\bar{X}_h$, are assumed to be unknown. Note that for any arbitrary weights w_{hj} of unit sum, the constraints (8.83) and (8.84) are equivalent to the following two constraints:

$$\sum_{h=1}^{L} \Omega_h \sum_{j \in s_h} w_{hj} = 1 \tag{8.86}$$

and

$$\sum_{h=1}^{L} \Omega_h \sum_{j \in s_h} w_{hj} x_{hj} = \bar{X} \tag{8.87}$$

For the optimal tuning weights $\bar{w}_{n_h}^*(j)$, find the estimator $\bar{y}_{\text{StTuned(cs)}}^{(c)}$. Suggest doubly jackknifed estimators of variance and a confidence interval estimator. Create a hypothetical pumpkin farm where the pumpkins can be divided into three mutually exclusive strata, say *Sumbo*, *Mumbo*, and *Jumbo* pumpkins. Select a sample of size n using the method of proportional allocation. Study the nominally 90%, 95%, and 99% coverages by the confidence interval estimators you suggested. Discuss the difficulties you encountered, if any.

Exercise 8.8 Consider a semituned estimator of the population mean $\bar{Y}$ in stratified random sampling defined as

$$\bar{y}_{\text{St(dell)}}^* = \sum_{h=1}^{L} \Omega_h \sum_{j \in s_h} w_{hj} \bar{y}_h(j) \tag{8.88}$$

where

$$\bar{y}_h(j) = \frac{n_h \bar{y}_h - y_{hj}}{n_h - 1} \tag{8.89}$$

is the hth stratum sample mean of the study variable obtained by removing the jth unit from the sample s_h, and w_{hj} is the semituned stratum weight such that across all strata the following two constraints are satisfied:

$$\sum_{h=1}^{L} \Omega_h \sum_{j \in s_h} w_{hj} = 1 \tag{8.90}$$

and

$$\sum_{h=1}^{L} \Omega_h \left[\sum_{j \in s_h} w_{hj} \bar{x}_h(j) - \bar{X} \right] = 0 \tag{8.91}$$

where

$$\bar{x}_h(j) = \frac{n_h \bar{x}_h - x_{hj}}{n_h - 1} \tag{8.92}$$

is the hth stratum sample mean of the auxiliary variable obtained by removing the jth unit from the sample s_h. It is named semituned because only sample means are jackknifed and weights are not jackknifed. Find a set of semituned weights w_{hj} such that across all the strata the log-likelihood distance function defined by

$$\sum_{h=1}^{L} \Omega_h \sum_{j \in s_h} \ln(w_{hj}) \tag{8.93}$$

is optimized, subject to the tuning constraints (8.90) and (8.91). For the semituning weights w_{hj} find the estimator $\bar{y}^*_{St(dell)}$. Suggest doubly jackknifed estimators of variance and a confidence interval estimator. Create a hypothetical pumpkin farm where the pumpkins can be divided into three mutually exclusive strata, say *Sumbo*, *Mumbo*, and *Jumbo* pumpkins. Select a sample of size n using the method of proportional allocation. Study the coverage of the nominally 90%, 95%, and 99% confidence interval estimators you suggested. Discuss the difficulties you encountered, if any.

Exercise 8.9 Consider a semituned estimator of the population mean $\bar{Y}$ in stratified random sampling defined as

$$\bar{y}^*_{StTuned(cs)} = \sum_{h=1}^{L} \Omega_h \sum_{j \in s_h} w_{hj} \bar{y}_h(j) \tag{8.94}$$

where

$$\bar{y}_h(j) = \frac{n_h \bar{y}_h - y_{hj}}{n_h - 1} \tag{8.95}$$

is the hth stratum sample mean of the study variable obtained by removing the jth unit from the sample s_h, and w_{hj} are tuned weights such that across all strata the following two constraints are satisfied

Tuning in stratified sampling

$$\sum_{h=1}^{L} \Omega_h \sum_{j \in s_h} w_{hj} = 1 \qquad (8.96)$$

and

$$\sum_{h=1}^{L} \Omega_h \sum_{j \in s_h} w_{hj} \bar{x}_h(j) = \bar{X} \qquad (8.97)$$

where

$$\bar{x}_h(j) = \frac{n_h \bar{x}_h - x_{hj}}{n_h - 1} \qquad (8.98)$$

is the hth stratum sample mean of the auxiliary variable obtained by removing the jth unit from the sample s_h. The name semituned comes that only sample means are jackknifed and weights are not jackknifed.

Consider tuning the weights w_{hj} such that across all strata the chi-square type distance function, defined by

$$\sum_{h=1}^{L} (2^{-1} n_h) \sum_{j \in s_h} q_{hj}^{-1} \left(w_{hj} - n_h^{-1} \right)^2 \qquad (8.99)$$

is optimized, subject to the tuning constraints (8.96) and (8.97), where q_{hj} is a choice of weights in the hth stratum. For the alternative tuning weights w_{hj}, find the estimator $\bar{y}^*_{\text{StTuned(cs)}}$. Suggest doubly jackknifed estimators of variance and confidence interval. Create a hypothetical pumpkin farm where the pumpkins can be divided into three mutually exclusive strata, say *Sumbo*, *Mumbo*, and *Jumbo* pumpkins. Select a sample of size n using the method of proportional allocation in each stratum. Study coverage by the nominally 90%, 95%, and 99% confidence interval estimators you suggested. Discuss the difficulties you encountered, if any. (*Hint*: See Chapter 9 to adjust semituning weights.)

Exercise 8.10 Consider a newly tuned estimator of the population mean $\bar{Y}$ for the combined stratified random sampling defined by

$$\bar{y}^{(g)}_{\text{StTuned(cs)}} = \sum_{h=1}^{L} \Omega_h \sum_{j \in s_n} \left\{ (n_h - 1)^2 \bar{w}_{n_h}(j) - (n_h - 2) \right\} \bar{y}_h(j) \qquad (8.100)$$

where

$$\bar{y}_h(j) = \frac{n_h \bar{y}_h - y_{hj}}{n_h - 1} \qquad (8.101)$$

is the hth stratum sample mean of the study variable obtained by removing the jth unit from the sample s_h, and $\Omega_h = N_h/N$ is the hth stratum weight. The tuned weights $\bar{w}_{n_h}^*(j)$ are computed using pooled information across all strata such that a new weighted chi-squared type distance function defined as

$$D_{\mathrm{c}} = \sum_{h=1}^{L} \Omega_h \left[\frac{n_h}{2} \sum_{j \in s_h} q_{hj}^{-1} \left\{ 1 - (n_h - 1)\bar{w}_{n_h}(j) - n_h^{-1} \right\}^2 \right] \qquad (8.102)$$

is optimized subject to the following two tuning constraints:

$$\sum_{h=1}^{L} \Omega_h (n_h - 1) \sum_{j \in s_h} \bar{w}_{n_h}(j) = \left(\sum_{h=1}^{L} n_h \Omega_h - 1 \right) \qquad (8.103)$$

and

$$\sum_{h=1}^{L} \Omega_h \left(\hat{G}_{xh}\right)^{n_h} (n_h - 1) \sum_{j \in s_h} \bar{w}_{n_h}(j) \left(\hat{G}_{xh}(j)\right)^{1-n_h}$$
$$= \sum_{h=1}^{L} \Omega_h \left(\hat{G}_{xh}\right)^{n_h} \sum_{j \in s_h} \left(\hat{G}_{xh}(j)\right)^{1-n_h} - \bar{X} \qquad (8.104)$$

where

$$\hat{G}_{xh}(j) = \left(\prod_{i \neq j=1}^{n_h} x_{hi} \right)^{\frac{1}{n_h - 1}} \qquad (8.105)$$

is the jth jackknifed geometric mean obtained from the hth stratum geometric mean $\hat{G}_{xh} = \left(\prod_{i=1}^{n_h} x_{hi} \right)^{\frac{1}{n_h}}$ after removing the jth unit. The combined population mean $\bar{X} = \sum_{h=1}^{L} \Omega_h \bar{X}_h$ of the auxiliary variable is assumed to be known. For the optimal tuning weights $\bar{w}_{n_h}(j)$, find the estimator $\bar{y}_{\mathrm{StTuned(cs)}}^{(g)}$. Suggest doubly jackknifed estimator of variance and confidence interval. Create a hypothetical pumpkin farm where the pumpkins can be divided into three mutually exclusive strata, say *Sumbo*, *Mumbo*, and *Jumbo* pumpkins. Select a sample of size n using the method of proportional allocation from each stratum. Study coverage by the nominally 90%, 95%, and 99% confidence interval estimators you suggested. Discuss the difficulties you encountered, if any.

Exercise 8.11 Consider the problem of estimating the geometric mean in stratified sampling defined as

$$G_y = \left[\prod_{h=1}^{L}\left(\prod_{i=1}^{N_h} y_{h_i}\right)\right]^{\frac{1}{N}} \qquad (8.106)$$

where N_h is the total number of units in the hth stratum and $N = \sum_{h=1}^{L} N_h$. Without loss of generality, a naive estimator of the geometric mean G_y is given by

$$\hat{G}_y = \left[\prod_{h=1}^{L}\left(\prod_{i=1}^{n_h} y_{h_i}\right)\right]^{\frac{1}{n}} \qquad (8.107)$$

where n_h is the number of units selected from the hth stratum using a simple random with replacement sampling scheme such that the total sample size $n = \sum_{h=1}^{L} n_h$. Then the jackknifed estimator of the geometric mean after dropping the jth unit from the hth stratum is given by

$$\hat{G}_y(h_j) = \left[\prod_{h=1}^{L}\left(\prod_{i(\neq j)=1}^{n_h} y_{h_i}\right)\right]^{\frac{1}{n-1}} \qquad (8.108)$$

A jackknifed estimator of the geometric mean is given by

$$\hat{G}_{\text{Jack}} = \frac{1}{n}\sum_{h=1}^{L}\sum_{j=1}^{n_h} \hat{G}_y(h_j) \qquad (8.109)$$

Assume an estimator of the variance of the average jackknife estimator $\hat{G}_{\text{Jack}}$ is given by

$$\hat{v}(\hat{G}_{\text{Jack}}) = \sum_{h=1}^{L}\sum_{j=1}^{n_h} c_{hj}\left(\hat{G}_y(h_j) - \hat{G}_{\text{Jack}}\right)^2 \qquad (8.110)$$

Determine, if possible, the values of weights c_{hj} such that $\hat{v}(\hat{G}_{\text{Jack}})$ can be considered as an estimator of the variance of the sample geometric mean in stratified sampling. Support your findings with simulation study. *Hint*: The weights c_{hj} can be a function of N_h/N, which may help to improve the estimator of variance.

Exercise 8.12 Consider the problem of estimating the harmonic mean in stratified sampling, defined as:

$$H_y = N\left(\sum_{h=1}^{L}\sum_{i=1}^{N_h} y_{hi}^{-1}\right)^{-1} \qquad (8.111)$$

where N_h is the total number of units in the hth stratum and $N = \sum_{h=1}^{L} N_h$. Without loss of generality, a naive estimator of the harmonic mean H_y is given by

$$\hat{H}_y = n \left(\sum_{h=1}^{L} \sum_{i=1}^{n_h} y_{hi}^{-1} \right)^{-1} \tag{8.112}$$

where n_h is the number of units selected from the hth stratum using the simple random with replacement sampling scheme such that the total sample size $n = \sum_{h=1}^{L} n_h$. The jackknifed estimator of the harmonic mean after dropping the jth unit from the hth stratum is given by

$$\hat{H}_y(h_j) = (n-1) \left(\sum_{h=1}^{L} \sum_{i(\neq j)=1}^{n_h} y_{hi}^{-1} \right)^{-1} \tag{8.113}$$

An average jackknifed estimator of the harmonic mean is given by

$$\hat{H}_{\text{Jack}} = \frac{1}{n} \sum_{h=1}^{L} \sum_{j=1}^{n_h} \hat{H}_y(h_j) \tag{8.114}$$

Assume an estimator of the variance of the average jackknife estimator $\hat{H}_{\text{Jack}}$ is given by

$$\hat{v}(\hat{H}_{\text{Jack}}) = \sum_{h=1}^{L} \sum_{j=1}^{n_h} c_{hj} \left(\hat{H}_y(h_j) - \hat{H}_{\text{Jack}} \right)^2 \tag{8.115}$$

Determine, if possible, the values of weights c_{hj} such that $\hat{v}(\hat{H}_{\text{Jack}})$ can be considered as an estimator of the variance of the sample harmonic mean in stratified sampling. Support your findings with a simulation study. *Hint:* The weights c_{hj} can be a function of N_h/N, which may help to improve the estimator of variance.

Exercise 8.13 Suppose that a sample of size n_h is selected from the hth stratum using the SRSWOR scheme for $h = 1, 2, \ldots, L$. Let r_h be the responses observed in the hth stratum. Assume the data are missing completely at random. Then consider the newly tuned estimator of the population mean $\bar{Y}$ in stratified nonresponse random sampling defined as

$$\bar{y}_{\text{StTuned(cs)}}^{\text{nr}} = \sum_{h=1}^{L} \Omega_h \sum_{j=1}^{r_h} \left\{ (r_h - 1)^2 \bar{w}_h(j) - (r_h - 2) \right\} \bar{y}_h(j) \tag{8.116}$$

where

$$\bar{y}_h(j) = \frac{r_h \bar{y}_h - y_{hj}}{r_h - 1} \tag{8.117}$$

is the hth stratum sample mean $\bar{y}_h = r_h^{-1} \sum_{i=1}^{r_h} y_{hi}$ of the study variable obtained by removing the jth unit from the mean of the responding units, and $\Omega_h = N_h/N$ are the hth stratum weights. The tuned weights $\bar{w}_h(j)$ are computed independently in the h strata such that the chi-squared type distance function defined as

$$D_h = \frac{r_h}{2} \sum_{j=1}^{r_h} q_{hj}^{-1} \left[1 - (r_h - 1)\bar{w}_h(j) - r_h^{-1} \right]^2 \tag{8.118}$$

is optimized, subject to the following two tuning constraints:

$$\sum_{j=1}^{r_h} \bar{w}_h(j) = 1 \tag{8.119}$$

and

$$\sum_{j=1}^{r_h} \bar{w}_h(j) \bar{x}_h(j) = \frac{\bar{x}_h^* - r_h(2 - r_h)\bar{x}_h}{(r_h - 1)^2} \tag{8.120}$$

where

$$\bar{x}_h(j) = \frac{r_h \bar{x}_h - x_{hj}}{r_h - 1} \tag{8.121}$$

is the jth jackknifed estimator obtained from the hth stratum sample mean $\bar{x}_h = r_h^{-1} \sum_{i=1}^{r_h} x_{hi}$ by removing the jth unit. The hth stratum sample mean of the auxiliary variable based on all sampled units in stratum h, $\bar{x}_h^* = n_h^{-1} \sum_{i=1}^{n_h} x_{hi}$, is assumed to be known. For the optimum tuning weights $\bar{w}_h(j)$, find the estimator $\bar{y}_{\text{StTuned(cs)}}^{\text{nr}}$. Suggest doubly jackknifed estimators of variance and a confidence interval estimator. Create a hypothetical pumpkin farm where the pumpkins can be divided into three mutually exclusive strata, say *Sumbo*, *Mumbo*, and *Jumbo* pumpkins, based on their known circumferences. Select a sample of size n using the method of proportional allocation to each stratum. Create an environment where the farmer may find some pumpkins lost or stolen from the sample due to some unavoidable circumstances. For various levels of nonresponse, study the nominally 90%, 95%, and 99% coverages by the confidence interval estimators you suggested. Use R or SAS. Discuss the difficulties you encountered, if any.

Exercise 8.14 Multistage stratified random sampling

Suppose that a population of N units is first subdivided into L mutually exclusive homogeneous subgroups called strata. Let the number of first-stage units (fsu) in the hth stratum ($h = 1, 2, ..., L$) be N_h, and let the total number of first stage units in all the strata taken together be $\sum_{h=1}^{L} N_h = N$. Let the number of second-stage units (ssu) in the ith fsu ($i = 1, 2, ..., N_h$) in the hth stratum be denoted by M_{hi}, the total number of ssu in the hth stratum and in all strata taken together being, respectively, $\sum_{i=1}^{N_h} M_{hi}$ and $\sum_{h=1}^{L} \sum_{i=1}^{N_h} M_{hi}$. Let Y_{hij} be the value of the study variable of the jth ssu in the ith fsu. We define the following totals: Total of the values of the study variable in the ith fsu:

$$Y_{hi} = \sum_{j=1}^{M_{hi}} Y_{hij} \tag{8.122}$$

Mean for the ith fsu:

$$\bar{Y}_{hi} = Y_{hi}/M_{hi} \tag{8.123}$$

Total for the hth stratum:

$$Y_h = \sum_{i=1}^{N_h} \sum_{j=1}^{M_{hi}} Y_{hij} = \sum_{i=1}^{N_h} Y_{hi} = \sum_{i=1}^{N_h} M_{hi} \bar{Y}_{hi} \tag{8.124}$$

Mean per fsu:

$$\bar{Y}_h = Y_h/N_h = \sum_{i=1}^{N_h} M_{hi} \bar{Y}_{hi}/N_h \tag{8.125}$$

Mean per ssu:

$$\bar{\bar{Y}}_h = Y_h \bigg/ \left(\sum_{i=1}^{N_h} M_{hi}\right) = \sum_{i=1}^{N_h} M_{hi} \bar{Y}_{hi} \bigg/ \left(\sum_{i=1}^{N_h} M_{hi}\right) \tag{8.126}$$

Total of the values of the study variable:

$$Y = \sum_{h=1}^{L} Y_h = \sum_{h=1}^{L} \sum_{i=1}^{N_h} Y_{hi} = \sum_{h=1}^{L} \sum_{i=1}^{N_h} \sum_{j=1}^{M_{hi}} Y_{hij} \tag{8.127}$$

Mean per fsu:

$$\bar{Y} = Y/N \tag{8.128}$$

Mean per ssu:

$$\bar{Y}_h = Y \bigg/ \left(\sum_{h=1}^{L} \sum_{i=1}^{N_h} M_{hi} \right) \tag{8.129}$$

From the hth stratum, out of N_h fsu's, n_h fsu are selected using an SRSWOR scheme. Out of the M_{hi} ssu's in the ith selected fsu ($i = 1, 2, ..., n_h$), m_{hi} ssu's are selected using an SRSWOR scheme. The total number of sampled ssu's in the hth stratum is $\sum_{i=1}^{n_h} M_{hi}$ and in all strata taken together is $\sum_{h=1}^{L} \sum_{i=1}^{n_h} m_{hi} = n$, the total sample size. Let y_{hij} ($h = 1, 2, ..., L; i = 1, 2, ..., n_h; j = 1, 2, ..., m_{hi}$) denote the value of the study variable in the jth selected ssu in the ith selected fsu in the hth stratum. From all combined strata, an unbiased estimator of the population total Y is given by

$$\hat{Y}_{\text{unbiased}} = \sum_{h=1}^{L} \frac{N_h}{n_h} \sum_{i=1}^{n_h} \frac{M_{hi}}{m_{hi}} \sum_{j=1}^{m_{hi}} y_{hij} \tag{8.130}$$

Consider the newly semituned estimator of the population total Y in multistage stratified random sampling, defined as

$$\hat{Y}_{\text{Tuned(ms)}} = \sum_{h=1}^{L} \frac{N_h}{n_h} \sum_{i=1}^{n_h} M_{hi} \sum_{j=1}^{m_{hi}} w_{hij} \bar{y}_{hi}(j) \tag{8.131}$$

where

$$\bar{y}_{hi}(j) = \frac{m_{hi} \bar{y}_{hi} - y_{hij}}{m_{hi} - 1} \tag{8.132}$$

Determine the semituned weights w_{hij} such that the chi-square type distance function,

$$D_1 = \frac{1}{2} \sum_{h=1}^{L} \frac{N_h}{n_h} \sum_{i=1}^{n_h} M_{hi} \sum_{j=1}^{m_{hi}} \frac{\left(w_{hij} - m_{hi}^{-1}\right)^2}{q_{hij} m_{hi}^{-1}} \tag{8.133}$$

is optimal, subject to the two tuning constraints

$$\sum_{h=1}^{L} \frac{N_h}{n_h} \sum_{i=1}^{n_h} M_{hi} \sum_{j=1}^{m_{hi}} w_{hij} = N \tag{8.134}$$

and

$$\sum_{h=1}^{L}\frac{N_h}{n_h}\sum_{i=1}^{n_h}M_{hi}(m_{hi}-1)\sum_{j=1}^{m_{hi}}w_{hij}\bar{x}_{hi}(j) = \left(\sum_{h=1}^{L}\frac{N_h}{n_h}\sum_{i=1}^{n_h}M_{hi}\bar{x}_{hi}-X\right) \quad (8.135)$$

where X is the known population total of the auxiliary variable, and where $\sum_{j=1}^{m_{hi}} w_{hij} = 1$, and other symbols have their usual meanings for two-stage sampling. Suggest a doubly jackknifed estimator of variance similar to the one suggested in the chapter, and write a program to study coverage by the 90%, 95%, and 99% confidence intervals. Discuss the difficulties you observed. Also, consider the log-likelihood distance function and study the properties of the resultant estimator. Report your findings in each case.

Tuning using multiauxiliary information

9.1 Introduction

In this chapter, we introduce a new semituned estimator of population mean and estimate its variance in the presence of multiauxiliary information. The new semituning methodology is illustrated by a model used to estimate the weight of pumpkins with the help of three known auxiliary variables viz., their circumferences, amount of fertilizer, and amount of farmyard manure (FYM) used. At the end of the chapter, exercises include fully tuned estimators in the presence of multiauxiliary information.

9.2 Notation

Let y_i and $x_{1i}, x_{2i}, \ldots, x_{ki}, i = 1, 2, \ldots, N$ be the values of the ith unit of the study variable and of the k auxiliary variables, respectively, in the population Ω. Here we consider the problem of estimating the population mean of the study variable

$$\bar{Y} = N^{-1} \sum_{i=1}^{N} y_i \tag{9.1}$$

by assuming that the population means of the auxiliary variables

$$\bar{X}_t = N^{-1} \sum_{i=1}^{N} x_{ti}, \quad t = 1, 2, \ldots, k \tag{9.2}$$

are known.

Let y_i and $x_{1i}, x_{2i}, \ldots, x_{ki}, i = 1, 2, \ldots, n$ be the values of the ith unit of the study variable and k auxiliary variables in the sample s drawn using a simple random sampling scheme.

Let

$$\bar{y} = n^{-1} \sum_{i=1}^{n} y_i \tag{9.3}$$

and

$$\bar{x}_t = n^{-1}\sum_{i=1}^{n} x_{ti}, \quad t=1,2,\ldots,k \tag{9.4}$$

be the sample means for the study and auxiliary variables, respectively.

9.3 Tuning with a chi-square distance function

We consider here a new semituned estimator of the population mean $\bar{Y}$, defined as

$$\bar{y}_{\text{MTuned(cs)}} = \sum_{j \in s}\left[(1-n)w_j + 1\right]\bar{y}(j) \tag{9.5}$$

where

$$\bar{y}(j) = \frac{n\bar{y} - y_j}{n-1} \tag{9.6}$$

is the sample mean of the study variable obtained by removing the jth unit from the sample s, and w_j is the semituned weight such that the following $(k+1)$ constraints are satisfied:

$$\sum_{j \in s} w_j = 1 \tag{9.7}$$

and

$$\sum_{j \in s} w_j \bar{x}_t(j) = \frac{(n\bar{x}_t - \bar{X}_t)}{n-1} \tag{9.8}$$

where

$$\bar{x}_t(j) = \frac{n\bar{x}_t - x_{tj}}{n-1}, \quad t=1,2,\ldots,k \tag{9.9}$$

is the sample mean of the auxiliary variable obtained by removing the jth unit from the sample s. The calibration constraint (9.8) is due to Deville and Särndal (1992), and constraint (9.7) is due to Owen (2001). Note that here we are not using jackknifed weights, so we will refer to this as a semituned method.

First, we suggest semituning the weights w_j so that the chi-square type distance function, defined as

$$(2^{-1}n)\sum_{j \in s} q_j^{-1}\left(w_j - n^{-1}\right)^2 \tag{9.10}$$

is minimized, subject to tuning constraints (9.7) and (9.8), where q_j is some choice of weights.

The Lagrange function is given by

$$L_1 = (2^{-1}n) \sum_{j \in s} q_j^{-1} (w_j - n^{-1})^2 - \lambda_0 \left\{ \sum_{j \in s} w_j - 1 \right\} \\ - \sum_{t=1}^{k} \lambda_t \left\{ \sum_{j \in s} w_j \bar{x}_t(j) - (n-1)^{-1} (n\bar{x}_t - \bar{X}_t) \right\} \quad (9.11)$$

where λ_0 and λ_t are Lagrange multiplier constants.

On setting

$$\frac{\partial L_1}{\partial w_j} = 0 \quad (9.12)$$

we have

$$w_j = \frac{1}{n} \left\{ 1 + q_j \lambda_0 + q_j \sum_{t=1}^{k} \lambda_t \bar{x}_t(j) \right\} \quad (9.13)$$

Using Equation (9.13) in Equations (9.7) and (9.8), a set of normal equations, to find the optimum values of λ_0 and λ_t is given by

$$A_{(k+1) \times (k+1)} \lambda_{(k+1) \times 1} = C_{(k+1) \times 1} \quad (9.14)$$

where

$$A_{(k+1) \times (k+1)} = \begin{bmatrix} \sum_{j \in s} q_j, & \sum_{j \in s} q_j \bar{x}_1(j), & \cdots, & \sum_{j \in s} q_j \bar{x}_k(j) \\ \sum_{j \in s} q_j \bar{x}_1(j), & \sum_{j \in s} q_j \{\bar{x}_1(j)\}^2, & \cdots, & \sum_{j \in s} q_j \bar{x}_1(j) \bar{x}_k(j) \\ \sum_{j \in s} q_j \bar{x}_k(j) \bar{x}_1(j), & \sum_{j \in s} q_j \bar{x}_k(j) \bar{x}_2(j), & \cdots, & \sum_{j \in s} q_j \{\bar{x}_k(j)\}^2 \end{bmatrix}$$

$$\lambda_{(k+1) \times 1} = \begin{bmatrix} \lambda_0 \\ \lambda_1 \\ \vdots \\ \lambda_k \end{bmatrix}_{(k+1) \times 1} \quad \text{and} \quad C = \begin{bmatrix} 0 \\ \left(\frac{n}{n-1}\right)(n\bar{x}_1 - \bar{X}_1) - \sum_{j \in s} \bar{x}_1(j) \\ \vdots \\ \left(\frac{n}{n-1}\right)(n\bar{x}_k - \bar{X}_k) - \sum_{j \in s} \bar{x}_k(j) \end{bmatrix}$$

The semituned weights w_j are given by

$$w_j = \frac{1}{n}\left[1 + \sum_{t=1}^{k} \Psi_{tj}\left\{\left(\frac{n}{n-1}\right)(n\bar{x}_t - \bar{X}_t) - \sum_{j\in s}\bar{x}_t(j)\right\}\right] \quad (9.15)$$

with

$$\Psi_{tj} = q_j A^{-1}_{(k+1)\times(k+1)} H_{(k+1)\times j}$$

where

$$H^t_{j\times(k+1)} = [1, \bar{x}_1(j), \bar{x}_2(j), \ldots, \bar{x}_k(j)]_{j\times(k+1)}$$

Thus, under the chi-square (cs) distance function, the newly semituned estimator (9.5) of the population mean becomes

$$\bar{y}_{\text{MTuned(cs)}} = \frac{1}{n}\left[\sum_{j\in s}\bar{y}(j) + \hat{\beta}_{\text{MTuned}}C\right] \quad (9.16)$$

where

$$\hat{\beta}_{\text{MTuned}} = A^{-1}_{(k+1)\times(k+1)} H_{(k+1)\times 1}\left[q_j \bar{y}(j)\right]_{(1\times n)} = (\hat{\beta}_0, \hat{\beta}_1, \ldots, \hat{\beta}_k)^t$$

Note that for the choice of weights $q_j = 1$, it can be easily seen that the estimator $\bar{y}_{\text{MTuned(cs)}}$ becomes a multiple linear regression type estimator of the form

$$\bar{y}_{\text{MTuned(cs)}} = \bar{y} + \sum_{t=1}^{k}\hat{\beta}_t(\bar{X}_t - \bar{x}_t) \quad (9.17)$$

9.3.1 Estimation of variance and coverage

We suggest an adjusted weighted estimator of the variance of the estimator $\bar{y}_{\text{MTuned(cs)}}$, defined by

$$\hat{v}_{\text{MTuned(cs)}} = n(n-1)^3 \sum_{j\in s} w_j^2 \left\{\bar{y}^{\text{MTuned(cs)}}_{(j)} - \bar{y}_{\text{MTuned(cs)}}\right\}^2 \quad (9.18)$$

Such adjustment may vary from survey to survey or according to the researchers' interest based on the type of data at hand. Note that each newly semituned doubly jackknifed estimator of the population mean is given by

$$\bar{y}^{\text{Tuned(cs)}}_{(j)} = \frac{n\bar{y}_{\text{Tuned(cs)}} - n\{(n-1)w_j + 1\}\bar{y}(j)}{n-1} \tag{9.19}$$

for $j = 1, 2, \ldots, n$.

The attained coverage by the nominally $(1-\alpha)100\%$ confidence interval estimates constructed using this newly semituned estimator of the population mean is estimated by observing the proportion of times the true population mean $\bar{Y}$ falls within the interval estimates given by

$$\bar{y}_{\text{MTuned(cs)}} \pm t_{\alpha/2}(\text{df} = n-k-1)\sqrt{\hat{v}_{\text{MTuned(cs)}}} \tag{9.20}$$

Here we generated a special type of population having one study variable Y, three auxiliary variables X_1, X_2, and X_3 where the ith population unit takes value for study variable y_i, and x_{1i}, x_{2i}, and x_{3i} for the three auxiliary variables. The value y_i for the study variable Y and the value x_{1i} of the first auxiliary variable X_1 are given by

$$y_i = \bar{Y} + \sigma_y y_i^* \tag{9.21}$$

and

$$x_{1i} = \bar{X}_1 + \sigma_{x_1} x_{1i}^* \sqrt{1 - \rho_{yx_1}^2} + \rho_{yx_1} \sigma_{x_1} y_i^* \tag{9.22}$$

where y_i^* and x_{1i}^* are independent standard normal variables each with mean zero and variance one, and with $\rho_{yx_1} = 0.8$. Similarly, we also generated two variables, x_{2i} and x_{3i}, whose correlations with the study variable were 0.3 and 0.6, respectively. The standard deviations of the three auxiliary variables were $\sigma_{x_1} = 10$, $\sigma_{x_2} = 2$, and $\sigma_{x_3} = 2$, and that of the study variable was $\sigma_y = 5$.

We studied coverage of the 90%, 95%, and 99% confidence intervals, formed as in Equation (9.20), by selecting 10,000 random samples from the population with one study variable and three auxiliary variables. The results obtained through the simulation are presented in Table 9.1.

Table 9.1 shows that the coverage by the newly semituned multiauxiliary estimator of the population mean performs very well for moderate sample sizes. We note in particular that the attained coverage is within 1% of the nominal coverage for sample sizes 14–40 for the 90% intervals, and within 1% of the nominal coverage for sample sizes 11–40 for 95% and 99% intervals.

9.3.2 R code

The following R code, **PUMPKIN101.R**, was used to study the coverage by the newly semituned multiauxiliary information estimator based on the chi-square type distance function.

Table 9.1 **The newly semituned multiauxiliary information estimator**

Sample size (n)	90% coverage	95% coverage	99% coverage
11	0.9123	0.9591	0.9933
12	0.9146	0.9602	0.9937
13	0.9131	0.9584	0.9935
14	0.9059	0.9535	0.9924
15	0.9059	0.9577	0.9929
16	0.9074	0.9525	0.9909
17	0.9078	0.9555	0.9914
18	0.9074	0.9532	0.9904
19	0.9038	0.9541	0.9906
20	0.9026	0.9525	0.9921
21	0.9038	0.9521	0.9899
22	0.9038	0.9512	0.9915
23	0.8996	0.9502	0.9900
24	0.8989	0.9510	0.9880
25	0.9009	0.9487	0.9906
26	0.9072	0.9512	0.9899
27	0.9021	0.9520	0.9907
28	0.9069	0.9536	0.9914
29	0.9031	0.9518	0.9911
30	0.9034	0.9522	0.9908
31	0.9035	0.9549	0.9916
32	0.8984	0.9477	0.9886
33	0.8997	0.9511	0.9912
34	0.9001	0.9508	0.9899
35	0.9000	0.9485	0.9898
36	0.9017	0.9510	0.9894
37	0.9029	0.9499	0.9904
38	0.8975	0.9496	0.9897
39	0.8984	0.9473	0.9903
40	0.9023	0.9504	0.9895

PROGRAM NAME: PUMPKIN101.R
```
set.seed(2013)
N<-10000
x1s<-rnorm(N, 0, 1)
x2s<-rnorm(N, 0, 1)
x3s<-rnorm(N, 0, 1)
y1s<-rnorm(N, 0, 1)
y<-1700 + 5*y1s
x1<-500 + 10*x1s*sqrt(1-0.8^2) + 0.8*10*y1s
x2<-50 + 2*x2s*sqrt(1-0.3^2) + 0.3*2*y1s
x3<-300 + 2*x3s*sqrt(1-0.6^2) + 0.6*2*y1s
XMEAN1 = mean(x1)
XMEAN2 = mean(x2)
```

```
XMEAN3 = mean(x3)
YM<-mean(y)
nreps<-10000
ESTP=rep(0,nreps)
ci1.max=ci1.min=ci2.max=ci2.min=ci3.max=ci3.min=vESTP=ESTP
for (n in seq(11,40,1))
  {
  for (r in 1:nreps)
    {
    us<-sample(N,n)
    XS1<-x1[us];XS2<-x2[us];XS3<-x3[us];YS<-y[us]
    X1M<-mean(XS1);X2M<-mean(XS2);X3M<-mean(XS3)
    YMJ = (sum(YS)-YS)/(n-1)
    X1MJ = (sum(XS1)-XS1)/(n-1)
    X2MJ = (sum(XS2)-XS2)/(n-1)
    X3MJ = (sum(XS3)-XS3)/(n-1)
    A<-matrix(ncol=4,nrow=4)
    A[1,1] = n
    A[1,2] = A[2,1] = sum(X1MJ)
    A[1,3] = A[3,1] = sum(X2MJ)
    A[1,4] = A[4,1] = sum(X3MJ)
    A[2,2] = sum(X1MJ^2)/(n^2)
    A[2,3] = A[3,2] = sum(X1MJ*X2MJ)
    A[2,4] = A[4,2] = sum(X1MJ*X3MJ)
    A[3,3] = sum(X2MJ^2)/(n^2)
    A[3,4] = A[4,3] = sum(X2MJ*X3MJ)
    A[4,4] = sum(X3MJ^2)/(n^2)
    B<-rep(0,4)
    B[2] = n*(n*X1M-XMEAN1)/(n-1)-sum(X1MJ)
    B[3] = n*(n*X2M-XMEAN2)/(n-1)-sum(X2MJ)
    B[4] = n*(n*X3M-XMEAN3)/(n-1)-sum(X3MJ)
    solve(A,B)->AL
    WI = (1/n)*(1+AL[1]+AL[2]*X1MJ+AL[3]*X2MJ+AL[4]*X3MJ)
    EST_I = n * ((1-n)*WI +1) * YMJ
    ESTP[r] = sum(EST_I)
    ESTP_J = (ESTP[r]-EST_I)/(n-1)
    ESTP[r] = ESTP[r]/n
    vj<-(WI^2)*(ESTP_J-ESTP[r])^2
    vESTP[r] = n*(n-1)^3*sum(vj)
    ci1.max[r]<- ESTP[r]+qt(0.95,n-4)*sqrt(vESTP[r])
    ci1.min[r]<- ESTP[r]-qt(0.95,n-4)*sqrt(vESTP[r])
    ci2.max[r]<- ESTP[r]+qt(0.975,n-4)*sqrt(vESTP[r])
    ci2.min[r]<- ESTP[r]-qt(0.975,n-4)*sqrt(vESTP[r])
    ci3.max[r]<- ESTP[r]+qt(0.995,n-4)*sqrt(vESTP[r])
    ci3.min[r]<- ESTP[r]-qt(0.995,n-4)*sqrt(vESTP[r])
    }
```

```
round(sum(ci1.min<YM & ci1.max>YM)/nreps,4)->cov1
round(sum(ci2.min<YM & ci2.max>YM)/nreps,4)->cov2
round(sum(ci3.min<YM & ci3.max>YM)/nreps,4)->cov3
cat(n,cov1,cov2,cov3,'\n')
}
```

9.3.3 Numerical illustration

In the following example, we explain the intermediate steps involved in the computation of the proposed semituned estimator in the presence of three auxiliary variables.

Example 9.1 Suppose that we took a sample of $n = 29$ pumpkins from the preceding population and measured y (the weights in lbs), x_1 (the circumference in inches), x_2 (the amount of special fertilizer used in lbs/plot), and x_3 (the amount of FYM used in lbs/plot).

Weight, y_i	Circumference, x_{1i}	Fertilizer, x_{2i}	FYM, x_{3i}
1695	496	53	297
1695	490	47	300
1702	509	43	299
1699	509	50	301
1699	500	49	299
1695	486	49	296
1698	501	49	298
1700	502	50	300
1690	479	48	296
1698	497	51	299
1703	512	50	300
1700	501	50	299
1691	481	49	298
1691	492	48	295
1694	493	49	301
1695	490	53	296
1698	492	51	298
1702	492	51	300
1703	511	49	302
1692	498	45	297
1693	485	48	299
1703	506	50	302
1696	493	51	298
1698	495	49	301
1704	510	50	300
1698	507	49	300
1702	501	49	302
1692	470	46	296
1695	485	48	299

Construct the 90% confidence interval estimate of the average weight $(\bar{Y})$ by assuming that the population means of the three auxiliary variables, $\bar{X}_1 = 500$, $\bar{X}_2 = 50$, and $\bar{X}_3 = 300$, are known.

Solution. An R program can be easily written to calculate the semituned weights and jackknife estimates of the average pumpkin weights:

w_j	$\bar{y}_{(j)}^{\text{MTuned(cs)}}$
0.03448270	1697.270
0.03448280	1697.275
0.03448285	1697.286
0.03448273	1697.277
0.03448276	1697.278
0.03448277	1697.273
0.03448276	1697.277
0.03448274	1697.278
0.03448279	1697.268
0.03448273	1697.275
0.03448273	1697.282
0.03448274	1697.278
0.03448277	1697.268
0.03448278	1697.269
0.03448276	1697.272
0.03448270	1697.270
0.03448273	1697.275
0.03448273	1697.280
0.03448275	1697.283
0.03448283	1697.272
0.03448278	1697.272
0.03448273	1697.283
0.03448273	1697.273
0.03448276	1697.277
0.03448273	1697.283
0.03448275	1697.276
0.03448275	1697.282
0.03448283	1697.273
0.03448278	1697.274

The overall tuned estimate of the average weight is $\bar{y}_{\text{MTuned(cs)}} = 1697.276$ lbs with a standard error of $SE(\bar{y}_{\text{MTuned(cs)}}) = 0.7088123$ lbs. Thus, the 90% confidence interval estimate of the average weight of the pumpkins on this farm is between 1696.065 and 1698.487 lbs.

9.3.4 R code used for illustration

The following R code, **PUMPKIN101EX.R**, was used in the preceding numerical illustration.

PROGRAM PUMPKIN101EX.R

```
n<-29
YS<-c(1695,1695,1702,1699,1699,1695,1698,1700,1690,1698,
1703,1700,1691,1691,1694,1695,1698,1702,1703,1692,
1693,1703,1696,1698,1704,1698,1702,1692,1695)
XS1<-c(496,490,509,509,500,486,501,502,479,497,512,501,
481,492,493,490,492,492,511,498,485,506,493,495,
510,507,501,470,485)
XS2<-c(53,47,43,50,49,49,49,50,48,51,50,50,49,48,49,53,
51,51,49,45,48,50,51,49,50,49,49,46,48)
XS3<-c(297,300,299,301,299,296,298,300,296,299,300,299,
298,295,301,296,298,300,302,297,299,302,298,301,
300,300,302,296,299)
XMEAN1 = 500;XMEAN2 = 50;XMEAN3 = 300
XMEAN1=mean(x1)
XMEAN2=mean(x2)
XMEAN3=mean(x3)
YM<-mean(YS)
X1M<-mean(XS1);X2M<-mean(XS2);X3M<-mean(XS3)
YMJ = (sum(YS)-YS)/(n-1)
X1MJ = (sum(XS1)-XS1)/(n-1)
X2MJ = (sum(XS2)-XS2)/(n-1)
X3MJ = (sum(XS3)-XS3)/(n-1)
A<-matrix(ncol=4,nrow=4)
A[1,1] = n
A[1,2] = A[2,1] = sum(X1MJ)
A[1,3] = A[3,1] = sum(X2MJ)
A[1,4] = A[4,1] = sum(X3MJ)
A[2,2] = sum(X1MJ^2)/(n^2)
A[2,3] = A[3,2] = sum(X1MJ*X2MJ)
A[2,4] = A[4,2] = sum(X1MJ*X3MJ)
A[3,3] = sum(X2MJ^2)/(n^2)
A[3,4] = A[4,3] = sum(X2MJ*X3MJ)
A[4,4] = sum(X3MJ^2)/(n^2)
B<-rep(0,4)
B[2] = n*(n*X1M-XMEAN1)/(n-1)-sum(X1MJ)
B[3] = n*(n*X2M-XMEAN2)/(n-1)-sum(X2MJ)
B[4] = n*(n*X3M-XMEAN3)/(n-1)-sum(X3MJ)
solve(A,B)->AL
WI = (1/n)*(1+AL[1]+AL[2]*X1MJ+AL[3]*X2MJ+ AL[4]*X3MJ)
EST_I = n *( (1-n)* WI + 1) * YMJ
```

```
ESTP = sum(EST_I)
ESTP_J = (ESTP-EST_I)/(n-1)
ESTP = ESTP/n
vj<-(WI^2)*(ESTP_J-ESTP)^2
vESTP = n*(n-1)^3*sum(vj)
L<-ESTP+qt(.05,n-4)*sqrt(vESTP)
U<-ESTP+qt(.95,n-4)*sqrt(vESTP)
cbind(WI,sum(WI),ESTP_J)
cat("Tuned estimate:", ESTP, "SE: ",vESTP^.5,'\n')
cat("Confidence Interval:"," ", L,"; ", U,'\n')
```

9.4 Tuning with empirical log-likelihood function

We define a semituned empirical log-likelihood (ell) estimator of the population mean $\bar{Y}$ as

$$\bar{y}_{\text{MTuned(ell)}} = \sum_{j \in s} w_j^* \bar{y}(j) \tag{9.23}$$

where w_j^* are the positive semituned weights such that the following $(k+1)$ constraints are satisfied:

$$\sum_{j \in s} w_j^* = 1 \tag{9.24}$$

and

$$\sum_{j \in s} w_j^* \left\{ \bar{x}_t(j) - \frac{(n\bar{x}_t - \bar{X}_t)}{n-1} \right\} = 0, \quad t = 1, 2, \ldots, k \tag{9.25}$$

Here we suggest semituning the weights w_j^* such that the log-likelihood distance function defined by

$$\sum_{j \in s} \frac{\log\left(w_j^*\right)}{n} \tag{9.26}$$

is optimized, subject to the tuning constraints (9.24) and (9.25).

The Lagrange function is then given by

$$L_2 = \sum_{j \in s} \frac{\log\left(w_j^*\right)}{n} - \lambda_0^* \left\{ \sum_{j \in s} w_j^* - 1 \right\} \\ - \sum_{t=1}^{k} \lambda_t^* \left\{ \sum_{j \in s} w_j^* \left\{ \bar{x}_t(j) - \left(\frac{n\bar{x}_t - \bar{X}_t}{n-1}\right) \right\} \right\} \tag{9.27}$$

where λ_0^* and λ_t^* are Lagrange multiplier constants.

On setting

$$\frac{\partial L_2}{\partial w_j^*} = 0 \tag{9.28}$$

we have

$$w_j^* = \frac{1}{n\left[1 + \sum_{t=1}^{k} \lambda_t^* \left\{ \bar{x}_t(j) - \frac{(n\bar{x}_t - \bar{X}_t)}{n-1} \right\} \right]} \tag{9.29}$$

Constraints (9.24) and (9.25) yield $\lambda_0^* = 1$, and λ_t^* is a solution to the single parametric equation

$$\sum_{j \in s} \frac{\sum_{t=1}^{k} \left\{ \bar{x}_t(j) - \frac{(n\bar{x}_t - \bar{X}_t)}{n-1} \right\}}{1 + \sum_{t=1}^{k} \lambda_t^* \left\{ \bar{x}_t(j) - \frac{(n\bar{x}_t - \bar{X}_t)}{n-1} \right\}} = 0 \tag{9.30}$$

Thus, under the ell distance function, the newly semituned estimator (9.23) of the population mean becomes

$$\bar{y}_{\text{MTuned(ell)}} = \frac{1}{n} \sum_{j \in s} \frac{\bar{y}(j)}{\left\{ 1 + \sum_{t=1}^{k} \lambda_t^* \left(\bar{x}_t(j) - \frac{(n\bar{x}_t - \bar{X}_t)}{n-1} \right) \right\}} \tag{9.31}$$

9.4.1 Estimation of variance and coverage

As before, we suggest here an adjusted estimator of the variance of the estimator $\bar{y}_{\text{MTuned(ell)}}$:

$$\hat{v}_{\text{MTuned(ell)}} = n(n-1)^3 \sum_{j \in s} \left(w_j^* \right)^2 \left(\bar{y}_{(j)}^{\text{MTuned(ell)}} - \bar{y}_{\text{MTuned(ell)}} \right)^2 \tag{9.32}$$

Note that each newly semituned ell doubly jackknifed estimator of the population mean is given by

$$\bar{y}_{(j)}^{\text{Tuned(ell)}} = \frac{n \bar{y}_{\text{Tuned(ell)}} - n w_j^* \bar{y}_n(j)}{n-1} \tag{9.33}$$

for $j = 1, 2, \ldots, n$.

Attained coverage by the nominally $(1-\alpha)100\%$ confidence interval estimates constructed using this newly semituned ell estimator of the population mean is estimated by observing the proportion of times the true population mean $\bar{Y}$ falls within the interval estimates given by

$$\bar{y}_{\text{MTuned(ell)}} \mp t_{\alpha/2}(\text{df}=n-k-1)\sqrt{\hat{v}_{\text{MTuned(ell)}}} \tag{9.34}$$

Here we generated a special type of population having one study variable Y and three auxiliary variables X_t, $t=1,2,3$. The values for the ith units y_i and three auxiliary variables x_{ti}, $t=1,2,3$ of the four variable are simulated as follows:

$$y_i = \bar{Y} + \sigma_y y_i^* \tag{9.35}$$

and

$$x_{ti} = \bar{X}_t + \sigma_{x_t} x_{ti}^* \sqrt{1-\rho_{yx_t}^2} + \rho_{yx_t}\sigma_{x_t}y_i^* \tag{9.36}$$

where y_i^*, and x_{ti}^*, $t=1,2,3$ are independent standard normal variables with means zero and variances one. We set $\rho_{yx_1}=0.8, \rho_{yx_2}=0.3$, and $\rho_{yx_3}=0.6$. The values of the standard deviations of study variable, and the three auxiliary variables are set at $\sigma_y=5$, $\sigma_{x_1}=10$, $\sigma_{x_2}=2$, and $\sigma_{x_3}=2$.

We studied coverage of the 90%, 95%, and 99% confidence intervals based on the newly semituned estimator of the population mean by selecting 10,000 random samples from the population above. The values of λ_1^*, λ_2^*, and λ_3^* are approximated, under certain assumptions, by solving the following system of linear equations:

$$\begin{bmatrix} \sum_{j\in s}\Phi_{1j}^2, & \sum_{j\in s}\Phi_{1j}\Phi_{2j}, & \sum_{j\in s}\Phi_{1j}\Phi_{3j} \\ \sum_{j\in s}\Phi_{1j}\Phi_{2j}, & \sum_{j\in s}\Phi_{2j}^2, & \sum_{j\in s}\Phi_{2j}\Phi_{3j} \\ \sum_{j\in s}\Phi_{1j}\Phi_{3j}, & \sum_{j\in s}\Phi_{2j}\Phi_{3j}, & \sum_{j\in s}\Phi_{3j}^2 \end{bmatrix} \begin{bmatrix} \lambda_1^* \\ \lambda_2^* \\ \lambda_3^* \end{bmatrix} \approx \begin{bmatrix} \sum_{j\in s}\Phi_{1j} \\ \sum_{j\in s}\Phi_{2j} \\ \sum_{j\in s}\Phi_{3j} \end{bmatrix} \tag{9.37}$$

where

$$\Phi_{tj} = \left\{ \bar{x}_t(j) - \frac{(n\bar{x}_t - \bar{X}_t)}{n-1} \right\} \tag{9.38}$$

for $t=1,2,3$ are three pivots.

The use of values of λ_1^*, λ_2^*, and λ_3^* from Equation (9.37) increases the speed of the program compared to using an iterative method for finding a nonlinear solution to this nonlinear system of equations in Equation (9.30). The results obtained through the simulation are presented in Table 9.2.

Table 9.2 **Newly semituned empirical likelihood estimator**

Sample size (n)	90% coverage	95% coverage	99% coverage
8	0.8212	0.8558	0.9084
10	0.8814	0.9063	0.9432
12	0.9031	0.9258	0.9555
14	0.9215	0.9415	0.9643
16	0.9364	0.9530	0.9747
18	0.9479	0.9630	0.9779
20	0.9518	0.9652	0.9818
22	0.9618	0.9740	0.9856
24	0.9660	0.9751	0.9862
26	0.9728	0.9799	0.9887
28	0.9729	0.9793	0.9891
30	0.9766	0.9840	0.9918
32	0.9783	0.9856	0.9919
34	0.9810	0.9872	0.9933
36	0.9839	0.9891	0.9949

Table 9.2 shows that the attained coverages by the newly semituned ell estimator of the population mean remain close to the coverage attained by the chi-square distance function. In particular, the nominal 90% coverage is estimated as 90.31% for a sample of 12 pumpkins, the nominal 95% coverage is estimated as 95.30% for a sample of 16 pumpkins, and the nominal 99% coverage is estimated as 98.91% for a sample of 28 pumpkins.

Thus, for estimating average pumpkin weights using known multiauxiliary information, the newly semituned ell estimator $\bar{y}_{\text{MTuned(ell)}}$ could perform better because it does not take negative values and does not show over coverage for moderate sample sizes.

9.4.2 R code

We used the following R code, **PUMPKIN102.R**, to study the coverage of intervals constructed using the newly semituned ell estimator.

PROGRAM NAME: PUMPKIN102.R
```
set.seed(2013)
N<-10000
x1s<-rnorm(N, 0, 1)
x2s<-rnorm(N, 0, 1)
x3s<-rnorm(N, 0, 1)
y1s<-rnorm(N, 0, 1)
y<-1700 + 5*y1s
x1<-500 + 10*x1s*sqrt(1-0.8^2) + 0.8*10*y1s
x2<-50 + 2*x2s*sqrt(1-0.7^2) + 0.7*2*y1s
x3<-300 + 2*x3s*sqrt(1-0.6^2) + 0.6*2*y1s
XMEAN1 = mean(x1)
```

```
XMEAN2 = mean(x2)
XMEAN3 = mean(x3)
YM<-mean(y)
nreps<-10000
ESTP=rep(0,nreps)
ci1.max=ci1.min=ci2.max=ci2.min=ci3.max=ci3.min=vESTP=ESTP
for (n in seq(8,36,2))
  {
  for (r in 1:nreps)
      {
      us<-sample(N,n)
      XS1<-x1[us];XS2<-x2[us];XS3<-x3[us];YS<-y[us]
      X1M<-mean(XS1);X2M<-mean(XS2);X3M<-mean(XS3)
      YMJ = (sum(YS)-YS)/(n-1)
      X1MJ = (sum(XS1)-XS1)/(n-1)
      X2MJ = (sum(XS2)-XS2)/(n-1)
      X3MJ = (sum(XS3)-XS3)/(n-1)
      PHI1 = X1MJ-(n*X1M-XMEAN1)/(n-1)
      PHI2 = X2MJ-(n*X2M-XMEAN2)/(n-1)
      PHI3 = X3MJ-(n*X3M-XMEAN3)/(n-1)
      A<-matrix(ncol=3,nrow=3)
      A[1,1] = sum(PHI1^2)
      A[1,2] = A[2,1] = sum(PHI1*PHI2)
      A[1,3] = A[3,1] = sum(PHI1*PHI3)
      A[2,2] = sum(PHI2^2)
      A[2,3] = A[3,2] = sum(PHI2*PHI3)
      A[3,3] = sum(PHI3^2)
      B<-rep(0,3)
      B[1] = sum(PHI1)
      B[2] = sum(PHI2)
      B[3] = sum(PHI3)
      solve(A/n,B)->AL
      WI = (1/n)*(1/(1+AL[1]*PHI1+AL[2]*PHI2+AL[3]*PHI3))
      EST_I = n * WI * YMJ
      ESTP[r] = sum(EST_I)
      ESTP_J = (ESTP[r]-EST_I)/(n-1)
      ESTP[r] = ESTP[r]/n
      vj<-(WI^2)*(ESTP_J-ESTP[r])^2
      vESTP[r] = n*(n-1)^3*sum(vj)
      ci1.max[r]<- ESTP[r]+qt(0.95,n-4)*sqrt(vESTP[r])
      ci1.min[r]<- ESTP[r]-qt(0.95,n-4)*sqrt(vESTP[r])
      ci2.max[r]<- ESTP[r]+qt(0.975,n-4)*sqrt(vESTP[r])
      ci2.min[r]<- ESTP[r]-qt(0.975,n-4)*sqrt(vESTP[r])
      ci3.max[r]<- ESTP[r]+qt(0.995,n-4)*sqrt(vESTP[r])
      ci3.min[r]<- ESTP[r]-qt(0.995,n-4)*sqrt(vESTP[r])
      }
```

```
round(sum(ci1.min<YM & ci1.max>YM)/nreps,4)->cov1
round(sum(ci2.min<YM & ci2.max>YM)/nreps,4)->cov2
round(sum(ci3.min<YM & ci3.max>YM)/nreps,4)->cov3
cat(n, cov1,cov2,cov3,'\n')
}
```

9.4.3 Numerical illustration

The following example illustrates the main steps taken in the computation of the semi-tuned ell estimator using three auxiliary variables.

Example 9.2 Suppose that we took a sample of $n = 29$ pumpkins from the preceding population and measured y (the weights in lbs), x_1 (the circumference in inches), x_2 (the amount of special fertilizer used in lbs/plot), and x_3 (the amount of FYM used in lbs/plot).

Weight, y_i	Circumference, x_{1i}	Fertilizer, x_{2i}	FYM, x_{3i}
1704	503	51	300
1708	522	54	303
1690	485	44	301
1704	501	48	302
1697	507	49	300
1700	504	47	301
1697	505	50	299
1703	509	49	303
1701	505	49	299
1707	508	51	301
1693	494	48	298
1703	502	53	299
1699	504	50	299
1698	502	48	301
1699	495	52	300
1692	497	50	299
1697	489	50	300
1698	485	50	297
1696	490	49	299
1690	482	49	296
1700	500	47	300
1703	509	53	301
1706	513	51	303
1704	509	53	302
1706	504	53	301
1700	494	50	298
1699	499	49	300
1701	499	51	300
1700	500	53	300

Construct the 90% confidence interval estimate of the average weight ($\bar{Y}$) by assuming that the population means of the three auxiliary variables, $\bar{X}_1 = 500$, $\bar{X}_2 = 50$, and $\bar{X}_3 = 300$, are known.

Solution. One can compose a program to calculate the tuning weights and the jackknife estimates of the average weights as

w_j^*	$\bar{y}_{(j)}^{\text{MTuned(ell)}}$
0.03384245	1701.097
0.03013577	1707.627
0.03805451	1693.663
0.03410544	1700.634
0.03202130	1704.295
0.03257111	1703.331
0.03262418	1703.234
0.03226295	1703.877
0.03232396	1703.767
0.03260040	1703.287
0.03517541	1698.737
0.03452480	1699.895
0.03292254	1702.711
0.03348761	1701.715
0.03697995	1695.567
0.03517437	1698.737
0.03854410	1692.811
0.03902799	1691.960
0.03735532	1694.902
0.03946780	1691.174
0.03349629	1701.702
0.03291326	1702.732
0.03170869	1704.856
0.03320083	1702.227
0.03448996	1699.960
0.03589429	1697.480
0.03447456	1699.978
0.03516481	1698.766
0.03552910	1698.123

The overall tuned estimate of the average weight of a pumpkin on such type of a farm is $\bar{y}_{\text{MTuned(ell)}} = 1699.96$ lbs with a standard deviation of $SE(\bar{y}_{\text{MTuned(ell)}}) = 630.7616$ lbs. Thus, the 90% confidence interval estimate of the average weight of the pumpkins on this farm is 622.5305–2777.3900 lbs.

9.4.4 R code used for illustration

We used the following R code, **PUMPKIN102EX.R**, to solve the preceding numerical illustration.

PROGRAM PUMPKIN102EX.R
```
n<-29
YS<-c(1704,1708,1690,1704,1697,1700,1697,1703,1701,1707,
1693,1703,1699,1698,1699,1692,1697,1698,1696,1690,1700,
1703,1706,1704,1706,1700,1699,1701,1700)
XS1<-c(503,522,485,501,507,504,505,509,505,508,494,502,
504,502,495,497,489,485,490,482,500,509,513,509,504,494,499,499,500)
XS2<-c(51,54,44,48,49,47,50,49,49,51,48,53,50,48,52,50,
50,50,49,49,47,53,51,53,53,50,49,51,53)
XS3<-c(300,303,301,302,300,301,299,303,299,301,298,299,
299,301,300,299,300,297,299,296,300,301,303,302,301,298,300,300,300)
XMEAN1 = 500;XMEAN2 = 50;XMEAN3 = 300
YM<-mean(YS)
X1M<-mean(XS1);X2M<-mean(XS2);X3M<-mean(XS3)
YMJ = (sum(YS)-YS)/(n-1)
X1MJ = (sum(XS1)-XS1)/(n-1)
X2MJ = (sum(XS2)-XS2)/(n-1)
X3MJ = (sum(XS3)-XS3)/(n-1)
PHI1 = X1MJ-(n*X1M-XMEAN1)/(n-1)
PHI2 = X2MJ-(n*X2M-XMEAN2)/(n-1)
PHI3 = X3MJ-(n*X3M-XMEAN3)/(n-1)
A<-matrix(ncol=3,nrow=3)
A[1,1] = sum(PHI1^2)
A[1,2] = A[2,1] = sum(PHI1*PHI2)
A[1,3] = A[3,1] = sum(PHI1*PHI3)
A[2,2] = sum(PHI2^2)
A[2,3] = A[3,2] = sum(PHI2*PHI3)
A[3,3] = sum(PHI3^2)
B<-rep(0,3)
B[1] = sum(PHI1);B[2] = sum(PHI2);B[3] = sum(PHI3)
solve(A,B)->AL
WI = (1/n)*(1/(1+AL[1]*PHI1+AL[2]*PHI2+AL[3]*PHI3))
EST_I = n * WI * YMJ
ESTP = sum(EST_I)
ESTP_J = (ESTP-EST_I)/(n-1)
ESTP = ESTP/n
vj<-(WI^2)*(ESTP_J-ESTP)^2
vESTP = n*(n-1)^3*sum(vj)
L<-ESTP+qt(.05,n-4)*sqrt(vESTP)
U<-ESTP+qt(.95,n-4)*sqrt(vESTP)
cbind(WI,ESTP_J)
cat("Tuned estimate:", ESTP, "SE: ",vESTP^.5,'\n')
cat("Confidence Interval:"," ", L,"; ", U,'\n')
```

9.5 Exercises

Exercise 9.1 Consider the problem of estimating the population mean squared error (or, say finite population variance) given by

$$S_y^2 = \{2N(N-1)\}^{-1} \sum_{i \neq j} \sum_{\in \Omega} (y_i - y_j)^2 \quad (9.39)$$

(a) Consider a semituned jackknife estimator of S_y^2 defined by

$$\hat{\sigma}_{T(cs)}^2 = \sum_{i \neq j} \sum_{\in s} [(1 - n(n-1))w_{ij} + 1] s_y^2(y_{(i)}, y_{(j)}) \quad (9.40)$$

where

$$s_y^2(y_{(i)}, y_{(j)}) = \frac{2n(n-1)s_y^2 - (y_i - y_j)^2}{2n(n-1) - 2} \quad (9.41)$$

where

$$s_y^2 = \{2n(n-1)\}^{-1} \sum_{i \neq j} \sum_{\in s} (y_i - y_j)^2 \quad (9.42)$$

The weight w_{ij} is the semituned weight such that the following $(k+1)$ constraints are satisfied:

$$\sum_{i \neq j} \sum_{\in s} w_{ij} = 1 \quad (9.43)$$

and

$$\sum_{i \neq j} \sum_{\in s} w_{ij} s_{x_t}^2(x_{t(i)}, x_{t(j)}) = \frac{n(n-1)s_{x_t}^2 - S_{x_t}^2}{n(n-1) - 1} \quad (9.44)$$

with

$$s_{x_t}^2(x_{t(i)}, x_{t(j)}) = \frac{2n(n-1)s_{x_t}^2 - (x_{ti} - x_{tj})^2}{2n(n-1) - 2} \quad (9.45)$$

$$S_{x_t}^2 = \{2N(N-1)\}^{-1} \sum_{i \neq j} \sum_{\in \Omega} (x_{ti} - x_{tj})^2 \quad (9.46)$$

and

$$s_{x_t}^2 = \{2n(n-1)\}^{-1} \sum_{i \neq j} \sum_{\in s} (x_{ti} - x_{tj})^2 \quad (9.47)$$

for $t=1,2,\ldots,k$. Consider semituning of the weights w_{ij} such that the chi-square type distance function, defined as

$$\frac{1}{2}\sum_{i\neq j}\sum_{\in s}\frac{(w_{ij}-1/(n(n-1)))^2}{q_{ij}/(n(n-1))} \qquad (9.48)$$

is minimized, subject to the $(k+1)$ constraints. Suggest a doubly jackknifed estimator of variance of the resulting newly semituned estimator. Write R or SAS code to study the properties of the resultant estimators.

(b) Consider a semituned ell estimator of S_y^2 defined by

$$\hat{\sigma}^2_{\text{T(ell)}} = \sum_{i\neq j}\sum_{\in s} w^*_{ij} s^2_y(y_{(i)}, y_{(j)}) \qquad (9.49)$$

Consider semituning the weights w^*_{ij} such that the weighted log-likelihood distance, function defined as

$$\frac{1}{n(n-1)}\sum_{i\neq j}\sum_{\in s}\log\left(w^*_{ij}\right) \qquad (9.50)$$

is optimized, subject to the $(k+1)$ constraints. Suggest a doubly jackknifed estimator of variance of the resulting newly semituned estimators. Write code in any programming language to study the properties of the resultant estimators.

Exercise 9.2 Estimating geometric mean
Consider the problem of estimating the population geometric mean defined by

$$G_y = \left(\prod_{i=1}^{N} y_i\right)^{1/N} \qquad (9.51)$$

Now consider a tuned jackknifed estimator of the population geometric mean G_y given by

$$\hat{G}_{\text{Tuned(cs)}} = \sum_{j\in s}\left[\left\{(n-1)^2 \bar{w}_n(j) - (n-2)\right\}\hat{G}_y(j)\right] \qquad (9.52)$$

where

$$\hat{G}_y(j) = \left(\prod_{i\neq j=1}^{n} y_i\right)^{1/(n-1)}, \quad j=1,2,\ldots,n \qquad (9.53)$$

is the jth jackknifed estimator of the geometric mean of the study variable obtained by dropping the jth unit from the usual estimator of the geometric mean given by

$$\hat{G}_y = \left(\prod_{i=1}^{n} y_i\right)^{1/n} \qquad (9.54)$$

The tuning weights $\bar{w}_n(j)$ in the estimator $\hat{G}_{\text{Tuned(cs)}}$ are obtained by optimizing the tuned chi-square type distance function

$$D = \frac{n}{2}\sum_{j\in s} q_j^{-1}\left[1-(n-1)\bar{w}_n(j)-n^{-1}\right]^2 \tag{9.55}$$

subject to the following three tuning constraints:

$$\sum_{j\in s}\bar{w}_n(j) = 1 \tag{9.56}$$

$$\sum_{j\in s}\bar{w}_n(j)\left[\hat{G}_x(j)\right]^{(1-n)} = \frac{1}{(n-1)}\left[\sum_{j\in s}\left(\hat{G}_x(j)\right)^{(1-n)} - \bar{X}\left(\hat{G}_x\right)^{-n}\right] \tag{9.57}$$

$$\sum_{j\in s}\bar{w}_n(j)\left[\hat{G}_z(j)\right]^{(1-n)} = \frac{1}{(n-1)}\left[\sum_{j\in s}\left(\hat{G}_z(j)\right)^{(1-n)} - \bar{Z}\left(\hat{G}_z\right)^{-n}\right] \tag{9.58}$$

where $\bar{X} = N^{-1}\sum_{i=1}^{N} x_i$ and $\bar{Z} = N^{-1}\sum_{i=1}^{N} z_i$ denote the known population arithmetic means of the two auxiliary variables. The usual estimators of the geometric means of the auxiliary variables are given by $\hat{G}_x = \left(\prod_{i=1}^{n} x_i\right)^{1/n}$ and $\hat{G}_z = \left(\prod_{i=1}^{n} z_i\right)^{1/n}$. Let $\hat{G}_x(j) = \left(\prod_{i\neq j=1}^{n} x_i\right)^{1/(n-1)}$ and $\hat{G}_z(j) = \left(\prod_{i\neq j=1}^{n} z_i\right)^{1/(n-1)}$ be the jth jackknifed estimators of the geometric means of the auxiliary variables X and Z, respectively, obtained by dropping the jth unit from the sample s. Suggest a doubly tuned jackknife estimator of variance of the tuned estimator of the population geometric mean G_y, and investigate coverage by the nominal 90%, 95%, and 99% confidence intervals by generating a population of at least 10,000 pumpkins from the Statistical Jumbo Pumpkin Model (SJPM) for sample sizes in the range of 10–100. Comment on your findings.

Exercise 9.3 Consider the problem of estimating the population mean $\bar{Y}$ with a tuned estimator defined by

$$\bar{y}_{\text{Tuned(cs)}} = \sum_{j\in s}\left[\left\{(n-1)^2\bar{w}_n(j) - (n-2)\right\}\bar{y}_n(j)\right] \tag{9.59}$$

where

$$\bar{y}_n(j) = \frac{n\bar{y}_n - y_j}{n-1} \tag{9.60}$$

is the jth jackknifed sample mean $\bar{y}_n$. Obtain the tuning weights $\bar{w}_n(j)$ by minimizing the tuned chi-squared distance function given by

$$D = \frac{n}{2}\sum_{j\in s} q_j \left[1 - (n-1)\bar{w}_n(j) - n^{-1}\right]^2 \qquad (9.61)$$

subject to the following three constraints:

$$\sum_{i\in s} \bar{w}_n(j) = 1 \qquad (9.62)$$

$$\sum_{i\in s} \bar{w}_n(j)\bar{x}_n(j) = \frac{\bar{X} - n(2-n)\bar{x}_n}{(n-1)^2} \qquad (9.63)$$

$$\sum_{i\in s} \bar{w}_n(j)\bar{z}_n(j) = \frac{\bar{Z} - n(2-n)\bar{z}_n}{(n-1)^2} \qquad (9.64)$$

where $\bar{X}$ and $\bar{Z}$ are the known population means of the two auxiliary variables. Also $\bar{x}_n(j) = (n\bar{x}_n - x_j)/(n-1)$ and $\bar{z}_n(j) = (n\bar{z}_n - z_j)/(n-1)$ are the jackknifed sample means $\bar{x}_n$ and $\bar{z}_n$, respectively. Suggest a doubly tuned jackknife estimator of variance of the tuned estimator of the population mean $\bar{Y}$ and investigate the attained coverage by the nominal 90%, 95%, and 99% confidence intervals by generating a population of at least 10,000 pumpkins from the SJPM for sample sizes in the range of 10–100. Comment on your findings. Extend the results to more than two auxiliary variables.

Exercise 9.4 Develop semituned and fully tuned methods of variance estimation using multiauxiliary information, by taking different kinds of distance functions.

Exercise 9.5 Develop a model-assisted tuned estimator of the population mean using multiauxiliary information for different kinds of distance functions.

Exercise 9.6 Estimating harmonic mean
Consider the problem of estimating the population harmonic mean defined as

$$H_y = N \left(\sum_{i=1}^{N} y_i^{-1}\right)^{-1} \qquad (9.65)$$

Consider a tuned jackknifed estimator of the population harmonic mean H_y given by

$$\hat{H}_{\text{Tuned(cs)}} = \sum_{j\in s} \left[\left\{(n-1)^2 \bar{w}_n(j) - (n-2)\right\} \hat{H}_y(j)\right] \qquad (9.66)$$

where

$$\hat{H}_y(j) = (n-1)\left(\sum_{i\neq j=1}^{n} y_i^{-1}\right)^{-1}, \quad j=1,2,\ldots,n \qquad (9.67)$$

is the *j*th jackknifed estimator of the harmonic mean of the study variable obtained by dropping the *j*th unit from the usual estimator of the harmonic mean given by

$$\hat{H}_y = n \left(\sum_{i=1}^{n} y_i^{-1} \right)^{-1} \tag{9.68}$$

The tuning weights, $\bar{w}_n(j)$, in the estimator $\hat{H}_{\text{Tuned(cs)}}$ are obtained by minimizing the tuned chi-square type distance function

$$D = \frac{n}{2} \sum_{j \in s} q_j^{-1} \left[1 - (n-1)\bar{w}_n(j) - n^{-1} \right]^2 \tag{9.69}$$

subject to the following three tuning constraints:

$$\sum_{j \in s} \bar{w}_n(j) = 1 \tag{9.70}$$

$$\sum_{j \in s} \frac{\bar{w}_n(j)\hat{H}_x(j)}{n\hat{H}_x(j) - (n-1)\hat{H}_x} = \frac{1}{(n-1)} \left[\sum_{j \in s} \frac{\hat{H}_x(j)}{n\hat{H}_x(j) - (n-1)\hat{H}_x} - \frac{\bar{X}}{\hat{H}_x} \right] \tag{9.71}$$

$$\sum_{j \in s} \frac{\bar{w}_n(j)\hat{H}_z(j)}{n\hat{H}_z(j) - (n-1)\hat{H}_z} = \frac{1}{(n-1)} \left[\sum_{j \in s} \frac{\hat{H}_z(j)}{n\hat{H}_z(j) - (n-1)\hat{H}_z} - \frac{\bar{Z}}{\hat{H}_z} \right] \tag{9.72}$$

where $\bar{X} = N^{-1} \sum_{i=1}^{N} x_i$ and $\bar{Z} = N^{-1} \sum_{i=1}^{N} z_i$ denote the known population arithmetic means of the auxiliary variables.

Let

$$\hat{H}_x(j) = (n-1) \left(\sum_{i \neq j=1}^{n} x_i^{-1} \right)^{-1}, \quad j = 1, 2, \ldots, n \tag{9.73}$$

and

$$\hat{H}_z(j) = (n-1) \left(\sum_{i \neq j=1}^{n} z_i^{-1} \right)^{-1}, \quad j = 1, 2, \ldots, n \tag{9.74}$$

be the *j*th jackknifed estimators of the harmonic mean of the auxiliary variables obtained by dropping the *j*th unit from the sample *s* and then forming the usual estimators of the harmonic means, namely

$$\hat{H}_x = n \left(\sum_{i=1}^{n} x_i^{-1} \right)^{-1} \quad \text{and} \quad \hat{H}_z = n \left(\sum_{i=1}^{n} z_i^{-1} \right)^{-1} \tag{9.75}$$

Suggest a doubly tuned jackknife estimator of variance of the tuned estimator of the population harmonic mean H_y. Investigate the coverage of the nominal 90%, 95%, and 99% confidence intervals by generating a population of at least 10,000 pumpkins from the new SJPM for sample sizes in the range of 10–100. Comment on your findings.

Exercise 9.7 Consider the problem of estimating the population mean $\bar{Y}$ with a tuned estimator defined as

$$\bar{y}_{\text{Tuned(cs)}} = \sum_{j \in s} \left[\left\{ (n-1)^2 \bar{w}_n(j) - (n-2) \right\} \bar{y}_n(j) \right] \tag{9.76}$$

where

$$\bar{y}_n(j) = (n\bar{y}_n - y_j)/(n-1) \tag{9.77}$$

is the jth jackknifed sample mean $\bar{y}_n$. Obtain the tuning weights $\bar{w}_n(j)$ by minimizing the tuned chi-squared distance given by

$$D = \frac{n}{2} \sum_{j \in s} q_j \left[1 - (n-1)\bar{w}_n(j) - n^{-1} \right]^2 \tag{9.78}$$

subject to the following three constraints:

$$\sum_{i \in s} \bar{w}_n(j) = 1 \tag{9.79}$$

$$\sum_{j \in s} \bar{w}_n(j) [\hat{G}_x(j)]^{(1-n)} = \frac{1}{(n-1)} \left[\sum_{j \in s} (\hat{G}_x(j))^{(1-n)} - \bar{X}(\hat{G}_x)^{-n} \right] \tag{9.80}$$

and

$$\sum_{j \in s} \frac{\bar{w}_n(j)\hat{H}_z(j)}{n\hat{H}_z(j) - (n-1)\hat{H}_z} = \frac{1}{(n-1)} \left[\sum_{j \in s} \frac{\hat{H}_z(j)}{n\hat{H}_z(j) - (n-1)\hat{H}_z} - \frac{\bar{Z}}{\hat{H}_z} \right] \tag{9.81}$$

where $\bar{X}$ and $\bar{Z}$ are the known population means of the two auxiliary variables. Let $\hat{H}_z(j) = (n-1)\left(\sum_{i \neq j=1}^{n} z_i^{-1}\right)^{-1}$ be the jth jackknifed estimator of the harmonic mean of the auxiliary variable obtained by dropping the jth unit from the sample s and then forming the usual estimator of the harmonic mean of the auxiliary variable, given by $\hat{H}_z = n\left(\sum_{i=1}^{n} z_i^{-1}\right)^{-1}$. Let $\hat{G}_x(j) = \left(\prod_{i \neq j=1}^{n} x_i\right)^{1/(n-1)}$ be the jth jackknifed estimator of the geometric mean of the auxiliary variable, X, obtained by dropping jth unit from the sample geometric mean $\hat{G}_x = \left(\prod_{i=1}^{n} x_i\right)^{1/n}$. Suggest a doubly tuned jackknife estimator of variance of the tuned estimator of the population mean $\bar{Y}$ and investigate coverage of the nominally 90%, 95%, and 99% confidence intervals by generating a population of at least 10,000 pumpkins from a modified SJPM with two auxiliary variables, and for various sample sizes in the range of 10–100. Comment on your findings. Extend the results to more than two auxiliary variables.

Exercise 9.8 Consider the problem of estimating the population mean $\bar{Y}$ with a tuned estimator defined as

$$\bar{y}_{\text{Tuned(cs)}} = \sum_{j \in s} \left[\left\{ (n-1)^2 \bar{w}_n(j) - (n-2) \right\} \bar{y}_n(j) \right] \tag{9.82}$$

where

$$\bar{y}_n(j) = (n\bar{y}_n - y_j)/(n-1) \tag{9.83}$$

is the jth jackknifed sample mean $\bar{y}_n$ of the study variable. Obtain the tuning weights $\bar{w}_n(j)$ by minimizing the tuned chi-squared distance function given by

$$D = \frac{n}{2} \sum_{j \in s} q_j \left[1 - (n-1)\bar{w}_n(j) - n^{-1} \right]^2 \tag{9.84}$$

subject to the following four constraints:

$$\sum_{i \in s} \bar{w}_n(j) = 1 \tag{9.85}$$

$$\sum_{i \in s} \bar{w}_n(j) \bar{x}_{1n}(j) = \frac{\bar{X}_1 - n(2-n)\bar{x}_{1n}}{(n-1)^2} \tag{9.86}$$

$$\sum_{j \in s} \bar{w}_n(j) \left[\hat{G}_{x_2}(j) \right]^{(1-n)} = \frac{1}{(n-1)} \left[\sum_{j \in s} \left(\hat{G}_{x_2}(j) \right)^{(1-n)} - \bar{X}_2 \left(\hat{G}_{x_2} \right)^{-n} \right] \tag{9.87}$$

and

$$\sum_{j \in s} \frac{\bar{w}_n(j) \hat{H}_{x_3}(j)}{n\hat{H}_{x_3}(j) - (n-1)\hat{H}_{x_3}} = \frac{1}{(n-1)} \left[\sum_{j \in s} \frac{\hat{H}_{x_3}(j)}{n\hat{H}_{x_3}(j) - (n-1)\hat{H}_{x_3}} - \frac{\bar{X}_3}{\hat{H}_{x_3}} \right] \tag{9.88}$$

where $\bar{X}_1, \bar{X}_2,$ and $\bar{X}_3$ are the known population arithmetic means of the three auxiliary variables, $\bar{x}_{1n}(j) = (n\bar{x}_{1n} - x_{1j})/(n-1)$ is the jth jackknifed sample mean of the first auxiliary variable, $\hat{G}_{x_2}(j) = \left(\prod_{i \neq j=1}^{n} x_{2i} \right)^{1/(n-1)}$ is the jth jackknifed sample geometric mean for the second auxiliary variable, and $\hat{H}_{x_3}(j) = (n-1) \left(\sum_{i \neq j=1}^{n} x_{3i}^{-1} \right)^{-1}$ is the jth jackknifed sample harmonic mean of the third auxiliary variable. Let $\bar{x}_{1n} = n^{-1} \sum_{i=1}^{n} x_{1i}$,

$\hat{G}_{x_2} = \left(\prod_{i=1}^{n} x_{2i} \right)^{1/n}$, and $\hat{H}_{x_3} = n \left(\sum_{i=1}^{n} x_{3i}^{-1} \right)^{-1}$ be the sample arithmetic mean, sample geometric mean, and sample harmonic mean of the first, second, and third auxiliary variables, respectively. Suggest a doubly tuned jackknife estimator of variance of the tuned estimator of the population mean $\bar{Y}$ and investigate the coverage by the nominally 90%, 95%, and 99% confidence intervals on this mean by generating a population of at least 10,000 pumpkins from a modified SJPM with three auxiliary variables, for sample sizes in the range of 10–100. Comment on your findings. Extend the results to more than three auxiliary variables.

A brief review of related work

10.1 Introduction

In this chapter, we review, in brief, work done on the topics of calibration and jackknifing in the field of survey sampling. This review is not exhaustive, but it will allow the reader to better grasp the ideas of calibration and jackknifing.

10.2 Calibration

In this section, we will discuss a few important publications in the area of calibration as a follow-up to the article by Särndal (2007). Särndal (2007) defines calibration in three parts. For finite populations, the calibration approach consists of three objectives: (1) computation of weights subject to some constraints on the sample values and parameters of auxiliary variables; (2) use of these weights to find a linearly weighted estimator of total or other parameters; and (3) creation of a nearly unbiased estimator in the absence of nonresponse and other nonsampling errors. Ardilly (2006) defines calibration as a method of reweighing the design weights in the presence of several qualitative or quantitative auxiliary variables. Kott (2006) defines calibration weights as those that result in nearly a design-consistent estimator of a population parameter.

Different researchers have different opinions. Let us consider the dictionary meaning of the word *calibration*: the adjustment of something for improvement. One calibrates a machine, calibrates a thermometer, calibrates a balance, and calibrates a stopwatch. Say that I have a stopwatch that is not correctly (or precisely) measuring time. How do I calibrate the stopwatch? I take it to the watchmaker, then he or she will adjust (or calibrate) its cog to improve the stopwatch's precision. The watchmaker will compare it to another watch that is known to be accurate. The word *calibration* means "improvement" or "tuning."

In the same way, a survey statistician has design weights $d_i = \pi_i^{-1}$, which are, in particular, reciprocals of the inclusion probabilities π_i, $i \in \Omega$ and are known for the entire population and in particular are known for a given sample s. No doubt the design weights $d_i = \pi_i^{-1}$ in the Horvitz and Thompson (1952) estimator:

$$\hat{Y}_{HT} = \sum_{i \in s} d_i y_i \tag{10.1}$$

are able to estimate the population total $Y = \sum_{i \in \Omega} y_i$. Is it possible to replace the design weights d_i in the Horvitz and Thompson (1952) estimator with new weights, say w_i, that improve the estimator's precision? If so, then this new estimator is called a calibrated estimator of the population total and is given by

$$\hat{Y}_{\text{Cal}} = \sum_{i \in s} w_i y_i \tag{10.2}$$

where the weights, w_i, $i \in s$, are called calibrated weights. Note that the calibrated weights are required only for units selected in the sample s and are not required for every unit in the population Ω. In the same way, a calibrated cog is required only for a stopwatch that does not show the correct time; calibrated cogs are not required for all stopwatches in the entire shop, or in the whole world.

We define calibration weights as design weights (or any other mechanism or cog of an estimator) that have been calibrated based on some standard (set of constraints) involving known parameters of auxiliary variables, in order to the precision and/or consistency of estimates of parameters of the study variable. Note that calibrated weights are computed at the estimation stage and not at the selection stage.

Deville and Särndal (1992) considered five distance functions between the calibrated weights w_i and the design weights d_i as shown in Table 10.1.

Among these five distance functions defined by Deville and Särndal (1992), the first one, DS_1, became popular because it leads to a generalized regression (greg) estimator. The main difficulty with the calibrated weights in greg is that these weights do not satisfy the desired nonnegativity constraint. The other distance functions, DS_2, DS_3, DS_4, and DS_5, can guarantee the nonnegativity of the resultant calibration weights. Farrell and Singh (2002a) have suggested a penalized chi-square distance function and have shown that the Searls (1964) estimator is a special case of their

Table 10.1 Deville and Särndal (1992) distances

Number	Distance function
1	$DS_1 = \dfrac{1}{2} \sum_{i \in s} \dfrac{(w_i - d_i)^2}{d_i q_i}$
2	$DS_2 = \dfrac{1}{2} \sum_{i \in s} \dfrac{1}{q_i} \left[w_i \log \left(\dfrac{w_i}{d_i} \right) - w_i + d_i \right]$
3	$DS_3 = 2 \sum_{i \in s} \dfrac{(\sqrt{w_i} - \sqrt{d_i})^2}{q_i}$
4	$DS_4 = \sum_{i \in s} \dfrac{1}{q_i} \left[-d_i \log \left(\dfrac{w_i}{d_i} \right) + w_i - d_i \right]$
5	$DS_5 = \dfrac{1}{2} \sum_{i \in s} \dfrac{(w_i - d_i)^2}{w_i q_i}$

proposal. Singh (2003) has shown that the Hartley and Ross (1954) estimator is also a special case of penalized distance function.

For one auxiliary variable, Deville and Särndal (1992) considered minimizing the distance function DS_1 subject to the calibration constraint

$$\sum_{i \in s} w_i x_i = X \tag{10.3}$$

The appropriate Lagrange function L is given by

$$L = \sum_{i \in s} (w_i - d_i)^2 (d_i q_i)^{-1} - 2\lambda \left(\sum_{i \in s} w_i x_i - X \right) \tag{10.4}$$

On differentiating Equation (10.4) with respect to w_i, and equating to zero, we have

$$w_i = d_i + \lambda \, d_i q_i x_i \tag{10.5}$$

On substituting Equation (10.5) into Equation (10.3) and solving for λ, we have

$$\lambda = \left(\sum_{i \in s} d_i q_i x_i^2 \right)^{-1} \left(X - \sum_{i \in s} d_i x_i \right) \tag{10.6}$$

On substituting Equation (10.6) into Equation (10.5), we have

$$w_i = d_i + \left(d_i q_i x_i \bigg/ \sum_{i \in s} d_i q_i x_i^2 \right) \left(X - \sum_{i \in s} d_i x_i \right) \tag{10.7}$$

Substitution of the value of w_i from Equation (10.7) in Equation (10.2) leads to the general regression estimator (greg) of total given by

$$\hat{Y}_{\text{greg}} = \sum_{i \in s} d_i y_i + \hat{\beta}_{\text{ds}} \left(X - \sum_{i \in s} d_i x_i \right) \tag{10.8}$$

where

$$\hat{\beta}_{\text{ds}} = \frac{\sum_{i \in s} d_i q_i x_i y_i}{\sum_{i \in s} d_i q_i x_i^2} \tag{10.9}$$

If $q_i = 1/x_i$, then the optimal weight w_i becomes

$$w_i = d_i X \bigg/ \sum_{i \in s} d_i x_i \tag{10.10}$$

and the resultant estimator reduces to the ratio estimator of population total, that is,

$$\hat{y}_R = \sum_{i \in s} d_i y_i \left(X \Big/ \sum_{i \in s} d_i x_i \right) \tag{10.11}$$

Singh, Horn, and Yu (1998) reported that there is no choice of q_i such that the resultant estimator (10.8) reduces to the product estimator of population total due to Murthy (1964).

Singh (2003, 2006) suggested an additional constraint:

$$\sum_{i \in s} w_i = \sum_{i \in s} d_i \tag{10.12}$$

He showed that the calibrated weights are then given by

$$w_i = d_i + \left\{ \frac{d_i q_i x_i \left(\sum_{i \in s} d_i q_i \right) - d_i q_i \left(\sum_{i \in s} d_i q_i x_i \right)}{\left(\sum_{i \in s} d_i q_i \right) \left(\sum_{i \in s} d_i q_i x_i^2 \right) - \left(\sum_{i \in s} d_i q_i x_i \right)^2} \right\} (X - \hat{X}_{HT}) \tag{10.13}$$

On substituting Equation (10.13) into Equation (10.2), the resultant calibrated estimator of the population total Y becomes an exact linear regression type estimator and is given by

$$\hat{Y}_s = \hat{Y}_{HT} + \hat{\beta}_{ols} (X - \hat{X}_{HT}) \tag{10.14}$$

where

$$\hat{\beta}_{ols} = \frac{\left(\sum_{i \in s} d_i q_i x_i y_i \right) \left(\sum_{i \in s} d_i q_i \right) - \left(\sum_{i \in s} d_i q_i y_i \right) \left(\sum_{i \in s} d_i q_i x_i \right)}{\left(\sum_{i \in s} d_i q_i \right) \left(\sum_{i \in s} d_i q_i x_i^2 \right) - \left(\sum_{i \in s} d_i q_i x_i \right)^2} \tag{10.15}$$

Following Stearns and Singh (2008), for a fixed sample design, the regression coefficient estimator $\hat{\beta}_{ols}$ in Sen (1953) and in Yates and Grundy (1953) can be written in the form

$$\hat{\beta}_{ols} = \frac{\sum_{i \neq j \in s} q_i q_j (d_i^2 x_i y_i - d_i x_i d_j x_j)}{\sum_{i \neq j \in s} q_i q_j (d_i^2 x_i^2 - d_i x_i d_j x_j)} = \frac{\frac{1}{2} \sum_{i \neq j \in s} q_i q_j (d_i y_i - d_j y_j)(d_i x_i - d_j x_j)}{\frac{1}{2} \sum_{i \neq j \in s} q_i q_j (d_i x_i - d_j x_j)^2} \tag{10.16}$$

Stearns and Singh (2008) suggested computing the weights q_i and q_j in pairs such that

$$q_i q_j = \left(\frac{\pi_{ij} - \pi_i \pi_j}{\pi_{ij}} \right) \quad (10.17)$$

where $\pi_{ij} = P(i \& j \in s)$ denotes the second-order inclusion probabilities. They suggested that although the choice of weight q_i may not be unique, for a fixed sample size design, the estimator of the regression coefficient will be unique and is given by

$$\hat{\beta}_{\text{ols}} = \frac{\hat{\text{cov}}(\hat{Y}_{\text{HT}}, \hat{X}_{\text{HT}})}{\hat{v}(\hat{X}_{\text{HT}})} \quad (10.18)$$

where $\hat{Y}_{\text{HT}}$ and $\hat{X}_{\text{HT}}$ stand for the Horvitz and Thompson (1952) estimators of the total for the study variable and for the auxiliary variable, respectively. The $\hat{\beta}_{\text{ols}}$ in Equation (10.18) is the estimator of the regression coefficient for probability proportional to size and without replacement sampling. The use of multiauxiliary information in the same setup can be found in Singh and Arnab (2011).

Berger, Tirari, and Tille (2003) also proposed a design-based simple alternative to the Montanari (1987) generalized regression estimator (greg) by implementing standard weighted least squares in their estimator. Wu and Sitter (2001) suggested another constraint

$$\sum_{i \in s} w_i = N \quad (10.19)$$

which also leads to a regression type consistent estimator of the population total for unequal probability sampling, and leads to an exact linear regression estimator for simple random sampling. The Wu and Sitter (2001) estimator is popular due to its model calibration. At almost the same time, Farrell and Singh (2002b) introduced the idea of model calibration. Montanari and Ranalli (2005) considered nonparametric model calibration to include auxiliary information at the estimation stage of a population parameter. Kim and Park (2010) suggested using the constraint (10.12) if the population size is unknown and using the constraint (10.19) if the population size is known. Singh and Sedory (2013, 2015) suggested a two-step calibration approach with a general constraint given by

$$\sum_{i \in s} w_i = \sum_{i \in s} k_i d_i \quad (10.20)$$

where k_i, $i \in s$ are constants to be determined at the second step while validating the resultant estimator to achieve minimum variance. In particular, there exist two choices of k_i, $i \in s$ such that calibration constraints (10.12) and (10.19) are special cases of Equation (10.20).

Rao (1994) and Singh (2001) considered the problem of estimating a general parameter of interest in survey sampling, which was later investigated by Singh

(2006) and Stearns and Singh (2008) in more depth. Rueda, Martínez, Martínez, and Arcos (2007) also considered the problem of estimating a distribution function with calibration techniques. Brewer (1999) introduced cosmetic calibration for unequal probability sampling designs. Martínez, Rueda, Arcos, and Martínez (2010) introduced optimum calibration points for estimating the distribution function that ultimately lead to an optimally calibrated estimator of the distribution function.

Hidiroglou and Särndal (1995, 1998) and Dupont (1995) considered a calibration technique in the two-phase sampling design. Calibrated weights are obtained for the second-phase sample by assuming that the first-phase sample information is known. Singh and Puertas (2003) considered the problem of estimating total, mean, and distribution function using two-phase sampling. Lundström and Särndal (1999) and Särndal and Lundström (2005) considered calibration as a standard method for the treatment of nonresponse in survey sampling. Rueda, Martínez, Arcos, and Muñoz (2009) considered the problem of estimating a population mean using a calibration technique for successive sampling. Särndal, Swensson, and Wretman (1992) have also derived calibration weights for two-stage sampling. Tracy, Singh, and Arnab (2003) considered the problem of calibration estimation in stratified random sampling. Kim, Sungur, and Heo (2007) also considered the problem of estimating a population mean in stratified sampling through a calibration technique. Tikkiwal, Rai, and Ghiya (2013) investigated several estimators obtained through a calibration technique for small domains. Arcos, Contreras, and Rueda (2014) introduced a novel calibration for social surveys. Dykes, Singh, Sedory, and Louis (2015) used a calibration technique to improve the Hansen and Hurwitz (1946) estimator in mail survey design in the presence of random nonresponse. This shows that calibration weights have been obtained by several researchers for various sampling schemes where the estimation of population total or a distribution function is concerned.

Owen (2001) suggested the use of a logarithm function; his resultant estimates are referred to as empirical log-likelihood estimators. Later, his idea of calibrating design weights in survey sampling was extensively investigated by several researchers, including Singh, Sedory, and Kim (2014), Singh and Kim (2011), Chen and Sitter (1999), Rueda, Muñoz, Berger, Arcos, and Martínez (2007), and Wu (2005), so no further detail is provided here.

Singh, Horn, and Yu (1998) introduced a method of higher-order calibration that could be used to obtain calibrated estimators of higher-order moments of the study variable, such as variance, by proposing calibration constraints on higher-order moments of the auxiliary variables. Later, the idea of higher-order calibration to estimate the variance of the linear regression estimator was widely accepted in the literature and was investigated by several researchers, such as Singh and Horn (1999), Singh, Horn, Chowdhury, and Yu (1999), and Singh (2001, 2004). The higher-order calibration approach has been applied to model calibration by several researchers, among them are Farrell and Singh (2002a, 2002b, 2005), Sitter and Wu (2002), Wu (2003), and Arnab and Singh (2003, 2005).

Also note that Singh and Sedory (2012) have derived calibrated maximum likelihood weights, as opposed to calibration weights, and Singh (2013) has suggested a dual problem of calibration of design weights. Both ideas could be further explored, but the discussion of that work is beyond the scope of this chapter.

10.3 Jackknifing

The statistical method of jackknifing is analogous to cutting an apple, a mango, or a pumpkin. Assume a person has an apple to eat. He can eat it without cutting it. He can make four slices and eat it. He can make eight slices and eat it. Ultimately the person eats the entire apple, and there is no benefit or loss caused by cutting it into four or eight pieces. Assume a person has 100 coins in a box. He takes one coin out of the box and counts the remaining 99 coins. He puts all 100 coins back into the box, removes another coin, and again finds that 99 coins remain. He can count several times, but will always find that there are 100 coins. It seems that when you jackknife anything, there is no gain or loss.

This is not the case when you jackknife a sample. Quenouille (1956) introduced the idea of jackknifing to reduce bias in the ratio estimator due to Cochran (1940). Upadhyaya, Singh, and Singh (2004) used it to construct nearly unbiased estimators when the auxiliary variable is negatively correlated with the study variable. Tukey (1958) was the first to use the jackknife technique to estimate variance of an estimator. The idea of variance estimation using the jackknife technique has gained popularity, due to its simplicity, and it has been widely used by survey statisticians for many types of survey designs. Wolter (1985) is a famous researcher who promoted the technique of jackknifing in estimating the variance of an estimator. Rao and Sitter (1995) used jackknifing while estimating the variance of the ratio estimator in two-phase sampling. Roy and Safiquzzaman (2003) applied the jackknife technique to a general class of estimators and estimated the variance of this general class of estimators for two-phase sampling. Arnab and Singh (2006) have proposed a new jackknife method for estimating variance from imputed data using the ratio method of estimation. Ramasubramanian, Rai, and Singh (2007) also applied the technique of jackknifing to estimate variance in two-phase sampling. Farrell and Singh (2010) considered the problem of jackknifing the calibrated estimator of a population mean in the two-phase sampling setup and estimated the variance of the chain ratio and chain regression type estimators. Berger and Skinner (2005) considered the problem of jackknifing for an unequal probability sampling scheme and examined the similarities between the jackknife and linearization methods. Their investigation was deeper than the estimator suggested by Campbell (1980). Singh, Kim, and Grewal (2008) suggested the problem of jackknifing scrambled responses when the data are collected on sensitive issues, making use of a randomization device. Jing, Yuan, and Zhou (2009) jackknifed empirical likelihood in the presence of nonlinear constraints. Singh and Arnab (2010) proposed the idea of adjusting bias in the estimator of variance of the ratio estimator in two-phase sampling, by combining calibration and jackknifing.

Though a short review of the literature on the topics of calibration and jackknifing has been made, an uninterrupted flow of publications on these topics seems to promise rapid progress in the foreseeable future. The work cited in this chapter is sufficient to put a researcher on a tack to do more work on the new concept of tuning design weights.

Bibliography

Amahia, G. N., Chaubey, Y. P., & Rao, T. J. (1989). Efficiency of a new estimator in PPS sampling for multiple characteristics. *Journal of Statistical Planning and Inference, 21,* 75–84.

Arcos, A., Contreras, J. M., & Rueda, M. (2014). A novel calibration estimator in social surveys. *Sociological Methods & Research, 43*(3), 465–489.

Ardilly, P. (2006). *Les techniques de sondage.* Paris: Editions Technip.

Arnab, R., & Singh, S. (2003). On estimation of population total using generalized regression predictor. In: *American Association for Public Opinion Research-Section on Survey Research Methods. Proceedings of the JSM, California,* 1–6.

Arnab, R., & Singh, S. (2005). A note on variance estimation for the generalized regression predictor. *Australian & New Zealand Journal of Statistics, 47*(2), 231–234.

Arnab, R., & Singh, S. (2006). A new method for estimating variance from data imputed with ratio method of imputation. *Statistics & Probability Letters, 76,* 513–519.

Bansal, M. L., & Singh, R. (1985). An alternative estimator for multiple characteristics in PPS sampling. *Journal of Statistical Planning and Inference, 11,* 313–320.

Bansal, M. L., Singh, S., & Singh, R. (1994). Multi-character survey using randomized response technique. *Communication in Statistics – Theory and Methods, 23*(6), 1705–1715.

Beale, E. M. L. (1962). Some use of computers in operational research. *Industrielle Organisation, 31,* 27–28.

Berger, Y. G., & Skinner, C. J. (2005). A jackknife variance estimator for unequal probability sampling. *Journal of the Royal Statistical Society, 67*(1), 79–89.

Berger, Y. G., Tirari, M. E. H., & Tille, Y. (2003). Towards optimal regression estimation in sample surveys. *Australian & New Zealand Journal of Statistics, 45*(3), 319–329.

Brewer, K. R. W. (1999). Cosmetic calibration with unequal probability sampling. *Survey Methodology, 25*(2), 205–212.

Brewer, K. R. W., & Hanif, M. (1983). *Sampling with unequal probabilities.* New York: Springer-Verlag.

Campbell, C. (1980). A different view of finite population estimation. In *Proceedings of the Survey Research Method Section of the American Statistical Association* (pp. 319–324).

Chen, J., & Sitter, R. R. (1999). Pseudo empirical likelihood approach to the effective use of auxiliary information in complex surveys. *Statistica Sinica, 9,* 385–406.

Cochran, W. G. (1940). Some properties of estimators based on sampling scheme with varying probabilities. *Australian Journal of Statistics, 17,* 22–28.

Cochran, W. G. (1977). *Sampling techniques* (3rd ed.). New York: John Wiley & Sons.

Das, A. K., & Tripathi, T. P. (1978). Use of auxiliary information in estimating finite population variance. *Sankhya C, 40,* 139–148.

Deville, J. C., & Särndal, C. E. (1992). Calibration estimators in survey sampling. *Journal of the American Statistical Association, 87,* 376–382.

Dupont, F. (1995). Alternative adjustments where there are several levels of auxiliary information. *Survey Methodology, 21,* 125–135.

Dykes, L., Singh, S., Sedory, S. A., & Louis, V. (2015). Calibrated estimators of population mean for a mail survey design. *Communications in Statistics – Theory and Methods, 18.* http://dx.doi.org/10.1080/03610926.2013.841932.

Farrell, P. J., & Singh, S. (2002a). Penalized chi-square distance function in survey sampling. In *Joint statistical meetings—section on survey research methods, NY*, pp. 963–968.

Farrell, P. J., & Singh, S. (2002b). Re-calibration of higher-order calibration weights. In *SSC annual meeting, proceedings of the survey methods section* (pp. 75–80).

Farrell, P. J., & Singh, S. (2005). Model assisted higher-order calibration of estimators of variance. *Australian & New Zealand Journal of Statistics, 47*(3), 375–383.

Farrell, P. J., & Singh, S. (2010). Some contribution to Jackknifing two-phase sampling estimators. *Survey Methodology, 36*(1), 57–68.

Finucan, H. M., Galbraith, R. F., & Stone, M. (1974). Moments without tears in simple random sampling from a finite population. *Biometrika, 61*(1), 151–154.

Gamrot, W. (2012). Estimation of finite population kurtosis under two-phase sampling for nonresponse. *Statistical Papers, 53*(4), 887–894.

Gupta, J. P., Singh, R., & Kashani, H. B. (1993). An estimator of the correlation coefficient in probability proportional to size with replacement sampling. *Metron, 51*(3-4), 165–178.

Hansen, M. H., & Hurwitz, W. N. (1943). On the theory of sampling from finite populations. *Annals of Mathematical Statistics, 14*, 333–362.

Hansen, M. H., & Hurwitz, W. N. (1946). The problem of non-response in sample surveys. *Journal of the American Statistical Association, 41*, 517–529.

Hansen, M. H., Hurwitz, W. N., & Madow, W. G. (1953). *Sample survey methods and theory*. New York: John Wiley and Sons.

Hartley, H. O., & Ross, A. (1954). Unbiased ratio estimators. *Nature, 174*, 270–271.

Hidiroglou, M. A., & Särndal, C. E. (1995). Use of auxiliary information for two-phase sampling. *Proceedings of the survey research method section of the American Statistical Association: Vol. 2* (pp. 873–878).

Hidiroglou, M. A., & Särndal, C. E. (1998). Use of auxiliary information for two-phase sampling. *Survey Methodology, 24*(1), 11–20.

Horvitz, D. G., & Thompson, D. J. (1952). A generalization of sampling without replacement from a finite universe. *Journal of the American Statistical Association, 47*, 663–685.

Isaki, C. T. (1983). Variance estimation using auxiliary information. *Journal of the American Statistical Association, 78*, 117–123.

Jing, B. Y., Yuan, J., & Zhou, W. (2009). Jackknife empirical likelihood. *Journal of the American Statistical Association, 104*(487), 1224–1232.

Kim, J. K., & Park, M. (2010). Calibration estimation in survey sampling. *International Statistical Review, 78*, 21–39.

Kim, Jong-Min, Sungur, E. A., & Heo, T. Y. (2007). Calibration approach estimators in stratified sampling. *Statistics & Probability Letters, 77*(1), 99–103.

Kott, P. S. (2006). Using calibration weighting to adjust for nonresponse and coverage errors. *Survey Methodology, 32*, 133–142.

Lahiri, D. B. (1951). A method for sample selection providing unbiased ratio estimates. *Bulletin of the International Statistical Institute, 33*(2), 133–140.

Lohr, S. L. (2010). *Sampling: Design and analysis* (2nd ed.). Pacific Grove, CA: Duxbury Press.

Lundström, S., & Särndal, C. E. (1999). Calibration as a standard method for treatment of nonresponse. *Journal of Official Statistics, 15*(2), 305–327.

Mangat, N. S., & Singh, R. (1992–1993). Sampling with varying probabilities without replacement: A review. *Aligarh Journal of Statistics, 12 & 13*, 75–106.

Martínez, S., Rueda, M., Arcos, A., & Martínez, H. (2010). Optimum calibration points estimating distribution function. *Journal of Computational and Applied Mathematics, 233*, 2265–2277.

Martinez, S., Arcos, A., Martinez, H., & Singh, S. (2015). Estimating population proportions by means of calibration estimators. *Revista Colombiana de Estadística, 38*(1), 267–293.

Midzuno, H. (1951). On the sampling system with probability proportional to sum of sizes. *Annals of the Institute of Statistical Mathematics*, *2*, 99–108.

Montanari, G. E. (1987). Post sampling efficient QR-prediction in large sample surveys. *International Statistical Review*, *55*, 191–202.

Montanari, G. E., & Ranalli, M. G. (2005). Nonparametric model calibration estimation in survey sampling. *Journal of American Statistical Association*, *100*(472), 1429–1442.

Murthy, M. N. (1964). Product method of estimation. *Sankhya A*, *26*, 69–74.

Neyman, J. (1934). On the two different aspects of the representative method: The method of stratified sampling and the method of purposive selection. *Journal of the Royal Statistical Society*, *97*, 558–606.

Owen, A. B. (2001). *Empirical likelihood*. Boca Raton, FL: Chapman & Hall.

Pearson, K. (1896). Mathematical contribution to the theory of evolution. III. Regression: Heredity and panmixia. *Philosophical Transactions of the Royal Society of London Series A*, *187*, 253–318.

Quenouille, M. H. (1956). Notes on bias in estimation. *Biometrika*, *43*, 353–360.

Ramasubramanian, V., Rai, A., & Singh, Randhir (2007). Jackknife variance estimation under two-phase sampling. *Model Assisted Statistics and Applications*, *2*, 27–35.

Rao, J. N. K. (1966). Alternative estimators in PPSWR sampling for multiple characteristics. *Sankhya A*, *28*, 47–60.

Rao, J. N. K. (1994). Estimating totals and distribution functions using auxiliary information at the estimation stage. *Journal of Official Statistics*, *10*(2), 153–165.

Rao, J. N. K., & Sitter, R. R. (1995). Variance estimation under two-phase sampling with application to imputation for missing data. *Biometrika*, *82*(2), 453–460.

Roy, D., & Safiquzzaman, M. (2003). Jackknifing a general class of estimators and variance estimation under two-phase sampling. *Metron – International Journal of Statistics*, *LXI*(1), 53–74.

Rueda, M., Martínez, S., Martínez, H., & Arcos, A. (2007a). Estimation of the distribution function with calibration methods. *Journal of Statistical Planning and Inference*, *137*, 435–448.

Rueda, M., Martínez, S., Arcos, A., & Muñoz, J. F. (2009). Mean estimation under successive sampling with calibration estimators. *Communications in Statistics – Theory and Methods*, *38*, 808–827.

Rueda, M., Muñoz, J. F., Berger, Y. G., Arcos, A., & Martínez, S. (2007b). Pseudo empirical likelihood method in the presence of missing data. *Metrika*, *65*, 349–367.

Särndal, C. E. (2007). The calibration approach in survey theory and practice. *Survey Methodology*, *33*(2), 99–119.

Särndal, C. E., & Lundström, S. (2005). *Estimation in surveys with nonresponse*. New York: John Wiley & Sons, Inc.

Särndal, C. E., Swensson, B., & Wretman, J. H. (1992). *Model assisted survey sampling*. New York: Springer-Verlag.

Searls, D. T. (1964). The utilization of a known coefficient of variation in the estimation procedure. *Journal of the American Statistical Association*, *59*, 1225–1226.

Sen, A. R. (1952). Present status of probability sampling and its use in estimation of farm characteristics. *Econometrika*, *20*, 130 (abstract).

Sen, A. R. (1953). On the estimate of the variance in sampling with varying probabilities. *Journal of the Indian Society of Agricultural Statistics*, *5*, 119–127.

Singh, H. P., Solanki, R. S., & Singh, S. (2015). Estimation of Bowley's coefficient of skewness in the presence of auxiliary information. *Communications in Statistics – Theory and Methods*, *43*(22), 4867–4880.

Singh, S. (2001). Generalized calibration approach for estimating the variance in survey sampling. *Annals of the Institute of Statistical Mathematics*, *53*(2), 404–417.

Singh, S. (2003). *Advanced sampling theory with applications: How Michael selected Amy: (Vols. 1 & 2)*. The Netherlands: Kluwer.

Singh, S. (2004). Golden and silver jubilee year-2003 of the linear regression estimators. In *Presented at the joint statistical meeting, Toronto, ASA section on survey research methods* (pp. 4382–4389).

Singh, S. (2006). Survey statisticians celebrate golden jubilee year 2003 of the linear regression estimator. *Metrika, 63*(1), 1–18.

Singh, S. (2012). On the calibration of design weights using a displacement function. *Metrika, 75*, 85–107.

Singh, S. (2013). A dual problem of calibration of design weights. *Statistics: A Journal of Theoretical and Applied Statistics, 47*(3), 566–574.

Singh, S., & Arnab, R. (2010). Bias-adjustment and calibration of Jackknife variance estimator in the presence of non-response. *Journal of Statistical Planning and Inference, 140*(4), 862–871.

Singh, S., & Arnab, R. (2011). On calibration of design weights. *Metron – International Journal of Statistics, LXIX*(2), 185–205.

Singh, S., & Grewal, I. S. (2012). Estimation of finite population variance using partial jackknifing. *Journal of the Indian Society of Agricultural Statistics, 66*(3), 427–440.

Singh, S., & Horn, S. (1998). An alternative estimator in multi-character surveys. *Metrika, 48*, 99–107.

Singh, S., & Horn, S. (1999). An improved estimator of the variance of the regression estimator. *Biometrical Journal, 41*(3), 359–369.

Singh, S., Horn, S., & Yu, F. (1998). Estimation of variance of general regression estimator: Higher level calibration approach. *Survey Methodology, 24*, 41–50.

Singh, S., Horn, S., Chowdhury, S., & Yu, F. (1999). Calibration of the estimators of variance. *Australian & New Zealand Journal of Statistics, 41*(2), 199–212.

Singh, S., Joarder, A. H., & King, M. L. (1996a). Regression analysis using scrambled responses. *Australian Journal of Statistics, 38*(2), 201–211.

Singh, S., & Kim, Jong-Min (2011). A pseudo-empirical log-likelihood estimator using scrambled responses. *Statistics and Probability Letters, 81*, 345–351.

Singh, S., Kim, Jong-Min, & Grewal, I. S. (2008). Imputing and jackknifing scrambled responses. *Metron – International Journal of Statistics, LXVI*(2), 183–204.

Singh, S., Mangat, N. S., & Gupta, J. P. (1996b). Improved estimator of finite population correlation coefficient. *Journal of the Indian Society of Agricultural Statistics, 48*, 141–149.

Singh, S., & Puertas, S. M. (2003). On the estimation of total, mean and distribution function using two-phase sampling: Calibration approach. *Journal of the Indian Society of Agricultural Statistics, 56*(3), 237–252.

Singh, S., Rueda, M., & Sanchez-Borrego, I. (2010). Random non-response in multi-character surveys. *Quality & Quantity, 44*, 345–356.

Singh, S., & Sedory, S. A. (2012). Calibrated maximum likelihood design weights in survey sampling. In *JSM: Section on survey research methods* (pp. 4000–4014).

Singh, S., & Sedory, S. A. (2013). Two-step calibration of design weights in survey sampling. In *Survey research methods section, presented at the JSM* (pp. 2928–2942).

Singh, S., & Sedory, S. A. (2015). Two-step calibration of design weights in survey sampling. *Communications in Statistics – Theory and Methods*. http://dx.doi.org/10.1080/03610926.2014.892137 (In press)

Singh, S., Sedory, S. A., & Kim, J.-M. (2014). An empirical likelihood estimate of the finite population correlation coefficient. *Communications in Statistics – Simulation and Computation, 43*(6), 1430–1441.

Sitter, R. R. (1997). Variance estimation for the regression estimator in two-phase sampling. *Journal of the American Statistical Association, 92*, 780–787.

Sitter, R. R., & Wu, C. (2002). Efficient estimation of quadratic finite population functions. *Journal of the American Statistical Association, 97*, 535–543.

Srivastava, S. K., & Jhajj, H. S. (1980). A class of estimators using auxiliary information for estimating finite population variance. *Sankhya C, 42*, 87–96.

Srivastava, S. K., & Jhajj, H. S. (1986). On the estimation of finite population correlation coefficient. *Journal of the Indian Society of Agricultural Statistics, 38*, 82–91.

Stearns, M., & Singh, S. (2008). On the estimation of the general parameter. *Computational Statistics and Data Analysis, 52*, 4253–4271.

Thompson, M. E. (1997). *Theory of sample surveys* (1st ed.). London: Chapman & Hall.

Tikkiwal, G. C., Rai, P. K., & Ghiya, A. (2013). On the performance of generalized regression estimator for small domains. *Communications in Statistics – Simulation and Computation, 42*, 891–909.

Tracy, D. S., & Singh, S. (1999). Calibration estimators in randomized response surveys. *Metron, 55*, 47–68.

Tracy, D. S., Singh, S., & Arnab, R. (2003). Note on calibration in stratified and double sampling. *Survey Methodology, 29*, 99–104.

Tukey, J. W. (1958). Bias and confidence in not-quite large samples (abstract). *Annals of Mathematical Statistics, 29*, 61–75.

Upadhyaya, L. N., Singh, H. P., & Singh, S. (2004). A family of almost unbiased estimators for negatively correlated variables using jackknife technique. *Statistica, 64*, 47–63.

Wakimoto, K. (1971). Stratified random sampling (III): Estimation of the correlation coefficient. *Annals of the Institute of Statistical Mathematics, 23*, 339–355.

Wolter, K. M. (1985). *Introduction to variance estimation*. Berlin: Springer.

Wu, C. (2003). Optimal calibration estimators in survey sampling. *Biometrika, 90*, 937–951.

Wu, C. (2005). Algorithms and R codes for the pseudo empirical likelihood method in survey sampling. *Survey Methodology, 31*(2), 239–243.

Wu, C., & Sitter, R. R. (2001). A model-calibration to using complete auxiliary information from survey data. *Journal of the American Statistical Association, 96*, 185–193.

Yates, F., & Grundy, P. M. (1953). Selection without replacement from within strata with probability proportional to size. *Journal of the Royal Statistical Society Series B, 15*, 253–261.

Author Index

Note: Page numbers followed by b indicate boxes and t indicate tables.

A

Amahia, G.N., 191–192, 195–197, 291b
Arcos, A., 17, 287–288, 291–293b
Ardilly, P., 283, 291b
Arnab, R., 8, 287–289, 291b, 294–295b

B

Bansal, M.L., 165–166, 178, 191–192, 195–197, 291b
Beale, E.M.L., 39, 291b
Berger, Y.G., 287–289, 291b, 293b
Brewer, K.R.W., 3, 287–288, 291b

C

Campbell, C., 289, 291b
Chaubey, Y.P., 191–192, 195–197, 291b
Chen, J., 288, 291b
Chowdhury, S., 288, 294b
Cochran, W.G., 3, 289, 291b
Contreras, J.M., 288, 291b

D

Das, A.K., 90, 291b
Deville, J.C., 28–29, 39, 258, 284, 284t, 291b
Dupont, F., 288, 291b
Dykes, L., 288, 291b

F

Farrell, P.J., 8, 49–50, 74, 156–159, 284–285, 287–289, 292b
Finucan, H.M., 22, 292b

G

Galbraith, R.F., 22, 292b
Gamrot, W., 21, 292b
Ghiya, A., 288, 295b
Grewal, I.S., 86, 289, 294b
Grundy, P.M., 286–287, 295b
Gupta, J.P., 139–140, 292b, 294b

H

Hanif, M., 3, 291b
Hansen, M.H., 31, 165–166, 175–176, 288, 292b
Hartley, H.O., 284–285, 292b
Heo, T.Y., 288, 292b
Hidiroglou, M.A., 288, 292b
Horn, S., 194, 286, 288, 294b
Horvitz, D.G., 199, 203, 205–210, 211t, 212–216, 283–284, 292b
Hurwitz, W.N., 31, 165–166, 175–176, 288, 292b

I

Isaki, C.T., 90, 292b

J

Jhajj, H.S., 90, 139–140, 295b
Jing, B.Y., 289, 292b
Joarder, A.H., 55–58, 294b

K

Kashani, H.B., 139–140, 292b
Kim, J.K., 287, 292b
Kim, Jong-Min, 140, 288–289, 292b, 294b
King, M.L., 55–58, 294b
Kott, P.S., 283, 292b

L

Lahiri, D.B., 170, 182, 292b
Lohr, S.L., 3, 292b
Louis, V., 288, 291b
Lundström, S., 288, 292–293b

M

Madow, W.G., 31, 292b
Mangat, N.S., 140, 191–192, 195–197, 292b, 294b
Martinez, H., 17, 287–288, 292b
Martinez, S., 17, 287–288, 292b

Midzuno, H., 204, 293*b*
Montanary, G.E., 74, 287, 293*b*
Muñoz, J.F., 288, 293*b*
Murthy, M.N., 286, 293*b*

N

Neyman, J., 220–221, 293*b*

O

Owen, A.B., 28, 42, 258, 288, 293*b*

P

Park, M., 287, 292*b*
Pearson, K., 139, 293*b*
Puertas, S.M., 288, 294*b*

Q

Quenouille, M.H., 8, 289, 293*b*

R

Rai, A., 289, 293*b*
Rai, P.K., 288, 295*b*
Ramasubramanian, V., 289, 293*b*
Ranalli, M.G., 74, 287, 293*b*
Rao, J.N.K., 166, 191–192, 195–197, 287–289, 293*b*
Rao, T.J., 191–192, 195–197, 291*b*
Ross, A., 284–285, 292*b*
Roy, D., 289, 293*b*
Rueda, M., 192–193, 287–288, 291–294*b*

S

Safiquzzaman, M., 289, 293*b*
Sanchez-Borrego, I., 192–193, 294*b*
Särndal, C.E., 28–29, 39, 258, 283–285, 288, 291–293*b*
Searls, D.T., 284–285, 293*b*
Sedory, S.A., 140, 287–288, 291*b*, 294*b*
Sen, A.R., 204, 286–287, 293*b*
Singh, H.P., 25, 289, 293*b*, 295*b*
Singh, R., 139–140, 165–166, 178, 191–192, 195–197, 291–293*b*
Singh, Randhir., 289, 293*b*
Singh, S., 3, 8, 17, 25, 31–32, 38, 49–50, 55–58, 74, 77, 86, 140, 156–159, 192–197, 199, 209, 284–289, 291–295*b*
Sitter, R.R., 11, 74, 287–289, 291*b*, 293*b*, 295*b*
Skinner, C.J., 289, 291*b*
Solanki, R.S., 25, 293*b*
Srivastava, S.K., 90, 139–140, 295*b*
Stearns, M., 77, 286–288, 295*b*
Stone, M., 22, 292*b*
Sungur, E.A., 288, 292*b*
Swenson, B., 288, 293*b*

T

Thompson, D.J., 199, 203–210, 211*t*, 212–216, 283–284, 292*b*
Thompson, M.E., 3, 295*b*
Tikkiwal, G.C., 288, 295*b*
Tille, Y., 287, 291*b*
Tirari, M.E.H., 287, 291*b*
Tracy, D.S., 55–58, 288, 295*b*
Tripathi, T.P., 90, 291*b*
Tukey, J.W., 8, 289, 295*b*

U

Upadhyaya, L.N., 289, 295*b*

W

Wakimoto, K., 139–140, 295*b*
Wolter, K.M., 289, 295*b*
Wretman, J.H., 288, 293*b*
Wu, C., 74, 287–288, 295*b*

Y

Yates, F., 286–287, 295*b*
Yuan, J., 289, 292*b*
Yu, F., 286, 288, 294*b*

Z

Zhou, W., 289, 292*b*

Subject Index

Note: Page numbers followed by *t* indicate tables.

A
Auxiliary variable, 2*t*

C
Calibration, 283–288
Census, 1–2
Chi-square type distance, 29, 37
Coefficient of variation, 20
Correlation coefficient, 51, 139–140

D
Displacement function, 208–214
Double suffix variable, 17
Doubly jackknifing, 14–15, 33, 44, 66, 72, 91, 181
Dual to empirical log-likelihood (dell), 41, 69–74, 98–106, 113, 141, 179–191, 233

E
Estimator, 3

F
Frequently asked questions, 17

G
Generalized regression (greg), 37–38
Geometric mean, 23, 58

H
Harmonic mean, 23, 60, 83
Horvitz-Thompson estimator, 199–200

J
Jackknifed geometric mean, 22–23, 58–59, 81–82
Jackknifed harmonic mean, 24, 60–61, 83–84
Jackknifed pooled mean, 22–23
Jackknifed pooled variance, 26
Jackknifed predicted values, 64
Jackknifed sample mean, 11
Jackknifed sample median, 25
Jackknifed sample proportion, 16–17
Jackknifed weights, 28, 87, 199–200
Jackknifing, 8–13, 289

K
Kurtosis, 20

L
Log-likelihood, 267–274

M
Midzuno-Sen sampling, 210–211
Model assisted, 63–69, 78
Multi-auxiliary information, 257
Multi-character surveys, 165–166
Multi-stage sampling, 254

N
Negative weights, 41
Nonresponse, 52, 78

O
Ordered moments, 4

P
Parameter, 2–3
Partial jackknifing, 49–50
Partially jackknifed correlation, 51
Partially jackknifed variance, 49–50, 86
Pooled mean, 22–23
Pooled variance, 25
Population mean, 3
Population mean squared error, 4
Population variance, 4, 50, 85–87, 275
Proportion, 16–17

R

Ratio estimator, 38, 155
R code, 7–8, 34–37, 45–46, 68–69, 73–74, 102–103, 110–111, 116, 148–149, 176–177, 188–190, 205–206, 212–213, 228–229, 235–239, 261–264, 270–272
Regression coefficient, 157–160, 168
Regression estimator, 31, 38

S

Sample mean, 3, 11–13
Sensitive variable, 55, 195
Skewness, 19
Statistical Jumbo Pumpkin Model, 5
Stratified sampling, 219
Study variable, $2t$

T

Total sum of squares, 21
Transformations, 165–166, 191–192
Tuning jackknife technique, 27
Tuning with chi-square, 28–41